Further praise for *Wicked Problems*

"In this wise, wide-ranging, and wonderful work, Guru Madhavan catalogs the kinds of engineering that have shaped the modern world—often not in the ways expected. And all too often such engineering is carried out by those who persist and persevere in the background to maintain our world from collapsing, and who are ignored without credit or reward, a point seldom appreciated by those who allocate funding and celebrate excellence."

—**Don Norman**, former vice president of Apple, author of *The Design of Everyday Things* and *Design for a Better World*

"*Wicked Problems* showcases Guru Madhavan's distinctive gifts as a soulful systems engineer and storyteller. This is a deep and delightful book and, ultimately, a hopeful one that illustrates how important systems engineering concepts are to addressing our most vexing societal problems."

—**Michele Gelfand**, John H. Scully Professor of Management, Stanford University; author of *Rule Makers, Rule Breakers*

"All too often, we treat problems the way we treat coffee—we separate what is easily soluble and ignore the rest. How much value could we unlock if we engineered a new approach? In this wonderful and energizing book, Guru Madhavan provides just such a new approach. It is an espresso machine for the imagination."

—**Rory Sutherland**, Vice Chairman of Ogilvy UK, author of *Alchemy: The Surprising Power of Ideas That Don't Make Sense*

"With a well-informed and novel narrative, Guru Madhavan provides a compelling portrait of the role engineers and engineering have played and will play in the societies we build, enjoy, and care for—a thoroughly enjoyable odyssey."

—**Sir Jim McDonald**, President, Royal Academy of Engineering

"Philosophical meditations are few and far between in the practice—and literature—of engineering. In this rich and reflective book, using Edwin Link's revolutionary flight simulator as an archetype to engage with wicked problems, Guru Madhavan provides us with a brilliant engineering vision for a just, sustainable, and civil society."
—**W. Bernard Carlson**, Joseph L. Vaughan Professor Emeritus of Humanities, University of Virginia; author of *Tesla: Inventor of the Electrical Age*

"*Wicked Problems* is a feast of ideas, insights, and inspiration. Using a riveting range of examples from history, philosophy, music, and culture, Guru Madhavan presents a compelling civic vision for engineering and an engineering vision for civic life."
—**Venkatesh Narayanamurti**, Benjamin Peirce Professor of Technology and Public Policy Emeritus and founding dean of the School of Engineering and Applied Sciences, Harvard University

"In this deeply insightful and inspired creation, Guru Madhavan provides a syntax and structure to expand our imagination for life's intractable problems—it should be a go-to guide for decision makers of all kinds."
—**Richard Meserve**, former president, Carnegie Institution for Science

"Well-told tales and a core commitment to engineering fundamentals permeate Guru Madhavan's *Wicked Problems*—it's a refreshing escape from the technology hype that surrounds us and a responsible call to action on leveraging the lessons of the past to guide a better future."
—**Chandrakant D. Patel**, chief engineer and senior fellow, Hewlett Packard Laboratories

"A much-needed and elegant account—and application—of the work of Edwin Link, one of the truly great pioneers in aviation history."

—**Paul Ceruzzi**, curator emeritus, Smithsonian National Air and Space Museum; author of *GPS* and *Beyond the Limits: Flight Enters the Computer Age*

"What an accomplishment! Guru Madhavan melds engineering with science, art, poetry, politics, history, and biography, and distills potent implications for making the world better using a flight-simulator mindset. A wonderful and hopeful read for anyone fascinated by failures and how to prevent them—that is, most all of us."

—**Randolph M. Nesse**, author of *Good Reasons for Bad Feelings* and coauthor of *Why We Get Sick*

Wicked Problems

Link Aviation Devices Inc.   June 19-1935

Wicked Problems

How to Engineer a Better World

Guru Madhavan

W. W. NORTON & COMPANY
Independent Publishers Since 1923

For information about permission to reproduce selections from this book, write to Permissions, W. W. Norton & Company, Inc., 500 Fifth Avenue, New York, NY 10110

For information about special discounts for bulk purchases, please contact W. W. Norton Special Sales at specialsales@wwnorton.com or 800-233-4830

Manufacturing by Lakeside Book Company
Book design by Lovedog Studio
Production manager: Lauren Abbate

ISBN 978-0-393-65146-1

W. W. Norton & Company, Inc.
500 Fifth Avenue, New York, N.Y. 10110
www.wwnorton.com

W. W. Norton & Company Ltd.
15 Carlisle Street, London W1D 3BS

1 2 3 4 5 6 7 8 9 0

The views expressed in this book are those of the author and of those individuals and authors of the publications consulted, and do not necessarily represent the views of the National Academies of Sciences, Engineering, and Medicine.

To Binghamton

There are mountains hidden in treasures.
There are mountains hidden in swamps.
There are mountains hidden in the sky.
There are mountains hidden in mountains.
There are mountains hidden in hiddenness.
This is complete understanding.

—Eihei Dōgen

To be an engineer and nothing but an engineer
means to be potentially everything and actually
nothing.

—José Ortega y Gasset

Contents

Preface: **Clocks and Clouds**   *xiii*

Prologue: **Airplane Mode**   *1*

Chapter One: **Wicked Efficiency**   *21*

  Refrain: Pitch   *47*

Chapter Two: **Wicked Vagueness**   *67*

  Refrain: Roll   *88*

Chapter Three: **Wicked Vulnerability**   *111*

  Refrain: Yaw   *133*

Chapter Four: **Wicked Safety**   *153*

  Refrain: Surge   *175*

Chapter Five: **Wicked Maintenance**   *195*

  Refrain: Sway   *221*

Chapter Six: **Wicked Resilience**   *241*

  Refrain: Heave   *264*

Epilogue: **Civicware**   *285*

Acknowledgments   *303*

Sources and Resources   *307*

Index   *351*

Sept. 29, 1931.
E. A. LINK, JR
1,825,462
COMBINATION TRAINING DEVICE FOR STUDENT
AVIATORS AND ENTERTAINMENT APPARATUS
Filed March 12, 1930   4 Sheets—Sheet 1

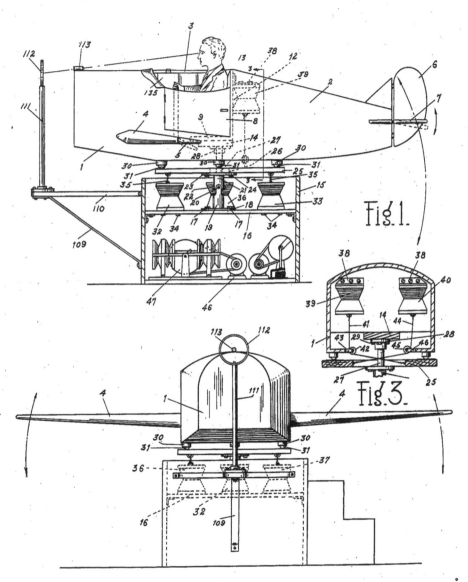

Fig.1.

Fig.3.

Fig.2.

INVENTOR
EDWIN A. LINK. JR.
BY
ATTORNEY

Clocks and Clouds

LIFE COMES WITH PERNICIOUS PROBLEMS, SOME WITH recognizable plots. We excel in solving the well-behaved ones, but intractable problems create more questions. In a 1965 lecture, the British philosopher Karl Popper classified problems using the metaphors of clocks and clouds. He compared systems that obey logic to clocks and those that defy logic to clouds. Unlike a timepiece's predictability, the shifting subtleties of clouds reside in their myriad forms and shadows. For William Shakespeare, clouds were simultaneously dragons, bears, lions, citadels, capes, and cliffs. And in John Constable's assiduous sky paintings, the cloudscapes were "the chief organ of sentiment." Their fluctuations and formlessness made it clear that "no two days are alike, not even two hours."

This book is about the cloudiest of problems, the "wicked" problems. Each of them is unique. With wicked problems, we tend to carve off pieces that invite rational solutions, leaving the rest for others to manage. In some cases, tackling part of a problem may deceive us into thinking that the whole problem is tamed. "Look, I've not tamed the whole problem," as scholar Charles West Churchman once observed, "just the growl; the beast is still as wicked as ever." Scholar John Warfield used the term "spreadthink," the opposite of groupthink, to describe how, in a wicked problem, individuals cannot focus and agree on its critical elements.

Look at a teakettle filled with hot water. You might ask why the water is hot. The first answer could be that the kinetic energy of the water molecules is high. The molecules whack one another and the wall of the kettle at higher speeds as a result. That's a scientific angle. A second answer might be that the kettle was on a stove with the flame turned on. That's an answer rooted in logic. A third answer could be that someone wanted hot water for tea. That's an answer based on intention. And a fourth answer, depending on one's belief, could be that it is part of god's plan for the universe. In these answers, all correct, we see how a teakettle is both a clock and a cloud. Clock systems construct our comforts. They direct electrons to send messages, summon shared rides, trade stocks, swipe for dates, and livestream K-pop. With cloud systems, we have only shadow-boxed the problems. Our conflicted response to climate change, a kind of global teakettle, offers a stark example of our failure.

A simple equation named "the monster" can help us think about wickedness: $I = 2^{n(n-1)}$, where I is the number of states a system could have with n number of elements. Suppose all the components interact with one another straightforwardly. Then, a two-element system results in 4 states, and a three-element system has 64 states. In a ten-element system, the number of states will exceed the number of stars in our galaxy. While wicked problems are inherently complex, we increase their complexity with our human bramble of beliefs and deficits, paradoxes and priors.

We spray around "aerosol words," easy to release but challenging to recapture. Think of some common terms: "society," "sustainability," "economy," "equity," "green," "liberty," "love," "peace," "progress," "innovation," "justice," "diversity," and "data." Each word encompasses multiple, muddled meanings. Every aerosol word and its sense becomes a compromise, barely reached. And worse, wicked problems are invoked so often, they are tainted by simple familiarity. As I'll argue in this book, engaging with wicked systems requires more than good intentions, creativity, and expertise. We need a communal code of conduct—or in an engineering sense, a *concept of operations* to train and treat our approaches to gain more profound improvements.

This book is double stranded. One strand follows a forgotten engineer; the other examines forgotten uses for engineering. Together, they weave an engineering vision for civics and a civic vision for engineering. While nonfiction, the book's aspiration may feel like fiction. Engineers, after all, aren't commonly invoked as pillars of democracy. Yet as we'll see, engineering does more than tech support. Engineering is a carrier of history, simultaneously an instrument and the infrastructure of politics. It's among the oldest cultural processes of know-how, far more ancient than the sciences of know-what. And through engineering, civics can gain a more structured, systemic, and survivable sense of purpose. By applying engineering concepts in a civic context, engineering can usefully grow the policy lexicon and enhance its cultural relevance. The usefulness of civics and engineering is often realized only in their breakdowns, much like trust, most longed for in their absence.

Democracy is the ultimate voice-activated application, defined by differences and decibels. Yet, there are no noise-canceling headphones. Unlike virtual assistants, democracies don't simply yield to our vocal commands. Democracy is a perpetual raw material, never a finished product. Developing a civic consciousness to achieve democracy's goals will fundamentally require a systems engineering approach, just as remedying the professional deficits of engineering will depend on civic participation. Indeed, civic engagement may exist without engineering, but no engineering can effectively exist without civic engagement.

This book will advance engineering concepts for cultural use. The driving desire is to responsibly appraise and engage with the inner nature and character of life's wicked problems. As the historian Arthur Schlesinger Jr. memorably said, we suffer today from much more *pluribus* than *unum*. Our world thrives on variety and diversity. Still, curtailing chaos requires sufficient cohesion. We need civic operating standards. Better living with wickedness will depend on better engineering of wickedness. Civics and engineering aren't just about what we choose to do; they are about what we choose to become. This book is about that once-and-future sensibility.

Wicked Problems

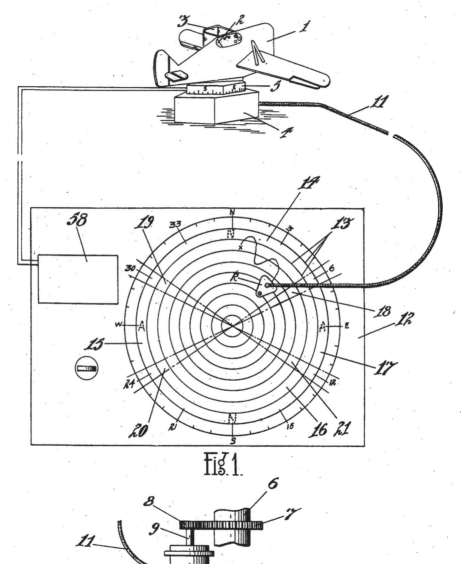

Fig.1.

Fig.2.

INVENTOR.
EDWIN A. LINK, JR.
BY Philip S. Hopkins.
ATTORNEY.

Airplane Mode

I change, but I cannot die . . .
Like a child from the womb, like a ghost from the tomb,
I arise and unbuild it again.

—Percy Bysshe Shelley

There's no love in a carbon atom, no hurricane in a water
molecule, no financial collapse in a dollar bill.

—Peter Dodds

Say, it's only a paper moon / Sailing over a cardboard sea
But it wouldn't be make-believe / If you believed in me

—Harold Arlen, Yip Harburg, and Billy Rose

IN LATE MAY 1941, 50 BRITISH CADETS SET OFF FOR THE
United States. Each had received a small blue book of instructions. "You are going to America as guests," the book opened, describing the "great, friendly, yet different nation." The trainees were warned not to speak out about what was wrong with their hosts and their country.

The group boarded *Pasteur*, a lice-infested liner once owned by the French. The Atlantic was rough. The seasick cadets slept on hammocks and bedraggled bunks, fearing German torpedo attacks. They arrived filthy and famished at the dispatching station in Halifax, Nova Scotia, and were deloused and fed. As they embarked on a long train ride, the cadets were issued thick long clothes, overcoats, and woolen underwear and socks. One problem: it was June, and they were going to Texas. Their destination was the No. 1 British Flying Training School.

The town of Terrell housed one of six schools created under the American Lend-Lease Act dedicated to Royal Air Force pilot training. The others were in Lancaster, California; Miami, Oklahoma; Mesa, Arizona; Clewiston, Florida; and Ponca City, Oklahoma, along with a short-lived summer school in Sweetwater, Texas. Each site contained an airfield, a ground school, a hangar, and a maintenance facility. The Royal Air Force set the course syllabus to train 500 novice students for flying badges at each site each year.

Six months earlier, President Franklin Roosevelt had sought public support for the lend-lease proposal. He viewed the American obligation to Great Britain as a good neighbor lending a hose to stop the fire of World War II. "Manufacturers of watches, farm implements, linotypes, cash registers, automobiles, sewing machines, lawn mowers and locomotives are now making fuses, bomb packing crates, telescope mounts, shells, pistols and tanks. But all our present efforts are not enough," Roosevelt said in his December 1940 fireside chat. "We must have more ships, more guns, more planes—more of everything." The United States, he stressed, must be "the great arsenal of democracy." Congress passed the bill in March 1941.

The lend-lease project eventually cost nearly $50 billion (approaching a trillion in today's dollars), with Great Britain receiving the lion's share of matériel. The US War Department authorized the British Air Ministry to operate the contract flying schools. The British Flying Training Schools rekindled a special military relationship between the countries, as captured in the inscription of Terrell's coat of arms: "The seas divide us, the skies unite us."

British cadets were captivated by open country and abundance: from breakfast eggs to Buicks and from Coca-Cola to city lights. They relished corn on the cob and grits, what they knew previously as animal feed and as a sand-gravel mix. They learned "squash" meant vegetable, not a game, and "gasoline" was petrol, as words ping-ponged between British accents and southern drawl. The locals gave cadets replacement cotton clothing to avoid heat rash from woolen garments, and sunglasses to fly in blinding brightness. As historian Tom Killebrew describes in *The Royal Air Force in Texas*, the cadets also witnessed the effects of segregation. Some had never seen Black people and didn't know whether to adhere to or rebel against their hosts' views. The cadets paid the Black custodial staff 50 cents weekly to make up their beds and a nickel to spit-shine shoes.

The Americans charged the British $21.60 per hour of primary training, which included ground classes on the basics of airmanship, armaments, and aircraft maintenance. Students confronted a mosaic of sensations as they learned takeoffs, turns, and touchdowns en route to their first solo flight. Advanced training offered them experience with instrument panels and flying in different wind conditions. Initially, the rate was $32.70 per hour for 180 hours in the 20-week training course. But British fighter aircraft such as the Supermarine Spitfire, an interceptor with elliptical wings, and Hawker Typhoon, a bomber for nighttime intrusions, flew at speeds exceeding 400 miles per hour. These new speeds posed new demands. In response, the Royal Air Force increased the training duration to 28 weeks.

While a Texas fog was no London fog, for aircraft, few conditions were more of a recipe for disaster than bad weather. Terrell was a "dust bowl when dry and a glue pot when wet," notes Killebrew. Flying in poor visibility was dangerous and even deadly. The weather often delayed commercial and cargo flights, and anyone worried about schedules was better off taking the train. Aviation couldn't become reliable until pilots were trained to achieve competency and confidence in any weather. As we'll see, overcoming these barriers required pilots to switch allegiance from their imperfect instincts to the aircraft's instruments, even when the latter contradicted their senses.

Training meant costs—in dollars and lives. Students named the Stearman, once famous as a crop duster, the "yellow peril." The fabric-covered Boeing biplane roared across the skies at 225 horsepower. Its open cockpit permitted a student and an instructor to sit in tandem. The Stearman was notorious for its "ground loop"; it circled out of control when landed improperly. Once, a student landed his Stearman on top of another, waiting on the tarmac for takeoff. In other cases, students who misused the controls overshot and crash-landed on the highway, in someone's backyard, or on trees. Some engines caught fire after landing, and some quit midair.

The wider aviation industry was shaking off turbulent experiences from the 1920s and 1930s and veering into a bold new age. Some years earlier, when the Navy and War Departments asked for a budget increase, President Calvin Coolidge had quipped: "Why don't they buy one airplane and let the aviators take turns flying?" By the 1940s, aircraft equipment was expensive and in demand. Inadequate maps and training manuals also accompanied the equipment shortage. The only way to learn how to fly at night and in bad weather was to fly at night and in bad weather. Still, a wartime footing demanded standards for professional training and proficiency. Aviation needed an approach that blended technology, psychology, and policy considerations and was realistic for initial and ongoing training.

The centerpiece of this lesser-known episode of the American and British military relationship was a general aviation trainer, the creation of Edwin Albert Link Jr., the chief protagonist of this book. Ed Link's pilot trainer consisted of a fuselage with mock wings and tail to give the impression of an airplane cockpit. Its accordion-like base bounced to imitate actual airborne motions, and its box circled 360 degrees, allowing infinite flight-like repetitions without injury. Under the hood were controls that replicated the cockpit experience.

Along with the stick and rudder pedals, an instrument panel with a clock and compass contained indicators for airspeed, altitude, turn and bank, oil pressure, and temperature. The pilots in this dark and cramped chamber had no outside horizon to reference, a precondition to train for instrument flight. The instructor sat at a desk with a map that traced the student's course using a special electrical recorder; the trainee and instructor communicated through two-way radio. The instructor sent electrical signals to the training unit at the turn of a switch. The trainer twisted and turned in a geometric ballet—puffing, wheezing, gasping, and groaning. The entire setup looked like a penny arcade ride. It didn't merely look like one; at one point, it *was* one. People dubbed it the Donald Duck, a winged armchair, and more memorably, the "blue box."

Ed Link asserted a systems approach for flight training. Even more significantly, though, his engineering brought women closer to aviation at a time when they experienced the barbs of discrimination. As more men enlisted for the war, the United States and United Kingdom faced a dwindling supply of instructors. Women became a natural fit as Link Trainer instructors. They initially worked as maintenance clerks at the Terrell flight school, fitting fabrics for planes and parachutes. Women were hired as instructors only after they agreed that they wouldn't date any cadets. Feelings, they were told, didn't have a place in an all-too-formal environment of "Mr." and "Miss," especially when women had to judge men's instrument readiness. The Link Trainer ultimately forged over half a million Allied Force pilots. With its synthetic environment, the Link Trainer didn't precisely portray the complexity of flying to the pilot but potently *implied* it. The Link Trainer was no illusion machine; it was an initiation. And much like it, this book aspires to a grounded approach to engage with the heights and depths of wicked problems.

✳

OUR WORLD IS FILLED with immortal problems. As with the struggles of early aviation, when "seat-of-the-pants" flying was the norm, we lack the know-how to gauge the size of such problems. Wicked problems have vexing staying power, working like franchises, with premieres and prequels, remakes, reissues, reboots, sequels, spin-offs, and streaming on demand. One can't opt out, log off, autocorrect, unsubscribe, block, mute, cancel, or shut down wicked problems. Nor can we click and clear them like browser history. As their forms change, so do their formulations.

In 1973, scholars Horst Rittel and Melvin Webber cataloged 10 properties of wicked problems. They argued that wicked problems contained "no definitive formulation." Each is unique, with no clear start and stop signs, and is usually a symptom of another problem with no clear entry and exit points. The so-called solutions aren't "true-or-false, but good-or-bad." What remains are opportunities to learn by trial and error.

The next time you drive across town, think about energy efficiency. You can technically express how efficient your car is in fuel use or how economical it is to get to your destination compared to other modes of transport. But the fact that you chose to drive could be behavioral, and having to go across town could result from poor urban planning, which may be political. Across these levels, engineers seek a baseline understanding of facts and assumptions, often shaping their products and services as an exercise in trade-offs. Certain factors and viewpoints are privileged over others, akin to a projector spotlighting the main actors, leaving the rest behind the scene.

Further, engineers often characterize themselves as "problem solvers," so much so that it's part of their primary professional identity. But for wicked problems, one must go beyond mere problem-solving. Engineers have succeeded in answering well-posed concerns partly because they influence how problems are posed, including how one thinks about energy efficiency. Scholar Russell Ackoff has observed that not all ways of viewing a problem are equally productive. Still, the most productive views are seldom evident. Problems should be approached from as many angles as possible before the measures are selected to address them. Effective problem formulation requires, first, awareness of the types of problems that lead to wickedness. Let's consider three types.

Call the first type "hard problems." They are bounded and boundable, and scientific principles, market pressures, and sponsor requirements neatly specify them. The outcomes are directed— even dictated—by customers, consumers, and clients. Have you used your smartphone to check in to a flight 24 hours earlier? Or, did you self-checkout at the grocery store or book a table at a chic gastropub and use a QR code to order tacos? These outcomes are products of hard problems. And as with standardizing production processes for Coca-Cola, cars, and canning crinkle-cut carrots, they can be *solved* by recombining and repurposing existing tools and techniques. Such problems can be mathematically manipulated, chemically configured, and materially improved. Ultimately, they can be "optimized" by applying available knowledge and experience with the idea that the best possible outcome exists and is achievable.

The second class, call it "soft problems," is in the arena of human behavior, which is complicated by political and psychological factors. Because their end points are unclear, and thorny constraints complicate their design, soft problems cannot be solved like hard problems; they can only be *resolved*. There are no easy fixes to a problem like traffic congestion. Adding more road capacity won't necessarily clear out bottlenecks on a throughway, nor can congestion pricing if the charges are unaffordable for most people who use a tunnel. Such interventions depend on the contexts for which they are designed: what works for Charing Cross Road may not work in Chengdu or Chennai. Even in the same city, what works for one bridge may not work for another. Notably, soft problems often involve how we assign value to time and how much we are willing to pay for it. They require political buy-in that may eventually pay for a service and behavior change that might incentivize more public transportation or teleworking arrangements. Since soft problems fuse technology, psychology, and sociology, resolving them yields an outcome that's not the best but only good enough. As Ackoff characterized it, the results are based not on optimizing but on "satisficing," an approach that satisfies and suffices.

The third class, "messy problems," emerges from differences and divisions created by our value sets, belief systems, ideologies, and convictions. A disease outbreak may involve hard problems with solutions such as barcode-tracked supplies or antibiotic deliveries. The outbreak's soft problems might require resolutions like mapping infectious disease spread or retooling the indoor environment to prevent the propagation of infection. Neither resolution is exact, but both are good enough. By contrast, a messy problem can involve a pathogen gaining antibiotic resistance or intersecting with delicate religious rituals, as we'll see with Ebola. How can we reframe centuries of tradition into safe, acceptable burial practices while respecting cultural sensitivities? Such messy situations may only be abstractly *dissolved* by transforming them into a different, possibly manageable state. Messy problems can be reframed out of existence not by optimizing or satisficing but by "idealizing." In Ackoff's words, this entails getting the matter "closer to an ultimately desired state, one in which the problem cannot or does not arise."

If they were works of art, hard problems would be photographs, offering clarity and directness. Soft problems are like blurry brushstrokes of impressionism, and messy problems are spilled and splattered abstractions. A wicked problem emerges when hard, soft, and messy problems collide. Think of them as a cubist collage where the truth is simultaneously sharp, shaky, and squiggly. All three are required for wickedness. Hardness is nestled in soft problems, and hardness and softness reside within messy problems. By extension, a solution can be within a resolution, and a dissolution might contain resolutions and solutions.

Of course, one might *absolve* the world's problems with a doomed shrug, but that's not the point of this book. The following chapters will vault across the terrains of hard, soft, and messy problems. We'll look at how hard problems become soft, how soft problems evolve into messy ones, and how hard, soft, and messy problems conspire to produce wickedness. We'll consider some departure points toward a balanced blend of hard solutions, soft resolutions, and messy dissolutions to wicked problems. To do so, we'll need a systems engineering approach.

$$*$$

THE WORD "SYSTEM" DENOTES a coming together to form a whole. Viewed this way, a screw or a sensor can be a system. An altitude meter is a system, inertial navigation is a system, a wing is a system, and so are the avionics for collision avoidance and a squadron of aircraft. Each system is best understood only through its interactions and not in isolation. Indeed, a system fails to function when its constituent units are separated; it loses vital properties that define its meaning and behavior in the first place. Ackoff reminded us that an automobile stripped apart couldn't transport people, and a disassembled person couldn't read, write, or live.

As this book will argue, systems engineering offers a versatile practice relevant to many modern problems. Systems engineering is not some specialty embedded in an AIE (acronym-intensive environment); it's a convening mindset. Systems engineers are always the learners, never the learned. A specialist engineer might flee from contradictions, but a systems engineer embraces them with up-front preparation, valuing prework over rework. In practice, systems engineers recognize *sensitivities* (when a commercial desire begins to degrade the environment), shape *synergies* (where multiple actions can achieve a benefit, a single activity cannot), and account for *side effects* (what influences what, beneficially and adversely).

Before specialization altered many engineering practices beyond recognition, a kind of "systems sensibility" pervaded early engineering. Consider the thinking behind the development of the Indus Valley from the Bronze Age. Sophisticated food production, grid-town planning, drains, dams, and dockyards were a testament not merely to civil engineering but to a form of *civic* engineering. They improved both standards of living and thinking. Systems planning to those engineers meant braiding commerce and culture, from recreation to conservation. Many Native traditions have applied such overarching awareness. Prominent successes of industrial systems engineering, including telecommunications and missile defense systems, were evident during World War II. The applications reached a broader scale between the 1950s and the 1970s in the defense, aerospace, urban-planning, and manufacturing sectors.

Systems engineering, like civic duty, occurs under the constraints of costs, schedules, and performance requirements. While systems engineers often focus on the needs, desires, end points, and context of a problem set, they know that local fixes will not produce a globally viable solution. An engineering process based on cost, schedule, and performance requirements works well for aircraft assembly. Once that plane begins to fly, however, it enters a broader social system with different levels of complexity.

Air travel began with aircraft developed for multiple uses, including military or civil applications. An interlinked network of testing, servicing, and scheduling supports each aircraft. But routing a flight cannot occur in a vacuum—multiple airlines use the same airports and air space. Centralized airspace evolved, overseen by the Federal Aviation Administration and internationally coordinated treaties. That system relies on the construction, maintenance, and ground operations of airports around the country, which in turn involve environmental impacts, noise control, ground transportation, and urban planning at the local and state level.

Should we add capacity at a near-city airport or build new facilities farther away? Each option now involves considerations far afield from aircraft yet central to the aviation system. That's why systems engineering often works best when it's not expected to produce an "engineering" solution.

Taking a systems engineering approach—duly considering all facets of a problem—with a far broader scope might sound unnatural, especially for the issues of business, government, and civic life. The idea of dividing a year into days or a circle into degrees and using clocks was unnatural and required an initial mental leap. Even flying was—and *is*—unnatural for humans. Still, we acquired the skills to do it, with iteration eventually shaping intuition. Similarly, a civic flight trainer can prepare us to better engage with our problems.

✳

CONCEPT DEVELOPMENT IS THE heartbeat of systems engineering. The first step in a systems design process is to prepare a concept of operations—or ConOps—that imagines the desired characteristics of a system from the perspectives of those who will engage with that system. In other words, ConOps are narratives about a situation we haven't necessarily experienced. In a more practical sense, ConOps are primarily about how a system would work, what functions it could responsibly fulfill, and what outcomes it should achieve. This book will adapt this foundational tool of systems engineering for appraising wicked problems. It will describe six attributes to inform a concept of operations: *efficiency, vagueness, vulnerability, safety, maintenance*, and *resilience*. There's no hierarchy among these attributes; each will produce a revelation and a relationship to the problem.

Consider how artists make complex visuals from the six elements of line, shape, form, space, color, and texture. And how good design comes from unity, balance, rhythm, pattern, variety, proportion, and emphasis. Or how taste is triggered by sweet, salty, bitter, astringent, and pungent, and how good composers braid melody, rhythm, texture, timbre, form, and harmony. Each element is crucial, akin to how an aircraft moves with six degrees of freedom, the position relative to Earth's horizon. The three rotational axes are *pitch* (tilting forward and backward), *roll* (pivoting side to side), and *yaw* (swiveling left and right); and the three translational axes—*surge* (moving forward and backward), *sway* (moving left and right), and *heave* (moving up and down). These motions together form a synergistic six.

As we saw earlier, the monster equation may intimidate us with astronomical problem states. But there are an equivalent number of possible approaches. A Rubik's Cube can have 43 quintillion, 252 quadrillion, 3 trillion, 274 billion, 489 million, 856 thousand possible positions, wrote its creator Ernő Rubik, with only one of these states as a starting point. Similarly, combining the features of efficiency, vagueness, vulnerability, safety, maintenance, and resilience can provide a rich set of possibilities.

The following six chapters will each profile a concept and discuss its hard, soft, and messy forms while considering its wicked formulations. The chapters will present signal tragedies and lesser-known tales, from the efficient design of battleships to how carpentry and a goat-footed god inform modern meanings of social resilience. We'll see how a city was coated with sugary sludge by overlooking a vulnerability, how vagueness surrounding a volcano curtailed global commerce, how a board game can provide insights for managing trash, and how a safety standard relates to a magnificent magician. We'll also see how maintenance as an integral engineering practice connects to ancient religious rituals and a maritime disaster worse than the *Titanic*. These narratives will demonstrate that engineering has always been a cultural choice. The key is to avoid traps laid by newness and novelty that claim to transform society; it's also about how we consciously care for our creations. Throughout these accounts, we'll absorb how best engineering occurs beyond its technical scope and what consequences await when it doesn't.

All too often, our cultural and commercial ideals have become such that governments or investors prefer a single action item to pursue. We make sense of the world with an ardent affection for narrow explanations—be it "monocausotaxophilia" ("the love of single causes that explain everything") or "monotheorism" ("preaching the one true theory's idealized models"). Yet true accomplishments reside in simultaneously understanding many facets of the problem, a feature the Link Trainer enabled with the goal of not crashing. Whether you are a chief executive or a citizen advocate, a block association or a blue-ribbon panel, none of our decisions can be optimized around a single concept: efficiency, safety, or resilience. Indeed, in a wicked problem, no single concept should win focus or attention. As in fitness, we can't sacrifice form for speed or core strength for convenience.

Engineering approaches to wicked problems can be realized only through multiple criteria. The friction among the six concepts— efficiency, vagueness, vulnerability, safety, maintenance, and resilience—can guide our solutions, resolutions, and dissolutions. "Without contraries is no progression," the Romantic poet William Blake wrote in *The Marriage of Heaven and Hell*. "Attraction and repulsion, reason and energy, love and hate, are necessary to human existence." Scholar Roger Martin has argued for contradictory proxies that work like opposable thumbs. For instance, an airline company's low-cost strategy can simultaneously include competing measures of customer satisfaction, employee satisfaction, and business profitability.

Wicked problems are not two-sided issues; they are six-sided issues. An acceptable solution, resolution, or dissolution should go through all six filters. What passes the efficiency filter may be trapped in the maintenance filter, and what clears the vagueness filter may not get through the resilience filter. We cannot mindlessly prioritize one concept over the other. Just as we ought not to look at the sun directly without filters, we shouldn't engage with wicked problems without a concept of operations. After all, a similar consciousness guided Ed Link's engineering from a dusty organ factory to deep space and deep-sea diving.

✳

"IN A DARK TIME, the eye begins to see," the poet Theodore Roethke wrote.

Ed Link's phone rang on the evening of February 10, 1934, and it was Charles "Casey" Jones, the world's airspeed record holder in 1931. The ace pilot, once taught by Wilbur Wright, was now the marketing man for Ed Link. Jones asked Ed Link to fly into Newark to demonstrate the pilot trainer to some army top brass. Two days earlier, President Roosevelt had ordered the Army Air Corps to take over mail delivery in the United States.

The morning of February 11 was drowsy and dark. In his leather helmet and goggles, the lanky 30-year-old Ed Link revved his Cessna and barked off the rain-soaked runway near Binghamton, New York. Frigid air whipped the windshield, and Ed Link was cold. He racketed over the Pennsylvania countryside, flying just above the tree level of the Endless Mountains with the Susquehanna River far below. The visibility deteriorated as thicker clouds swaddled him, but Ed Link pushed on. He turned left toward the Poconos, flying below the gray cotton-candy clouds. With hands stiff on the controls and eyes fixed on the instrument panel, Ed Link pressed ahead with a faint drone over the Appalachian ridges molded by the meandering Delaware River. The more tolerable terrains of northern New Jersey appeared but no signs of New York City through the thick blankets of gray. He was late for his meeting.

In Newark, the army officers got fidgety even as Jones assured them of Ed Link's arrival. As they decided to leave, Ed Link's steel screamed out of the fog and landed on the runway in a precisely timed fashion. The army officials could barely see Ed Link in the mist, even as he waved his hands from the cockpit. The landing was all the demo the officers needed on how to fly in bad weather. Gaining confidence in Ed Link's training system, they alerted Washington. Soon came an order for six Link Trainers at $3,400 each.

Aviation was different from other modes of transportation. It was the first to pursue "perfect all-weather operations, at least in any organized way," notes scholar Erik Conway. Operators of cars and trucks, or even trains and ships, didn't need to be concerned about blind driving—when required, they could stop and wait the weather out. But airplanes didn't have that luxury when airborne. Pilots couldn't circle for hours until the weather cleared because fuel was, after all, limited. Under these circumstances, as we'll see, delivering airmail was a perilous pursuit.

For most pilots in the 1920s and early 1930s, "contact flying"— keeping track of landmarks and railroads while flying close to the ground—was the norm. In limited visibility—or "blind flight"— lacking reference to the horizon, the pilots were disoriented, with their senses utterly unreliable. Their instincts contradicted their instruments—they felt they were turning in one direction while doing the opposite. They spun and stalled with vertigo. Accidents in which weather was a crucial factor "were dismissed simply as 'pilot error,' without further explanation," noted General Jimmy Doolittle, a pioneer of instrument flying. "This was true, but the error was too often the pilots' refusal to believe their instruments instead of their senses."

Before the army's abrupt interest in pilot training, almost no one was interested in Ed Link's flight trainer, and there was no market for it. As a teenager, Ed Link had spent a fortune on flying lessons. He thought that the overall expense—and the time—could be reduced if the bulk of the training could be done on the ground. Ed Link's hand-built invention, using the bellows, valves, and actuators of organs and pianos, coincided with the rise of instrument flying and enabled reliable and routine landings. A principal benefit of Ed Link's approach was that if people learned how to fly airplanes without a real airplane, then the real airplanes could be freed up for profitable use. His planeless instruction led to a psychological insight that not everything in the airplane had to be duplicated, only core parameters that gave pilots the essential flight experience.

The Link Trainer didn't merely apply an instrumental attitude to piloting. It generated a new coherence between those instruments and the person flying. That coherence grew beyond the confines of the trainer; it extended into the pilot's mind. This "synthesis of senses" has been compared to preparing a meal. Chefs can't decide whether a dish is ready through sight and smell alone and often use taste to confirm. Similarly, pilots employed sight, touch, sound, and even smell to understand flight conditions and guide their response. Now they added feedback from their instruments.

Ed Link's consequential landing in Newark is virtually unknown to the world. Soon after, more people discovered that airborne flight-training time could be slashed in half or more by simply learning how to fly on the ground. A central revelation in the story of the Link Trainer was, as one airmail pioneer paradoxically put it: "An airway exists on the ground, not in the air." That is, the development of ground-based training and organizations that support it was as crucial as the invention of the airplane itself. Ground-based training is now the default approach for cost and safety reasons. The Federal Aviation Administration requires all commercial pilots to gain their training for qualifications in flight simulators. And without simulators to train astronauts, there wouldn't be a space program. One observer wrote that it's "doubtful that any piece of equipment has been cursed more than the ego-busting Link Trainer" in the annals of aviation. The sweat of World War II pilots within the bounds of the blue box "would probably have floated an aircraft carrier." It's one thing to mass-produce 10,000 planes, but producing 10,000 qualified pilots to operate them in different conditions was an altogether different problem.

If the Wright brothers generated a "passion for wings," making airplanes a cultural symbol, Ed Link presented a principled protocol whose prime purpose was the preparation. Flight training proved as valuable as flight itself. In this book, we'll use Ed Link's work as a substrate to explore the cultural consequence of a systems engineering approach that should be recognized in the same league as the Wright brothers. Chapters are interspersed with alternating refrains—named after the six degrees of freedom: pitch, roll, yaw, surge, sway, and heave—that are an homage to Ed Link's engineering and the world it made. The refrains relate the flight trainer's development from carnival attraction to today's immersive environments.

Ed Link was a person in multiple mediums—land, space, and water. He connected them with his boundary-blurring belief that good problem formulation was a prerequisite to success. Ed Link seamlessly crossed over problem sets unhindered by specialty credentials, like a cubist ushering art into its modern era. Ed Link was thoroughly conversant in technology, behavior, and policy, what his device trained, and how it ordered and organized the experience. That's why ground-based training has resonated well beyond aviation and can now offer value to engineering our wicked problems in business, health, education, and policy. The confinement in Ed Link's blue box gave trainees the consciousness of the many angles to approaching a problem. In a way, it's an invitation to think *inside* the box.

✳

THE PHOTOGRAPHER DOROTHEA LANGE said that a camera is an instrument that teaches people how to see without a camera. In its journey from an exhibition entertainer to a competency creator, Ed Link's work may seem an unlikely symbol to inform civic engagement. But as this book will show, the Link Trainer was never a solo solution for a hard problem. It can serve as a mindset to identify the problem qualities, engage with them on their terms, play the future out, and be open to corrections.

We tend to pursue civic decisions as if they're about comparing one option to another. Civics isn't a blind taste test; it's more like a blind flight test. It requires assembling instrumental, sensory, and environmental awareness in our minds. In the parlance of flight training, civics is often in a state of unstable equilibrium, prone to changes, and sensitive to disturbances, with stability arising from tugs and pulls. Just as citizenship isn't about passing a civics test, civics is more than approved behaviors, observes scholar Peter Levine. Civics, in its ordinary meaning, is about how "concrete communities of people should decide and act together and develop the rules, values, resources, and habits necessary to succeed." Even if all talk around civics is civil and clued in on data, it isn't enough in this framing. "Deliberation without work is empty," Levine notes, "but work without deliberation is blind."

Engineering and civic life are only as functional as the people in them. And while both can create colossal returns with collective action, those returns are unattainable without common components of coordination. Engineering relies on standards often at the expense of diversity, and civic life thrives on diversity with frequently no standards for coordination. Many presume civics is innate and nothing extra is needed to be a good citizen. This book maintains that our civic fundamentals can be relearned and reestablished through an engineering concept of operations; call it stick-and-rudder civics. Like engineering, civic competencies are trainable traits. As the Link Trainer showed, training in confinement and under constraints can offer a path to freedom. But civics doesn't always mean freedom; it's made of degrees of freedom.

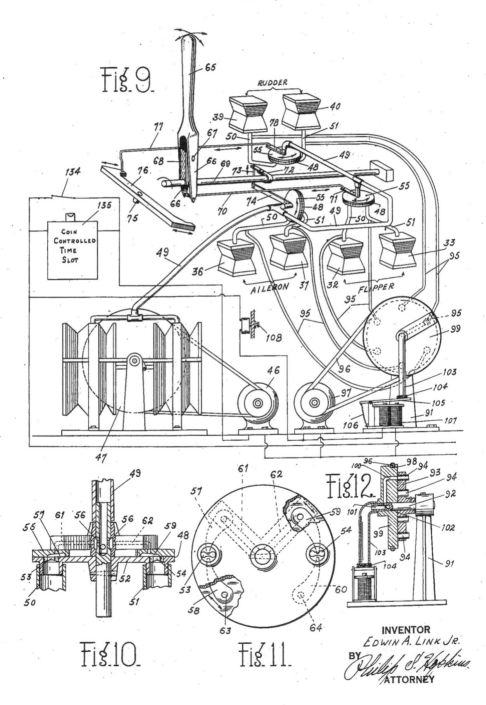

INVENTOR
Edwin A. Link Jr.
BY
Philip S. Hopkins
ATTORNEY

Chapter One

Wicked Efficiency

F EW KNEW WHY THE UNITED STATES HAD A NAVY IN THE late 19th century. America's marine doctrine contained no strategy, only a few tactical rules of thumb, and many ill-trained personnel. The navy was in a period of stasis, "run by faith and by habit," as historian Elting Morison observed.

In 1885, an American naval officer visited European shipyards to glean ideas for improving iron ships. His design cobbled together construction tips from Italian, Chilean, and British warships. The resulting vessel sailed for only five years and never sailed smoothly. The navy's combat vessels were engineered for multiuse, cleverly borrowing from European models. A ship was an arsenal, a barrack, a bank, a church, a fort, a hotel, a hospital, a home, a prison, a warehouse, a workshop, and a weapon. "Every portion of the warship and her equipment should be adapted to secure efficiency with the men available: the human factor cannot be neglected," wrote one shipbuilder. This shipwright argued that standardization was the "most obvious means of securing simplicity" and that an all-in-one efficiency should combine armor, expedience, and readiness. However, this drive for efficiency hid an embarrassing reality: many warships couldn't shoot straight; the success of that crucial element depended entirely on the human factor.

The navy required each ship to pass a quarterly target practice. The drills were distressing and left the vessels dirty. Cleanup was a pain, but luckily, these drills occurred only once every three months for the crew members. According to the *Army, Navy Journal*, these sessions were all show and no substance: "A brilliant display of gunnery. All the targets were left untouched." In one unsettling instance, an officer completely fabricated the data for the target-practice tests and mailed those papers in a box with cockroaches. As cunningly intended, the insects had ravaged the report by the time it reached the department.

In the 1890s, the navy's experiments had varying degrees of success in testing its warships. For instance, the "efficiency" tactic to boost power and range by mounting multiple guns on the same platform backfired. The turrets for five-inch guns were superimposed on the turrets for the heavier, eight-inch guns. When the turrets turned, they failed to compensate correctly for the effects of wind velocity on various calibers. The setup was useless for enabling the different guns to hit a target simultaneously. Moreover, an ill-placed ammunition hoist couldn't support more than one gauge at a time, quickly becoming an expensive hindrance. This "miracle of imagination," Morison wrote, "simply complicated the gunnery task enormously."

In 1898, during the Spanish-American War, Commodore George Dewey led the American Asiatic Squadron to triumph without casualties in the Battle of Manila Bay, ending Spain's colonial era in the Philippines. The easy American victory in the Pacific theater wasn't necessarily because of superior ships or strategy but due to the "complete inefficiency" of the Spanish Navy and their "shameful neglect of gun practice." However, the American rate of hits, "though greater than the enemy's, was nevertheless very small," wrote scholar Philip Alger in the *Proceedings of the United States Naval Institute*. The flagship USS *Olympia*, with four 8-inch guns and ten 5-inch guns, and USS *Baltimore*, with four 8-inch guns and six 6-inch guns, fell "at least 500 yards short of the enemy." And, of the 9,500 projectiles fired, only 123—or a mere 1.3 percent—made their mark on the Spanish vessels.

Alger identified two elemental errors of gunfire at sea. First, a "personal error" emerged between a gunner's intent to fire a gun and actually firing it. When pulling the trigger, the ship's normal motion on the seas displaced the gun and the line of sight on the target, leading to the second, a "pointing error." The gunners couldn't instantaneously correct their shots during a ship's roll because they couldn't clearly distinguish different objects at different distances. Consequently, any successful shot was not from skill but by sheer accident. The appalling displays of incompetence in American warships seemed everywhere, with shots missing even near-range targets. To fix the personal and pointing errors, gunners needed to estimate firing intervals promptly and precisely. They were also concurrently required to compensate for sight. These skills would transform naval gunnery and the overall efficiency of maritime warfare.

William Sowden Sims, a 40-year-old lieutenant, was seasick from this dismal state of marksmanship. In 1898, after his return from Paris as an attaché at the American embassy, the navy assigned Sims to brand-new *Kentucky*. The 368-foot, 11,000-ton battleship carried a crew of 500, a library, a gym, a piano, and other luxuries. Despite the pathetic flop of its 70-odd guns in firing practice, a 1900 report declared *Kentucky* an "unqualified success." One magazine praised it as one of the "Greatest Fighting Machines Afloat" and as agile as a thoroughbred. But while passing through the Suez Canal toward the Asian waters, Sims's experiments proved that *Kentucky* was far inferior to its European counterparts. "*Kentucky* is not a battleship at all. She is the worst crime in naval construction," he wrote. *Kentucky*'s specialized design may have seemed economically efficient, but Sims knew it was hazardous. The people who built the guns weren't the same engineers who made the turrets, and those who made the carriages didn't speak to the engineers making the guns. Years later, Sims compared these ship-crippling vulnerabilities to pitting a blind gladiator against a heavily armed adversary. "We can send as many ships as we wish . . . and they will all be destroyed," he wrote in his Pulitzer-winning memoir, *The Victory at Sea*.

In Hong Kong, Sims met and was tutored by Sir Percy Scott, a pioneering captain and later admiral. Scott had significantly improved British marksmanship by devising a method of fixing and maintaining targets irrespective of a ship's roll. Sims would capitalize on these improvements for the US Navy. With an irascible intellect, a deadpan gravitas, and a Vandyke beard, Scott was notorious for his rum consumption and rough manners. Judged as "dreadfully backwards in all useful subjects," Scott joined the Royal Navy at 13 in 1866. At a young age, he noticed the deplorable state of gunnery, recalling that in the preceding years, "cleanliness of the ship and the state of men's bedding were still regarded as the most important factors of efficiency." A decade earlier, another illustrious fleet commander lamented that gunnery was "quite a secondary thing." Beauty beat out the brutal realities, and practicality bowed to prestige. Knowledge of artillery was considered merely an elective rather than an essential part of officer training. Seamen who pointed out these deficiencies were dismissed as faultfinders. Even the inspection reports neglected to examine the efficiencies—and ultimately, the effectiveness—of target drills, a practice absent until 1903.

In 1878, Scott was selected as a gunnery lieutenant for HMS *Excellent*, a disheveled triple-decker destroyer unworthy of its name. Everything about the ship was antiquated, and even firing at close range required extensive maneuvering. On each gun were two V-shaped metal pieces separated by 4 feet. From 6 feet behind the gun, the sailor would fire by pulling the trigger string when the V's and the target aligned. The optical antics required the alignment of both V's and target: one at 6 feet, the other at 10, and another at 3,000. "This called upon the eye to do more than any camera will do," Scott wrote. Even when the ship rolled the same amount, some guns fired earlier—and more erratically—than others. These problems continued for two more decades.

In the summer of 1896, as captain of the cruiser HMS *Scylla* on the Mediterranean, Scott was once again astonished that the effi-

ciency of housekeeping outshined the efficiency of maritime weaponry. "The state of the paintwork was the one and only idea. To
be the cleanest ship in the Fleet was still the objective for every
one; nothing else mattered," Scott wrote. With the ship's roll, poor
sights, shoddy gun quality, and pitiable firing rates, he condemned
the situation as "worse than useless."

Scott focused on overcoming the problem of waiting for a ship's
roll to align the target. He devised a telescope as a gun sight so the
shooter had only one point to focus on distinctively. Scott's crew
had little success in early practice rounds with one-inch guns. The
crew blamed the telescopes, but Scott knew the fault was in the
guns themselves. He installed newer guns using an electromagnetic
rig that provided excellent stability. The successive loading-and-
firing procedure with improved telescopic sight measure "proved to
be a great success," but the Admiralty rejected the idea. Had they
accepted it in 1898, "the Navy would have had an efficient instruc-
tional weapon and the country would have been saved £40,000 a
year" (about £7 million today) in ammunition.

Scott also demonstrated that men often fired by keeping one
eye open and the other on the telescope. Any quick adjustments
to the sight required an extra eye. Scott's solution was to bring in
a "sight setter," another person to elevate or divert the gun per the
shifting target. Early demonstrations of this continuous-aim firing
method were successful. "The Admiralty hauled me over the coals
for the innovation," Scott said. Instead of implementing his idea,
the higher-ups reprimanded him. "It was the same with all my proposals. They were all boycotted."

On a particularly rough day at the sea, Scott spotted a trainee
with a far superior firing record than others. This sailor nimbly
adjusted his telescope and the gun-elevating gear to keep the sight
steady according to his ship's speed and rolling of the target. Scott's
entire idea of continuous-aim firing was being put into action by
the remarkable trainee. "What one man could do intuitively the
others could be taught to do," Scott wrote. He codified the trainee's

act of dexterity into a disciplined procedure for the entire corps. He also fitted a contrivance called a "dotter" to ensure the gun aligned with the bull's-eye and, if it deviated from the target, to indicate by how much.

In 1899, Scott returned to England. Call it *naval* gazing—the Admiralty sat on Scott's training concept for three more years. They demanded more data for the dotter from the frustrated captain and temporarily slashed his salary. The Admiralty transferred Scott to HMS *Terrible*, a long-range cruiser with a seemingly apt name that was commissioned to China via the Suez Canal. Scott found that the gun sights on *Terrible* were also unusable. He adapted his ideas and instruments from *Scylla*, improvising telescope crosswires by plucking long hairs from a midshipman. Now that the sighting appliances were dependable, firing speed became more crucial. It was better to have someone who hit six targets out of eight shots in one minute than someone who scored four out of four over an extended period. The firing accuracy in this new training regime increased by an order of magnitude, and the firing rate quadrupled. Even with these transformative results, the "Admiralty remained immovable," Scott lamented. "Like all authors of new ideas I was laughed at."

Sims, the American, was far less patient than his British mentor. He imported Scott's continuous-aim firing innovation to the US Asiatic Squadron by regearing the guns and crew training. He reported his results to the Bureau of Ordnance, which received Sims's claims with disbelief and promptly relegated them to a basement file cabinet. Over the next two years, Sims wrote 13 more reports, and the bureau finally decided to conduct its own tests at the Washington Navy Yard. Unfortunately, the bureau ran the tests on the banks of the Anacostia River instead of on the rolling deck of a ship. This meant that gunners did not need to change their gun gears. The physics and the results were naturally different, and the navy concluded that Sims's claims were extreme and continuous-aim firing was impossible.

In late 1901, the irritated lieutenant broke the entire chain of command and fired off a letter to President Theodore Roosevelt detailing the moribund state of the navy. The knowledgeable commander in chief, a former assistant secretary of the navy, was perturbed and immediately ordered a target practice for the North Atlantic Squadron. Annoyed by their abysmal firing record, Roosevelt elevated Sims to the head of target practice and demanded that his reports be distributed to every naval officer in service. Sims mandated gun practices for all ships, initiated firing contests between fleet ships, and continually refined the gunnery techniques. By 1905, the firing speed and accuracy in the US Navy had soared under Sims's command. The result was improved marksmanship, with the metric of hits per gun per minute in target practice increasing at least 3,000 percent in three years, a significant step in perfecting fire control. Perhaps inspired by his protégé's radical move (and success), Scott sent a letter to the Lordship in late 1904. Scott demanded an overdue upgrade on all the gun sights across the British Navy. After years of ignoring the crushing weight of facts, the Admiralty finally promoted Scott as the inspector of naval target practice.

Scott and Sims were celebrated as "gun doctors" for making the UK and US gunners the world's best. The "hard efficiency" that Scott and Sims engineered required identifiable targets and well-defined training. After years of struggle, they achieved the rapid and reliable marksmanship that helped propel the US and UK navies to superpower status. Yet, efficiency is such that it's always susceptible to being replaced by greater efficiency. The two men who bonded over a specific form of efficiency soon diverged on another. In 1906, the Royal Navy introduced HMS *Dreadnought* under Admiral Jackie Fisher, which abruptly revolutionized naval technology. The 20,000-tonner was formidable with its turbine-propelled speed; one historian said, "It could sink the whole German navy." The capital ship's superior main battery, range-finding, and fire-control capabilities outmoded the "pre-dreadnoughts," averting the need for long-range marksmanship.

In the United States, Sims was convinced that a few all-big-guns would have a decisive cost advantage over maintaining several smaller ships. "We have, indeed, no choice in the matter, if we are to remain a world power." In 1916, when Congress asked the navy to report on the "largest battleship which can be undertaken in the United States," the agency unhesitantly proposed a design for five 80,000-ton, nearly 1,000-foot ships with fifteen 18-inch guns, $50 million apiece. But with size comes sacrifice; supreme strength has bills to pay.

In the UK, after years of opposition to the *Dreadnought*-type projects, Scott reached a boiling point in late 1920. He lashed out at the Admiralty and Westminster in a series of letters to the *Times* of London. " 'What is the use of a battleship?' Why will you not tell me? I am very much annoyed with you," he wrote. He groaned that the dreadnoughts were doomed. Instead, a fleet of aircraft and submarines would make naval warfare efficient, making battleships useless "just as the motor vehicle has driven the horse from the road."

Whenever pursuing high performance, we must develop an intimate relationship with efficiency. One philosopher declared engineers to be "efficiency animals." The intent was to show how engineers nurture this alpha concept as an economic activity and cultural sensibility. One can optimize for efficiency to make our trains run on time, to pack our potato chips without powdering them, to direct-deposit our paychecks, and to generate quadrillion-times speed advantage for smartphones and stock trading. One can also optimize hard efficiency by trading off a customer's emotional connection to gain convenience in the selection, like a fast-food joint presenting only three combos. Optimization means sacrifice: higher accuracy may come with a lower range. But Scott and Sims balanced these variables by making trade-offs, typically how engineers design cars against the competing criteria of comfort, speed, style, and fuel economy.

We crave hard efficiencies just as the routine processes of cellular metabolism and maintenance use efficiency to enable our survival. But biological evolution isn't always coherent in an engineering sense and "selects" for efficiency in open-ended ways and schedules. In contrast, engineering for efficiency is consciously constrained. The obtained optimum isn't the absolute best but a relative best under local constraints. "If a nation feels that a funeral pyre should be aligned in a north-south direction to aid the dead's journey into heaven, the model to be optimized will incorporate this consideration as a design criterion irrespective of the truth of the claim," writes the scholar Billy Vaughn Koen. "In a society of cannibals, the engineer will try to design the most efficient kettle."

Hard efficiencies are objective, while soft efficiencies are subjective and a moving target, much like the shooting struggles before the continuous-aim technique. Soft efficiencies are conditional and contextual, with more nuances than numbers. We'll now visit this in the works of some 19th-century French engineers who profoundly influenced the economic theory of public welfare.

✳

IN THE EARLY 1800S, before the term "scientist" existed, engineers were the high priests of social arithmetic. In France, because engineers mixed mechanical insights with municipal idealism, one could boast: "Engineers *do* economics while others talk about it." Following the French Revolution, polytechnics emphasized an engineering education that would simultaneously design for state efficiency and social equality. These goals clashed when state engineers built roads, ports, and canals. Simply pursuing the efficiency of generating benefits from public infrastructure was one thing. It didn't necessarily mean those benefits spread equally among the constituents who paid for and used them. How can a project be efficient while maximizing its social benefits? This "soft efficiency" denoted multiple subjective perspectives and lacked precise targets. Public welfare problems hinged on "utility," a psychological factor that informed how French engineers thought about puzzling projects. Their practical insights would provide a critical breakthrough in the long-running quest for a coherent economic theory of costs, benefits, and behavior from individuals to institutions.

Let's first differentiate between technical efficiency and economic efficiency. As engineers often practice, technical efficiency can be as simple as achieving an output with input. Understanding how dependent an output is on its input can increase hard efficiencies, like improving the hit rate of guns. The technical efficiency is higher if fewer inputs are required to produce an output. We saw how Scott and Sims minimized demands on the soldier to scale up accuracy, from a ship to a squadron. With economic efficiency, the question of substituting one input or output with another often arises. A farm producing soybeans on an acre is technically efficient, but optimizing the acreage with high-yield fertilizers to reduce the production cost can make it economically efficient. Suppose the farm can produce rice in addition to soybeans, or the crops can be readily rotated with corn. The money invested and recouped becomes a yardstick for managing resources, relating cost and price, be it for profitably substituting land use with fertilizers or maximizing revenues with soybeans versus rice and corn.

Hard technical efficiency may exist within the problems of soft economic efficiency. But soft efficiencies cannot be solved like hard efficiencies; they can only be resolved. What one farmer values and what's suitable for one village may not be the case for another. These practical conditions point to satisficing: one can satisfy and suffice to achieve a resolution without gaining a solution. Economists would argue that increasing one or both efficiencies would increase the overall utility for a given input. But the struggle to understand the nebulous notions of utility had persisted since economic principles belonged to moral philosophy.

Adam Smith, the "Father of Capitalism," followed the steps of Plato, Copernicus, and John Locke and crystallized his views on utility as an essential property of commerce in 1776. He probed a commodity's value established in its use compared to its exchange. Water has much use value but little exchange value: it's beneficial everywhere with little or no price. Diamonds are the reverse, with little use and a high price. An old Indian parable captures this paradox. In the tale, a wealthy person instructed his servant to find what people would pay for his diamond. A vegetable vendor who had never seen a diamond offered 20 pounds of eggplants in exchange for the shiny rock. A mystified clothing merchant, with the idea of making an ornament out of it, proposed a few rupees. Finally, a judicious jeweler jumped to offer the maximum. In the eyes of different people, the diamond took on various incarnations. So, as Smith would realize later, a commodity's "relative utility" based on the value added is subjective. But he couldn't work out an efficiency metric for utility considerations in use and exchange.

The British philosopher Jeremy Bentham, known as the Father of Utilitarianism, writing in 1823, addressed utility as a property that produced "benefit, advantage, pleasure, good, or happiness." The French economist Jean-Baptiste Say, expanding Adam Smith's work, later linked utility to the intensity of a service's demand. It was again a matter of how people perceived its usefulness, like a jeweler appraising a diamond. And Say felt utility was "not confined to one human being, but applies to a whole class of society." Individual values could explain why some people pay for designer duds and champion cheeses. And collective value may explain why society prioritizes the development of drugs, devices, vaccines, and investments for national security and social security. But economists have often inferred utility in money, specifically in a "willingness to pay," even if it means converting complex factors like natural resources, well-being, or carbon emissions into dollars. Economic trade-offs are unlike engineering, where effects are usually known. And engineering trade-offs are different from purely monetary ones, knowing human preferences don't unthinkingly follow physical laws. In the economics of public welfare, it's not always clear who bears the costs and who receives the benefits. This problem perplexed the French polytechnic-trained engineers eager to maximize efficiencies in public infrastructure.

Cost-benefit analyses have been handy tools for approximating value since antiquity. However, the French engineers were frustrated by the unscientific approaches to measuring utility and value. They began refining them using their real-world cases of nation-building. As engineering became more mathematical in the 1820s, some technocrats felt that quantification "isolated engineers in a world of abstraction" instead of practical deeds. In one view, the aridity of numbers "killed off engineers' inventive capacity." Others saw numbers as an instrument of reason for engineering. Quantitative objectivity wasn't merely "an aid to private decisions, but a deeply public act," notes historian Theodore Porter, so decisions were "routine, almost automatic." Most people found the technical formulas credible but not necessarily comprehensible; technical analyses were beyond their competence. Numbers became "a tribute to democracy, an appeal to objectivity made necessary" to overcome the authority of position and privilege. And the large economic stakes of public projects decisively shaped this engineering transformation in France.

Claude-Louis Navier was a staunch servant of logic. He laid the engineering groundwork to analyze structures and their stability under different loads. Navier also described the motions of viscous fluids, informing studies of ocean currents, weather patterns, and aerodynamics. His name is engraved on the Eiffel Tower, along with those of 71 other men. He wrote a mathematical expression of utility so that public projects could produce greater gains. In 1830, his formula calculated the minimum threshold for canal use under which the project was a taxpayer boon or burden. He argued that projects resulting in savings from lower recurring costs were in the state's best interest.

Charles Joseph Minard, an inspector general of roads and bridges, improved on Navier's innocuous idea. While teaching a course on transportation, Minard confronted the same problem of characterizing utility. Like Navier, he pursued a financial metric for utility but zeroed in on personal elements of the benefits. He realized that income inequality was a prime problem of the public economy. Even with equal incomes, people may find greater use for a road connecting different towns than a community theater for evening entertainment.

Minard's chief advance was to consider opportunity costs explicitly. He assigned a price to the time individuals saved by taking one route over another. For instance, customers may create their own efficiencies during high traffic by taking a nearby bridge that charges twice as much as the tunnel located five miles north. As the context changes, our behavior changes, and with it comes a shift in how we experience efficiency. But ultimately, Minard focused on how a society collectively valued a project over time by studying the likes of Roman aqueducts. And he believed such assessments were strictly an engineer's job. Minard's most-remembered legacy was from his retirement years: how to present information visually. Minard drew up Napoleon's dire losses in the 1812–13 campaign against Russia. The Grande Armée was whipped and weakened by lousy weather and battle tactics. The Minard Map doubled as a flow graph, packing information about the army's course, direction, and head count in relation to time, temperature, and terrain. Although specific to a military context, this map remains a notable template for statistical graphics.

In the 1830s, Jules Dupuit's economic theory sought to avoid moral wrongs in favor of market gains, an idea that seems far more relevant today even if sometimes ethical reasoning is framed as "economically illiterate." Dupuit's mind worked like a can opener for problems vacuum-sealed with values, and he was "cold, reserved, cutting all at the same time." As a custodian of public funds, he repeatedly faced a simple question: Using what criteria should he approve a public program? A former student of Navier and later chief engineer of Paris, Dupuit's utility assessments began with maintaining roads.

Dupuit studied the impacts of road degradation on public transportation. The gravel roads were shallow and shaky and had to be swept periodically for carriage traffic. His fellow engineers argued that they could solve the problem by spreading new material over the road. Still, in a move that signaled his transition from a builder-engineer to a policy engineer, Dupuit made a case for economic efficiency. The road-maintenance problem involved frequency of use, degree of wear, choice, cost, and quality of materials, as well as the available labor to build it. Years earlier, Navier noted that investments were sensible only if their savings outweighed the costs, which would have worked for a hard efficiency problem with fixed circumstances. Customers use the roads under different conditions for different reasons. Dupuit could only summarize their motivations as an indirect proxy rather than judge them against some arbitrary standard.

Dupuit then used substitution calculus, an engineering method used to study whether new technology investments led to comparable savings in labor. He identified when a new investment turned profitable against the substitute options for road making and maintenance. Even if both costs and benefits were monetary, Dupuit knew that how people weighed the benefits of the road-maintenance program affected how they paid for it. These dynamics, typical of soft problems, aren't amenable to a simple technical fix. By understanding the central role of preferences and how they shaped policy choices, Dupuit could now walk the tightrope between market incentives and fairness.

Writing in 1844, he compared budget-planning problems to the situation of a person in pitch dark measuring the height of a wall. If the wall is at most six feet, the person can touch the top, but not if it's 50 feet. Calculus provided the ladder; and individual preferences, the lighting. But, for the edgy engineer, economics had to "come out of the 'clouds of philosophy'" and become more technical through experiments, models, and testing. The road-maintenance problem spurred him to regulate horse-drawn carriages with tolls, a pricing predicament he encountered in different forms during his career.

In 1840s Paris, Dupuit noticed that the government built centralized transportation systems to achieve political goals rather than efficiency, which called back the prickly problem of measuring utility. Dupuit invented a measure that depended on the sacrifice individuals were ready to make for the service. In other words, how much one was *willing* to pay rather than what one does pay. Utility, seen this way, determines a product's cash value. If you purchase a bottle of wine, how much are you willing to pay for a second bottle to gain extra satisfaction? This *marginal utility* proved critical for Dupuit in advancing two concepts.

The first, *consumer surplus*, is about how people experience satisfaction. You have invested in this book with your money to buy it and time to read it. You may have done so because you judged it worthwhile when many books were competing for your attention. The difference between what you invested in this book and what you gained from it, or the relative utility, is called consumer surplus. The publisher also needs to recover its costs for this book, ideally with a profit, which is producer surplus. Similarly, Dupuit's roads also needed to be worth people's investments and remain profitable for the state. Does the book enrich your perspectives, and does a road save you time, even if it's an expensive toll? Personal preferences and impressions of gains dictate those answers. So, one needs to simultaneously increase consumer and producer surplus, achievable with the second concept: *differential pricing*.

The idea of charging a different price for the same service to different people at different times and conditions is now a familiar situation, but Dupuit codified it mathematically. Cinemas charge different rates for the same movie with military, student, matinee, or senior citizen discounts. This is known as first-degree price discrimination. Producers might lower their rates on a scale for bulk purchases of office products, which is second-degree discrimination. And third-degree discrimination is based purely on a customer's ability to pay for first-class, business-class, or economy fare for trains and planes.

Dupuit placed utility maximization at the core of economic debates with consumer surplus and differential pricing. "The only real utility is that which people are willing to pay for," Dupuit declared. With France's intensive urbanization in the mid-19th century, Dupuit applied differential pricing for water use, a commodity with utility and scarcity. French engineers used preferential costs and benefits to plan railroads and tariffs for freight and passengers. These projects demonstrated that value and values were inseparable, and sound public policy depended on both.

Dupuit's findings are not collected in hefty volumes like those of the traditional thinkers in economics before and after him. But his work had widespread implications on the nature of "institutions, technology, and human nature," writes scholar Robert Ekelund. Dupuit's concepts informed the development of the "theory of the firm," in which companies focus on maximizing profits, and the "theory of consumer behavior," which focuses on how people spend their money, both informed by soft efficiency. Dupuit is now a distant star in economic history. Still, his insights as an engineer were at the frontier of cents and sensibility. His work wasn't about absolute financial freedom or market power but the constraints and motivations that create them.

If economic trade-offs are inevitable in setting policy priorities, so are the notions of privilege. Whose hunger matters more? How does one life compare to another? Such questions balance messy efficiencies and critical cultural choices. Making decisions based solely on cost-benefit efficiencies is akin to deciding on a home based on dollars over square feet without considering neighborhood safety and other necessities. Similarly, a cost-benefit calculus may suggest that dictatorship is more "efficient" than democracy and that lifestyle drugs have better incentives than neglected lifesaving vaccines. Indeed, most forms of cost-benefit calculations have ignored productive and unquantifiable considerations. Economists once debated the value of unpaid labor like housework and left it out of the national income statistics. Then for years, it was pointed out that "if a man married his maid, GDP would fall." We'll see next with "messy efficiency" that without wisdom, even sharp ideas such as cost-benefit can lead to wrong decisions. That's the risk of privileging numbers without realizing that they, like words, can have multiple contested meanings.

Scott and Sims achieved epic technical efficiency in naval cannons because their shooting targets were clear and straightforward, even in shaky circumstances. Dupuit's economic efficiencies comprised organizing public preferences based on what the "market can bear." But with messy efficiencies, even the concept of what a market is, whom it serves, and what it can bear is often morally dubious. Hence, messy problems and their efficiencies can't be solved by optimizing or resolved through satisficing. They can only be dissolved by idealizing a desired state of improvement. "As with all sciences," Dupuit wrote, "it is necessary to pass from axioms to elementary principles, and then gradually progress to the most complicated consequences."

✳

NOBEL-WINNING POLYMATH ALBERT SCHWEITZER once said pain is "a more terrible lord" than death. An economist may refer to diminishing health as a drop in "utility," but that should prompt us to think of scholar James Hartle's metaphor. Suppose one drops a cat, a cannonball, and an economics textbook from the same height. They hit the ground at the same acceleration due to gravity. These phenomena make sense from the laws of physics. Still, those insights reveal nothing about the qualities of cats, cannonballs, or economics. The kinds of efficiencies displayed in hard and soft scenarios don't necessarily make sense for messy conditions and may even create them. The politics of health provides a test case for messy efficiencies with its unmatched disputes because there's much—life itself—at stake.

As life spans have productively increased, so have discussions about increasing fairness and justice in health care. The demands to decrease disparities in access to care are also rising. People can now live longer with medicines for their morbidities. This new biology has altered the sociology of caregiving, for those in need and those able to provide. Health is also integral to an individual's identity and can foster a community's identity. Consider how the HIV/AIDS crisis of the 1980s spawned a global movement of committed activists whose work helped make clinical trials more inclusive. Now we have thousands of disease-specific lobbying groups to boost awareness and funding. While life-and-death discussions occur in labor, environmental, and defense policies, in health, they occupy center stage, reminds scholar Daniel Carpenter. "Health and illness shape who we are politically."

Let's look at some messy problems of health financing. Cost-effectiveness has been a preferred yardstick to consider the economic efficiency of medical interventions. It's a ratio of cost over health-adjusted life years (HALYs); the smaller this ratio, the better the gains per dollar spent. Two commonly used HALYs summarize the disease burden and compare how effective an intervention is in a population: quality-adjusted life years (QALYs) and disability-adjusted life years (DALYs).

QALYs are general proxies for the years lived and how much good health an individual can likely accrue from an intervention. They aren't linked to specific diseases but to how people value their health and how a community perceives a condition. People are expected to express what they are willing to trade off in terms of time or risk of death to regain perfect health, such as freedom from discomfort, pain, depression, or anxiety. The opinions are scored as a binary in this "feelings thermometer," with a score of one denoting perfect health and zero, death. When combined with costs, QALYs point to an economic utility that guides investments. If the QALYs are about healthy years gained, then DALYs are about healthy years lost. DALYs score the diminished life expectancy and quality of life also on a scale of zero and one, but inversely, where zero is perfect health, and one is death. Unlike QALYs, defined by preference-based weights from surveys, the disability weights of DALYs are determined by expert opinion. DALYs are standardized for studying different disease burdens worldwide, as in comparing type 2 diabetes in Sri Lanka to cancers in Spain or tuberculosis in Indonesia and India.

The dollars-over-QALYs gained, or the dollars-over-DALYs reduced, have become standard measures of cost-effectiveness. These efficiency ratios compare a new intervention to an older or no treatment, specifically by comparing the dollars saved for the health benefits produced. If an intervention is more costly and less effective, it's rejected. If it's more effective and less expensive, it's accepted. Suppose the intervention is more effective and costly; cost-effectiveness would be guided by a threshold of desired dollars-over-health benefits set by the government to finance it. While a health ministry, a finance ministry, and a commerce ministry within the same government may have different emphases, cost-effectiveness is considered a common standard for competing concerns.

However, ethics significantly complicate the Dupuit doctrine of economic efficiency. Older adults with lower HALYs may be viewed as a "bad investment" for those who pay for their care, just as people with disabilities who have less likelihood of recovery or who are more socially disadvantaged are discriminated against. Because HALYs combine health states across different communities between the spectrum of perfect health and death, the resulting efficiencies may not differentiate a life-enhancing outcome from a lifesaving one. Plus, QALYs use discounting to compare the costs and outcomes at different times, the idea being that people prefer to gain health now rather than sometime in the future. A healthy person is expected to discount the future differently than someone requiring health assistance. Finally, pure economic efficiency of deciding who is "worthy" can sideline equity considerations.

Those who believe that "a QALY is a QALY is a QALY" might equate economic efficiency with social efficiency. But does that mean that all QALYs are of equal social value, and who is to judge?

Does it mean that only younger and more productive individuals create a productive and prosperous society? And is efficiency all about what people might be "willing to pay" for their health? Pursuing cost-effectiveness without considering our values can create a condition called "rigor distortis," in which outcomes may be precisely efficient but generally catastrophic. Messy health and climate problems cannot be optimized or satisfied but can only be idealized as individuals transform a problem for the greater collective good. With many things in life, an addition could mean a benefit. However, that's not the case with the economic theory of utility maximization. In the following two examples, we'll see how messy efficiencies can create startling side effects.

The first unforeseen complication relates to the pricing of prescription drugs. Fair pricing has been a matter of morality since ancient Greek philosophy. The 13th-century theologian and "Doctor of the Church" Thomas Aquinas condemned usury practices when lenders earned unfair money with exorbitant loan interests. Aquinas called price gouging on precious materials and services after a disaster fraud. "No one ought to sell something that doesn't belong to him," he argued in *Summa theologica*. Some egregious examples in prescription drugs include a company more than quadrupling the price from $160 for an epinephrine injector to treat potentially fatal allergic reactions. One firm boosted the price of a nonpatented drug in a small market for treating severe infections from $13.50 to $750 per tablet. Another firm launched treatment for a neuromuscular disease at $750,000 for the first year's dosage, renewing at half that price for subsequent years. Such extraordinary amounts have rightly generated public outrage.

Some view the cost-effectiveness-based approach as an invitation to raise prices that are currently low relative to the medication's "perceived value." One can, after all, loosely interpret these "gold-standard" ratios. A government may reasonably want the most cost-effective outcome for its investments on the public's behalf. At the same time, patients and their families might wish for the most effective outcome that could lengthen their life span, no matter the cost. An insurer paying for the treatment may demand the least costly option, irrespective of the benefit. Thus, a value mania prevails in health policy, with contradicting notions about what value means, to whom, and why.

The US biopharmaceutical markets are unique. Drug pricing is a function of both health insurance and the legal rights of patent holders. However, unlike in other nations, when a patented product enters the US market, the buyers are often shielded from prices through the insurance industry. All other industrialized countries face a similar issue with their laws that grant patent rights for their products. But only the United States has no price controls to counteract the price pressure drug insurance creates, which doesn't exactly square with the idea of the government wanting to use public money efficiently. Thus, unsurprisingly, US residents pay more for branded prescription drugs than residents of other nations. Hence, two hard policy solutions effective on their own—health insurance and patent law—can collide to create a messy arms race.

To return to Dupuit's domain, consider a railroad serving thousands of businesses. Customers may differ in their willingness to pay for those services based on the available competition. A metals factory can choose from the pool of freight-rail and barge operators to transport its product from origin to destination. Comparatively, a farmer in the countryside with no trucking or barge options may rely on the railroad. The freight company would naturally wield greater power over the farmer than over the metals factory. Differential pricing, as Dupuit's work showed, can help the railroad recoup its investments, reinvest in maintenance, and ensure returns to its investors. But a pure efficiency approach thwarts all the best intentions if one misses the context sensitivities of health. We'll see this now in prioritizing new and improved vaccines.

Which vaccine candidate should get more attention and funding, and which less, and why? An analysis based strictly on cost-effectiveness would conclude that spending significant money to develop an Ebola vaccine would make little sense. Ebola might seem like a minor threat to people living in most of the world, even across Africa, the continent with the highest number of cases. Ebola-related deaths are far fewer than those from malaria, influenza, or tobacco use. Still, Ebola has dominated headlines due to the fear it evokes. The rare but severe disease has gruesome effects, such as uncontrolled bleeding and high death rates. So, people might more likely prioritize developing an Ebola vaccine.

In the 2014–16 epidemic, about 29,000 individuals caught Ebola. Nearly 12,000 people died in Guinea, Liberia, and Sierra Leone, the nations with the broadest transmission, 8 in Nigeria, 6 in Mali, and 1 in the United States. The accompanying panic extensively transmitted on news and social media shot Ebola to the top of global health priorities. But the vaccine candidates for Ebola developed years earlier had been shelved for cost-effectiveness. The dollars-over-QALYs-and-DALYs approach can bypass a plethora of interacting issues: contact tracing; public dread; hygiene; herd immunity; disproportionate impacts on low-income populations; disparities and inequities in health outcomes; new engineering platforms; storage and supply systems; losses in productivity; diplomatic and foreign policy measures; the evolving understanding of the disease, including the possibility of eradication, effective prevention alternatives, quarantines, lockdowns, and travel bans; and potential side effects of drugs and vaccines. Even with the "pure, liquid hope" of immunization, these complexities cannot be distilled into a clean ratio.

Unlike hard and soft efficiencies, messy efficiencies require a multicriteria approach. Multicriteria models allow one to consider various factors that go into a decision, featuring those not included in cost-effectiveness analysis—for example, traditional funeral rites and the struggle to curtail the spread of Ebola. In Guinea and Sierra Leone, in particular, the traditional custom of touching, bathing, and kissing those who died became a messy problem during the Ebola pandemic. Governments neglected to consider disease transmission through cultural practices. Thus, these burial rituals were an undetected source of spread that impeded early disease detection and containment. The World Health Organization noted that about 60 percent of the transmission in Guinea, and a higher percentage in Sierra Leone, was attributed to burial rituals in the first six months of the epidemic.

Communities resisted the blanket policy for mandated cremations due to distrust of politicians and public health officials who dismissed their practices as "backward." Sometimes, people performed secret burials to avoid contact tracing and disposal methods that they found offensive. In this case, the same disease had different impacts in the more religiously observant West Africa than in equatorial Africa. Relatives of the deceased also developed new customs for burial that were sensitive to their faith and attentive to the public health protocol: they used banana tree trunks to substitute for the corpses. They bathed, dressed, and hugged the symbolic trunks before burial. Multicriteria approaches can enable communities to include the most relevant dimensions and build the trust unachievable with hard cost-effectiveness ratios. In a multicriteria model, one can adjust the weight of attributes a community deems vital to create more dutiful efficiency measures.

Writing in 1956, two decades before Ebola was identified, scholar V. F. Ridgway cautioned that performance measurements one often sees with efficiency could have logical and lethal consequences, and they required judicious use. He knew that a cure could sometimes be worse than the disease.

✳

ADMIRAL SIMS WAS A lifelong iconoclast. He rocked the boat by jingling silver on dinner tables at formal events. Despite his contribution to the US Navy's modernization, the restless reformer refused institutional honors and wore no ribbons or medals. The biggest bangs for the buck he and Scott boosted were an emblem of rationality; the hard efficiencies standardized the superiority of

American and British ships. Now try that template with soft and messy problems. The shapes and shadows of efficiency will shift spryly, creating value as much in the frontal lobe as on the factory floor. These qualities become part of what's colloquially called "the effing debate," the tendency to prioritize efficiency over efficacy at all costs.

Society accepts efficiency for what it produces without knowing what it is. Efficiency has been called the chief "imperative of engineering," a "constitutive value," and a "holy grail." Yet, it's also about gaining power and exploiting resources. A 1628 book, *The Art of Logic*, assigned divine status in describing God, who joins form into matter, as the "efficient cause of man." Efficiency is time, money, speed, strength, productivity, behavior, misbehavior, willingness to pay, unwillingness to budge, market shares and what the markets won't share, what the markets can bear and what the markets *should* bear, gross domestic product and gross moral failure, all at once.

Efficiency isn't about the greatest benefit for the cost; "it means the greatest measurable benefit for the measurable cost," wrote scholar Henry Mintzberg. "In other words, efficiency means demonstrated efficiency, proven efficiency, above all, calculated efficiency. A management obsessed with efficiency is a management obsessed with measurement. The cult of efficiency is the cult of calculation. And therein lies the problem." With its many forms, efficiency itself can be an elusive idea. An "efficient restaurant" could mean fast service and so-so food. By analogy, an efficient society's promarket orientation can be antisocial.

Maximizing utility is one thing, but how about maximizing efficiency as the dominant value? The case of energy efficiency is worth exploring; whether it's pure good or something to aim for actively remains an open debate. Energy efficiency isn't merely a technical specification but is also a value statement in the world of wicked climate problems. Almost without exception, energy efficiency functions as a hard ratio assuming that its market motivations are separable from social equity. Miles per gallon, watts per square foot, lumens per watt, and kilowatt-hours per square meter can convey useful information without doing anything with justice and equity. "Energy efficiency" is more meaningful when discussing energy vagueness, vulnerabilities, safety, maintenance, and resilience.

Where do hard efficiencies end and soft efficiencies begin? When does messy turn into wicked? Scholar Byron Newberry has drawn upon the biological example of "ring species" as a helpful metaphor for engineering identity, which is also useful for the changing nature of efficiency. It relates to the genus of salamanders called *Ensatina* that encircle California's Central Valley, between the western Coast Ranges and the eastern Sierra Nevada. The 40-to-60-mile-long valley is hostile to the amphibians. The *Ensatina* don't move across the valley but around it, where the salamanders can interbreed with neighbors and maintain gene flow in the ring. However, in a zone of overlap on the mountains south of the valley are two distinct species of salamanders that don't interbreed. One is spotless brown, from the valley's west side; the other, with yellow and black blotches, is from the east. The plain and spotted salamanders are distinct enough to pronounce them two separate species, but that's wrong. In their full evolutionary context, it's difficult to tell where one salamander type ends and the other begins, slowly turning into each other. As with *Ensatina*, Newberry notes, engineers, too, cross the line "from being an engineer to being something else." Navier, Minard, and Dupuit were such kinds of engineers. We'll see many more in this book who have become something other than the traditional types of Scott and Sims. Similarly, the kinds of efficiencies they sought were also on a ring where one could see hard and soft problems as distinct but part of messy and, ultimately, wicked problems.

In wicked systems, not everything can be pursued as an efficiency measure, nor should it be. It's undoubtedly unpopular to seek inefficiency. But look at the flimsy functions of quick quantification, super specialization, and comfort capitalism. Efficiency can produce vast uselessness and costly consequences. In 2020, during the earlier stages of Covid-19, one analysis looked at a midsize American manufacturer making masks. The company was able to source melt-blown polymer, the main ingredient for the mask, more quickly than the elastic strings for the ear loops. The other available option was to use a spiral variety. Still, the elastic form was incompatible with the assembling device. The company had to manually unspool the coil, slowing the production process at a time of urgent need. The efficiency of an elastic-manufacturing system itself needed to be more elastic.

Global supply networks cheaply and successfully manufactured vaccines in India, medical-grade gloves in Malaysia, and nasal swabs in Italy. As with the mask maker, steady productions became fragile and incompatible when firms couldn't reproduce and repurpose their competencies during atypical times. Usually planned and queued for efficiency, intensive-care beds became unavailable when most needed. Moreover, companies were restless to order their employees back to "in-person" work during the prime of the pandemic. Their business case centered on the perceived need for efficiency from physical collaboration and the return to a previous state. When policies can't thread the needle between efficiency and resilience, they create new vulnerabilities. We can see how the notion of "efficient" risk dispersal across financial markets triggered the 2008 global economic meltdown. And so on—in the name of the supposed good of efficiency, rigorous intents have produced calamities.

If human ingenuity can harvest such industrial efficiency, it can also engineer more inclusive and morally necessary efficiency. This is only possible when efficiency is combined with other concepts of operation, such as vagueness, vulnerability, safety, maintenance, and resilience, the subjects of further chapters. That's why, to make the essential considerations of a wicked problem apparent, efficiency should be a departure point, not the destination.

Refrain

Pitch

I N AUGUST 1927, LIEUTENANT NORMAN GODDARD WAS
hell-bent for Hawaii. So much so that he painted the letters
H.B.H. on the rudder of *El Encanto*, the glistening, all-metal
monoplane favored to win the open-ocean race from California to
Hawaii.

Earlier that May, Charles "Slim" Lindbergh had soloed his
450-gallon, single-engine *Spirit of St. Louis* nonstop for 33½ hours
from New York to Paris. Until then an unknown airmail pilot,
Lindbergh's fame soared as he clinched the $25,000 purse put up
by hotelier Raymond Orteig. James Dole, a charismatic Harvard-
educated farmer turned pineapple magnate, saw an opportunity
here. By hosting a flight contest of his own, Dole wanted to produce the Pacific Lindbergh. The first prize for the nonstop feat was
$25,000; the second, $10,000.

Soon after the contest announcement, two army pilots pioneered
the transpacific flight in a plywood-skinned Fokker Trimotor, *Bird
of Paradise*. Favoring fuel more than food in payload, on June 28,
1927, Lester Maitland and Albert Hegenberger winged their way
from the San Francisco Bay into the broad blue expanse. After
25 hours and 50 minutes, they landed in paradise. However, the
civilian glory to end Hawaii's isolation was still up for grabs. The
San Francisco Examiner called the 2,400-mile contest the "greatest air race" in world history, comparing its life-and-death spectacle to Roman gladiators entering the arena. Fifteen competitors

signed up. But tragic crashes the week before the derby shrank the final race list to eight contenders. A ninth entrant, *Spirit of Peoria*, remained a possible contestant. Its pilot planned to carry a caged owl christened Colonel Pineapple to stare at him and his navigator to keep them awake. But the biplane was disqualified for inadequate fuel capacity.

On Tuesday, August 16, Dole's race day, Goddard and his navigator, Kenneth Hawkins, arrived in suit-and-tie, calf socks over pants, to set off to Honolulu from Oakland. They tested their radio adapted from high-end navy equipment, the detachable landing gear, and the inflatable rubber wing float. "When we get over to Honolulu we're going to reserve rooms for the rest of the racers," Hawkins said. "After that I'm going to rent a surf board and an outrigger canoe and practice up so I can show the boys a few tricks when they arrive." After days of preparation and photo ops, *El Encanto* was ready, with Wright Whirlwind engine, salmon-like fuselage, and three gas tanks, topped at 350 gallons.

The daredevil racers and nearly 100,000 enthusiasts who took the day off waited for a fog bank to clear at the Oakland field. Spectators showed up with their picnic chairs, snacks, and ukeleles. Those in San Francisco thronged every hill and high ground with binoculars. Around noon, the starter waved the checkered flag—a sharp silence set in.

First up, blue-and-yellow *Oklahoma* thundered down the runway and lifted off to a loud shout. Two minutes later, the crowd cheered anew as Goddard throttled *El Encanto*. The silver machine roared toward a 120-mile-per-hour takeoff but clumsily swerved left and right. The plane's wheels caved and toppled in smoke and dust, skidding Goddard's hopes and dreams 4,800 feet from the start line. Sirens wailed, and fire trucks rushed to the wreck; the announcers were relieved that the gas tanks didn't blow up. Goddard and Hawkins scrambled out, swearing at their plane. "I would rather have crashed in midocean than to have had this happen," Hawkins said.

Next up, *Pabco Pacific Flyer* rumbled, rattling into the nearby marshland, 7,000 feet away. The pilot and navigator, Livingston "Lone Eagle" Irving, crawled out in rage; a tractor towed his shining orange plane back to the field. Then, at 12:30 p.m., the sleek, cigar-shaped Lockheed Vega *Golden Eagle* whizzed off for the journey west as Goddard and Hawkins kept bickering over whose mistake it was. A minute later, the red-white-and-blue biplane *Miss Doran*, flown by a circus pilot accompanied by the aircraft's namesake, a 22-year-old Michigan schoolteacher, barreled down the runway. It barely rose, and circled back. The remaining three contestants—a cadmium-colored *Aloha* piloted by Martin Jensen, the yellow-blue *Woolaroc* with the Hollywood stunt flier Art Goebel, and the silver-green *Dallas Spirit* flown by the combat veteran Bill Erwin—cleared the runway.

But within minutes, *Dallas Spirit* returned with its fabric fuselage ripped loose. Then, the third plane to return, *Oklahoma*, hovered low over the crowd with a trail of dark smoke, strangled by a malfunctioned motor. *Miss Doran* sprinted off swiftly on its second attempt. "There it is; there's the Pacific Ocean. We're on our way for sure this time," said Mildred Doran, the young teacher on board. *Pabco* was ready for another try. It bounced uncontrollably for a slow climb, toppling over its wing. "Let's go home, you've had enough for one day," said Irving's wife. The crowd started to drift away, betting on the four planes that took off in the grand contest, leaving the rest behind, disabled and damaged.

Woolaroc steadily hummed dots and dashes after a day of flying. Art Goebel grew worried that he should have seen the islands by now. He was almost deaf with the continuous engine noise. His navigator, William Davis, assured him they were on course and Hawaii was within three hours. Goebel panicked, as the remaining fuel would last only two hours. Water landing seemed inevitable. But Davis boosted his radio range to develop a better fix and double-checked his compass. Soon they spotted a dim shore in the distance and erupted in joy. "Here we are," Goebel said. "By gosh, my job's done." Goebel and Davis started shooting pistols and dropping smoke bombs above Koko Crater and Diamond Head. *Woolaroc* reached Honolulu in 26 hours and 17 minutes, winning the first prize from the Pineapple King.

Two hours behind, Martin Jensen and his navigator, Paul Schluter, couldn't avoid the clouds. They flew *Aloha* a hundred feet above the water for many hours, eventually spotting Oahu. "For God's sake don't do any stunts," Schluter told Jensen. "We are lucky to be here." *Aloha* landed second as crowds screamed louder. Hours later, Dole grew agitated waiting for the other two planes yet to arrive: *Miss Doran* and *Golden Eagle*. And they would never come. Three days later, the mended *Dallas Spirit* flew a rescue mission with a salvaged radio from the wrecked *Pabco Pacific Flyer* to search for the two missing planes. "I believe with my whole heart that we will make it. . . . We will win because Dallas Spirit always wins," the pilot Bill Erwin wrote in his departure note. "It is the last and most wonderful adventure of life." The *Dallas Spirit*, too, disappeared. One poet wrote:

> The days crawl by—no hopeful words
> Are wafted back of the lost Dole-birds,
> Tell us, oh Gods of the Air and Sea,
> Where are the seven who soared so free
> Into the West?

The Dole Derby was a Pyrrhic pursuit, with two victors and 10 victims. Newspapers declared it "aviation asininity" and an "orgy of reckless sacrifice." Even *Aloha*'s pilot, Jensen, said, "The expenses turned out to be a lot higher than the winnings." Success in this mission required velocity and vision, reliability and recovery, fuel and faith. But the race fundamentally exemplified the danger of blind flight, in which pilots flew through treacherous conditions without the know-how to navigate clouds, fog, and poor visibility. These fliers needed the skills to interpret and trust the instruments that enabled them to gauge their relationship with the horizon. They often stalled, spiraled, or spun out when disoriented and dizzy. *Aloha*'s navigator, Schluter, vowed never to fly to Hawaii again. The winning Art Goebel, who was the first to sign up for the Dole race and also dreamed of spiraling around the Eiffel Tower and the Brooklyn Bridge, gave up stunt flying. Jensen was on a job with a movie studio to fly its mascot, a lion, across the country when he crashed his plane in poor visibility over the Arizona wilderness. Lost in the desert with a 400-pound cat, Jensen trekked for three days until he got to a ranch to call the studio, which was more interested in the lion.

"America has found her wings," Lindbergh, who declined to participate in the Dole race, wrote, "but she has yet to learn to use them." It took a few more years—and far more lives—before pilots had access to technology that could teach them flight skills without risking a crash. This technology would eventually take shape in Binghamton, a city in upstate New York with an unexpected legacy of engineering.

THE BINGHAMTON REGION WAS a vital trading post in the 17th and 18th centuries. During the American Revolution, the land belonged to the Haudenosaunee, whose name means to "form a cabin," connoting the strength and loyalty in the alliance of six nations of the Iroquois Confederacy. The crooked Susquehanna River and a calmer Chenango, blending in Binghamton, brought commerce, travel, and trade.

In 1786, a British barrister turned real estate mogul, William Bingham from Philadelphia, co-purchased about 33,000 acres of land at 12½ cents each. The owners divided the lot in 1790, and Bingham's 15,000 acres included today's Broome County. Ten years later, having never visited the area, Bingham, primarily working through an astute agent, Joshua Whitney, sold the lots for an extraordinary 10 to 15 dollars an acre.

In 1817, the engineering of the Erie Canal began. To some, the nearly $8 million, 363-mile project was the "Eighth Wonder of the World," and to others, it was the "Big Ditch." The towpath from Binghamton, a nearly 100-mile-long reservoir-fed waterway known as the Chenango Canal, opened in 1837. It had been hand-dug through mires and marshes by hundreds of migrant laborers.

In the years that followed, the lush rural economy became more extractive. Plow shops took root, and logging and rafting operations sprang up. Dairy farms swelled. The "iron horse" of the railroads led to fresher produce and faster businesses. Factory furnaces were commonplace in the 1850s. In the 1870s, some firefighters christened Binghamton "the Parlor City" as a homage to its hospitality that far surpassed their hometown, Scranton, Pennsylvania. Years later, a popular song celebrated Binghamton as "clean, fair, and grand."

✈

IN 1898, THE FIRST coin- and roll-operated piano was launched in the United States by Peerless Piano Player Company of St. Johnsville, a small town a hundred miles northeast of Binghamton. Their products featured controls for automatic roll rewinding at the tune's end. Such machines played masterpieces on demand—one could loop Schubert's *Unfinished* Symphony no. 8 in B Minor, Chopin's *Heroic* Polonaise in A-flat Major, and Dvořák's *New World* Symphony in E Minor. But then came a central ambiguity, in the words of an attendee at a player piano concert: "Should one applaud? For nobody is there. It is only a machine."

Between 1910 and 1925, about 85 percent of pianos made in the United States were player pianos, or pianos that played themselves. Their popularity in the early Jazz Age created a lucrative industry rivaling real estate and automobile businesses. The pianos brought forth "every touch in technique, every subtlety of expression and tone color; in fact, you hear the actual playing of a master musician," one company claimed. In 1911, one Niagara Falls company advertised a console with a three-hour, four-course dinner selection. The opening course was accompanied by the harmonies of Richard Wagner, the second course with the tunes of Ignace Leybach, and the third with Albert Gumble and Giuseppe Verdi, and the dessert concluded with ragtime. The player pianos were pneumatic virtuosos, powered primarily by suction. The basic design came from Alexandre-François Debain, a French instrument maker who honed the harmonium in the 1840s, whose portable versions, common in the Indian subcontinent, require hand pumping. When pressed, the player piano's foot pedals controlled airflow into the bellows and through it to the machinery of tubes, gaskets, hoses, inside valves, outside valves, pallet valves, flap valves, slide valves, and crankshafts. The air ran through the pipes and produced the piano's tones.

The music was "programmed" on holes spaced across a spool of rolled paper that turned in one direction at the desired speed and, in turn, defined the tempo. The pianist's pedaling accented the melody. After 1908, player pianos used standardized music rolls better suited for complex classical scores. The 88-note rolls had 88 holes spaced at 9 holes per inch on a standard 11¼-inch paper width and ran up to 100 feet in length. The application of perforated holes for music notation came from textile manufacturing, courtesy of French weaver Joseph Marie Jacquard, who, in 1801, engineered a loom attachment to automate patterning. The "Jacquard loom" followed earlier innovations by the machinist Jacques de Vaucanson, who automated silk rolling using pierced metal cylinders and launched the lathe. This momentous machine spun off the Industrial Revolution, and the same concept later revolutionized punched cards for computing.

Now, with player pianos, picture a conveyor belt or, loosely, a cassette tape. The information stored on the roll was read as the paper (instead of a magnetic film) moved over a tracker bar with holes arranged in a row perpendicular to the roll. The opening activated the pneumatic apparatus when the tracker and paper holes aligned. Compressed air from the bellows traveled through the pipes and valves to raise or lower the piano keys. Some units, called orchestrions, came with even more elaborate arrangements. They were attached to a complete wind orchestra, mandolins, organ pipes, traps, timpani, and triangles to ring out a medley of classical and popular tunes. And some player pianos spontaneously played when one deposited a coin.

Some argued that "programmed" music machines alienated artists from the stage and from themselves. "The nightingale's song is delightful because the nightingale herself gives it forth," one critic noted, sternly resisting "the menace of mechanical music" in 1906. "Music teaches all that is beautiful in this world. Let us not hamper it with a machine that tells the story day by day, without variation, without soul, barren of the joy, the passion, the ardor that is the inheritance of man alone." But mechanization also brought benefits: "Automatic instruments were always ready to play, day and night. They never went on strike, never were late, did not take breaks, and never came to work drunk." Competition led to a manic mechanical music market and an industry in flux, frenzy, and fight over design tweaks, which scholar Brian Dolan characterized as "chronic mechanitis."

One Binghamton-based firm, the Automatic Musical Company, purchased high-end pianos from the Schaff Brothers Piano Company of Huntington, Indiana, and added wooden casing and glass frames to Schaff's products. The Automatic Musical Company improved on the self-playing mechanism by using "endless rolls" with superior rewinding and replay features. However, expenses for a bitter patent dispute with a competitor edged the company toward bankruptcy in 1913. So, George T. Link at Schaff Brothers dispatched his son Edwin from Indiana to Binghamton to take over and operate the Automatic Musical Company. Edwin Link Sr. eventually purchased Automatic and incorporated it as Link Piano Company in 1916.

At its high point during the "golden age" of automatic instruments, Link Piano Company had 125 employees. The firm focused on expanding the reach of coin-operated pianos and orchestrions, competing with Peerless and the "mighty Wurlitzers." They also made repeating xylophones and some units without keyboards. The endless rolls in the Link piano units—which now lasted up to 15 tune lengths—provided greater variety for silent pictures: songs to accompany romance and fighting, dancing and pathos, humor and tragedy. Their versatile units were "distinctly a delight" in producing special effects, from screaming spirits to galloping horses. They eliminated technical annoyances and ensured constant music speed as harps and flutes swiftly harmonized with the push of a button.

The Link Company's pianos weren't a success for just theaters. Their nickelodeons, or nickel-operated pianos, drove traffic—and business—at soda fountains, ice-cream parlors, corner drug stores, skating rinks, ballrooms, and cabarets. High-end clubs and restaurants sought quarter and half-dollar slots for the affluent clientele. By 1925, the Link Company produced about 300 nickelodeons, 12 pipe organs, and many music rolls each year. The firm even launched its first automatic theater organ. With three keyboards and 11 ranks of pipes, the Link C Sharp Minor organ's mechanical operations were virtually noise-free, even with its electrical motors and generators.

One leading New York organist, fittingly named C. Sharpe Minor, wrote in a 1925 letter that the instrument was the "most pleasant surprise. The quality of the tone was beautiful, and the response of the action superb. I had never known before such instantaneous response from any pipe organ." The Link organ, "tonally and mechanically, was far ahead of any organ on the market today," he observed. Installations of this "architectural symphony" with elaborate jewelry and velvet curtains spread nationwide, from Elk lodges to opulent residences and mausoleums. In 1927, the organ became the centerpiece of Binghamton's majestic Capital Theater, well known for its variety shows. However, the silent-era popularity of Link theater pianos and organs was short lived. With the rise of the "talkies" and the economic downturn of the Great Depression, player pianos and organs were relegated to cultural obscurity.

But the technologies within the Link theater pianos would not remain backstage for long. Ed Link's inventiveness soon found them an unexpected second life in aviation.

HISTORIAN JOSEPH CORN DESCRIBED America's affection for aviation in the early 20th century as the "winged gospel." For some Americans, an airplane was a mechanical god, and their pilots, divine heroes. Aviation symbolized the promise of a peaceful future in the heavens. Like traditional faiths, aviation imbued people's lives with meaning. And like secular social visions, aviation became a technological reform movement. The resulting "air-mindedness" evoked utopian images of prosperity where each person could one day own an airplane. In Europe, Pablo Picasso oil-painted *Our Future Is in the Air* on canvas for a 1912 pamphlet promoting aviation. Picasso and Georges Braque compared themselves to the Wright brothers since their cubist creations were similarly trailblazing. The same year, the French dramatist Romain Rolland wrote: "Take possession of the air, submit the elements, penetrate the last redoubts of nature, make space retreat, make death retreat."

Despite the dangers of aviation, many pilots contributed to the romanticized notions of flight. In a 1924 issue of *Cosmopolitan*, the one-time pilot Winston Churchill, who had survived a take-off crash, wrote that "air is an extremely dangerous, jealous, and exacting mistress. Once under the spell, most lovers are faithful to the end, which is not always old age." Even the safety features of 1920s aviation seemed more like fashion for a debonair bachelor than protection. "Goggles, gloves, a heavy leather coat, and a cork-and-leather helmet were de rigueur," writes scholar Peter Pigott, and the Wright brothers dressed in their Sunday best to fly.

But, as the Dole Derby and other disasters demonstrated, flying was highly dangerous. Pilots faced enormous demands with maintaining complex coordination between what they saw clearly when they could and how they felt about it through their internal sensations of ears, nerves, and muscles. Before radio and other navigation aids, airmail pilots relied on tracking what was visible on the ground. In this "contact flying approach," fliers focused on a familiar landmark. They flew toward it, then fixed their sights on another landmark to track, repeating the process until they reached their destination. The transcontinental air system grew from the Union Pacific Railroad route to San Francisco by siting 15 landing fields, spaced 200 miles apart. Decades before satellite navigation, airmail pilots used Rand McNally road maps they could pick up at gas stations or the "iron compass" of the railroads, water towers with painted town names, and lonely ranch houses.

Pilots memorized the heights of hills, steeples, water tanks, and telephone poles so they didn't crash into them. And if pilots flew long distances, it was often at low altitudes, about 50 to 300 feet from the ground, always ready for emergency landings. Occasionally, aircraft got snarled in power lines, as was the case with a Canadian pilot in 1918 who was suspended above a store building for hours. One pilot said he got around "with a few drops of homing pigeon" in his veins. When pilots needed timely wind data over the farms, they looked to the cows below as a proxy since the animals always grazed with the wind behind them. If the bovine compass failed, they looked down for direction hints from rural outhouses, which always faced south. While this method may have worked in local landscapes, contact flying over enemy territories was extremely unsafe. The challenges of wartime flight demanded better pilot skills and higher flight.

In 1920, the US Army began a survey of "sky roadways" to improve airmail service. They mapped 44,000 miles with data from 90 flights. This effort resulted in a comprehensive listing of relevant landmarks, rooftop markings, and possible landing fields and patches, where excited crowds sometimes interfered with aircraft landing. A 1921 follow-up survey provided a log of coast-to-coast distances, landmarks, and flying directions for a transcontinental air network. The routes stretched from New York City to Bellefonte, Pennsylvania, then to Cleveland, then to Chicago, then to Omaha, and the longest leg to Cheyenne, then to Salt Lake City, then to Reno, and finally the shortest leg to San Francisco. Pilots preempted modern-day voice-guided navigation by writing down almost turn-by-turn directions every few miles. Their notes pointed out lakes, trees, railroads, telephone cables, and hangars.

One popular survey book, *Landing Field Guide and Pilot's Log Book*, often contained colorful commentary: "Following Michigan Central double track railway . . . Good fields but country is getting more rolling." "Plenty of fields large enough but one would have to be very careful to avoid granite boulders which are there in large numbers and sizes." "Absolute wilderness . . . Still flying compass course. Have not seen landmark house, field, town or sign of human life for one hour and ten minute until just now."

The book also contained helpful hints. "Don't fly if you can't walk straight; Don't invent new stunts; Don't be a flying fool a damn fool is safer." And then some caution: "Be conservative . . . Feel your Machine; don't let it fly you . . . It is better to be a Sky-Pilot than a Crack-Ace. Be conservative." And some bonus instructions: Don't drop ballast other than fine sand or water from the aircraft. Should the engine fail on takeoff, land straight ahead no matter the obstacle. A plane with a lifeless motor had the landing right-of-way, and no aircraft was to taxi faster than humans walking.

To ensure the mail crews were oriented correctly, in 1923, the US government created the "lighted airway" by drawing hundreds of yellow concrete arrows, each about 70 feet long, for fair-weather guidance. Each arrow had an adjacent 50-foot rotating acetylene searchlight to guide night flights. A single beacon—at 500 watts with 18-inch doublet lenses—provided 10 miles of visibility under clear conditions and automatically turned off during the day. About 250 of these beacons illuminated the Chicago-to-Cheyenne routes.

In 1924, the US cross-country flight service was born due

to lighted airways. The New York–to–San Francisco route had 10 radio stations, and the US Navy dispatched weekly weather reports. In these early days of cross-country flight service, a trip from east to west took about 34 hours, and the reverse, 29 hours. Subsequently, NOTAM, or Notice to Airmen (now called Notice to Air Missions), sent regular hazard updates along flight routes, brimming with codes and acronyms. By 1925, there were about 100 commercial planes in service. The following year, with approval from Congress, the postmaster general contracted out mail delivery to private companies. As we'll see, this move created a spate of complications, including a political debacle.

As cross-country flight became increasingly popular, one daredevil acrobat and airmail pilot, Elrey Borge Jeppesen, realized the need for standardized pilot directions. Bad weather could blur even the bright beacons. "Sometimes you couldn't see a mile, so what the hell good was a beacon?" Jeppesen said. So he purchased "little black books" at 10 cents each to start compiling directions to amend and improve on government aerial charts. While piloting aerial survey planes and photographing terrains, Jeppesen made notes of his surroundings that would prove lifesaving to other pilots. Jeppesen bought an altimeter to estimate the heights of landmarks—from mountains to smokestacks. He approximated safe landing distances for emergency strips and scouted alternative travel routes. His notes contained practical details such as landing locations, runway lengths, and simplified terrain views and altitudes.

Jeppesen included information on where to get the best weather report and fuel and what to do when successive storms gripped pilots. "He knew, for example, that when approaching a certain airfield, he had to keep the grain silo to his right and the road with its line of telephone poles to his left," Jeppesen's biographers note. "Or, when taking off from another field, he had to pull up enough to clear the barn and windmill and grove of trees at the end of the runway. Simple enough, but what about when the weather was bad or the visibility restricted? Exactly how far from the runway are those poles, that barn, and those trees? Just as important, how tall were they?" Jeppesen's scribbles and sketches—which he developed into standardized flight maps and called the Jepp charts—became a sought-after resource for pilots. "I didn't start out to chart the skies," Jeppesen said. "I did it to keep myself from getting killed."

By 1930, five major airlines existed, together carrying 400,000 people per year. As air travel became more popular, air-traffic problems soared. To ensure that planes didn't try to land simultaneously or fly toward each other at the same altitude, eastbound aircraft began flying at odd-thousand-foot altitudes and westbound flights at even altitudes. And while in 1923 only 250 beacons helped guide pilots, by 1933, 1,500 beacons lit American airways. Pilots relied on Morse-code dots and dashes in their cockpit AM radio receivers to differentiate the beacon lights they tracked to their destinations. They were flying on course if they heard a steady tone; otherwise, they were off course and needed to correct. Further, the beacons relayed their course with single Morse letter flashes through the dark: 1 for W, 2 for U, 3 for V, 4 for H, 5 for R, 6 for K, 7 for D, 8 for B, 9 for G, 10 for M. The letters formed the mnemonic "When Undertaking Very Hard Work Keep Direction by Good Methods."

As we'll see, a high-profile crash involving a US senator in 1935 goaded the government to create a national air-traffic management system, which fully emerged only in the early 1940s. But until then, the problem of blind flying remained, and lousy weather cut short journeys and lives. The successful solution, however, required Ed Link's ingenious engineering—to make pianos fly.

BORN IN 1904, ED LINK was mechanical minded. Eager for solitude, young Ed found his fun deconstructing and reassembling clocks. His other pastimes included tinkering with gunpowder at a friend's farm and blowing the door off the barn. There were other projects: building toy cannons that shot kerosene bombs, assembling a mechanical printer to produce a comic newspaper he sold for a penny, and designing an inflatable slide from his bedroom window to the street. Whenever he walked down the school hallway and entered the classroom, his jingling pants pockets, bulging with keys, chains, and knives, drew attention with every step.

At 12, Ed Link drew a pencil sketch of a submarine. "Unfortunately, it wasn't very practical; it most assuredly would have sunk," he confessed years later. His technical aptitude aside, Link was an academic disappointment. Ed Link was "just dumb enough to be a genius," an associate recalled. He was expelled from classes and dropped out of junior high school. Later in life, Ed Link maintained that formal education taught all one couldn't do.

In 1920, when living in California with his mother, Katherine, who was by then separated from Edwin Link Sr., young Ed Link caught the "air-mindedness" that acted on the imaginations of adventurous kids. During Ed Link's youth, many out-of-work pilots who had flown in World War I became acrobatic showmen. These nomadic barnstormers traveled county to county and coast to coast, stopping in cow pastures and at state fairs to entertain the locals. They organized flying circuses and showed off inversions, loops, and snap rolls at 100 miles per hour. These bold "birdmen" charged spectators 5 or 10 dollars for a few minutes' ride in their Curtiss Jennys—wood-framed surplus war stock described as "a bunch of parts flying in formation." The Jennys tended to spit oil and fumes on the pilot's face. They occasionally caused a fire in the plane's flammable fabric. Daredevil solo fliers walked on wings, changed planes in motion, and even played tennis while aloft. Some vaulted from aircraft to running boats or trains.

One morning in 1920, young Ed Link borrowed a motorcycle to drive to an airfield on Wilshire Boulevard in Los Angeles to "Fly with the Stars." A year earlier, Sydney Chaplin—Charlie Chaplin's brother—became a proprietor of a short-lived airline company. Ed Link had saved money from his work repairing motorcycles. He signed up for three lessons from Chaplin for $50 an hour (close to $750 an hour in today's value). In aviator gear replete with a silk scarf, Chaplin offered Ed Link a seat in the back of his Curtiss Oriole. In a death-defying experience, Ed Link got a lightning tour punctuated with loops and spins, where he hardly touched the controls. "That's a hell of a way to teach someone to fly," Ed Link later recalled, pointing out that most "old-time aviators, like Chaplin, started teaching their students by scaring them half to death."

Ed Link's family initially disapproved of his interest in aviation. When he returned to Binghamton in 1922, after short-lived jobs in Illinois and West Virginia, Ed Link joined his father's piano and organ firm. A grease-under-fingernail tinkerer, Ed Link revived and restored decrepit player units as his brother George oversaw the company's operations. In 1924, Ed Link rigged a vacuum cleaner–style device to remove lint from the piano tracker bars, which degraded the endless-music rolls. This fix solved a frequent nuisance that made the units dysfunctional. A year later, he directed the design and rollout of the Link C Sharp Minor organ, which the company installed in the hundreds. Undeterred by his parents' disapproval, in 1926, Ed Link completed his first solo flight, with lessons from a local aviator. Edwin Link Sr. was displeased with his son's passion for flying and fired Ed Link from the factory in response, but he quickly rehired him. Ed Link's inventive instincts clicked into gear during his lessons, perhaps spurred by the necessity of his penny-pinching to afford flight school. He began to imagine new ways to slash the high costs of pilot training. His idea sparked a new era.

WILBUR WRIGHT TRAINED HIS pilots to gauge the effects of wind by opening and closing the hangar doors. He sat at the controls and told his students to "imagine" air disturbances and the necessary aerodynamic response as they maneuvered the parked plane's rudder and wing controls. The idea of "flightless flight" defined many inaugural efforts for flight training.

One contemporary outlier of the Wrights was Glenn Curtiss, a motorbiking pioneer from Hammondsport, a Finger Lakes village 100 miles west of Binghamton, who embraced a different method. Because his planes were all single-seaters, there was no way students could learn from their instructor on the flight. Learning from others' airborne demonstrations and grading the flight from the ground was the only way. Under Curtiss's instruction, students translated what they observed from the ground into flight maneuvers. In contrast, the Wrights emphasized the "aerial" experience the students cultivated on the ground.

In 1910, the Wrights developed a "kiwi-trainer" in Huffman Prairie, Ohio, named after the flightless bird. In this trainer, students sat in a Wright Model B pusher biplane mounted on a motor-driven platform. They were introduced to basic flight movements and sensations. The Wrights charged an extraordinary one dollar a minute for these elementary lessons (approaching $2,000 an hour in today's value), and their courses were usually four hours minimum.

Around the same time, engineers developed several other "motion trainer" concepts. A British automobile engineer, Eardley Billing, came up with the idea of an oscillator. This rudimentary biplane sat on a rotatable undercarriage. The wobbling gave students a "feel" for open-air flight. At the Antoinette Flying School grounds, a French military camp, students learned simple balancing maneuvers while seated in a wooden barrel split lengthwise. When instructors disturbed the ropes in the mounting arrangement, as in tug-of-war, the student had to control the barrel's action and align with the horizon by adjusting pulleys.

In a similar effort to reconstruct balancing motions experienced in flight, after crashing his biplane on a telephone pole, British captain Haydn Sanders recovered the wreck, stripped out the motor, and sat the plane on a metal pivot. As the contraption rocked, the students sitting in it learned to operate the rickety actions of the elevator, rudder, and ailerons. But this device failed because it required steady wind.

World War I brought on new attempts to design flight trainers.

One wacky-seeming design came from New Yorker William Guy Ruggles. He affixed a chair in a gimbal-ring assembly powered by electric motors. The unit spun on all three axes, creating comic disorientation and nausea in students. Its inventor was confident that with blindfolding, one experienced intense illusions of airsickness. Ultimately the Ruggles Orientator—or the "bathtub," as some called it—only assessed a candidate's fitness to fly rather than training them how to fly.

To further improve on the Ruggles Orientator, one visually impaired Philadelphia native, Colonel William Charles Ocker—later called the "Father of Instrument Flying"—proposed a hooded version to create the nonvisual conditions. With his partner, Colonel Carl Crane, Ocker taught aviators to rely on instruments when they could not see the earth's surface. Ocker and Crane's conception of blind flight didn't mean flying blindly. In darkness, fog, and other poor-visibility conditions, "some artificial means must be used in order that flight may be possible," they wrote in their influential book, *Blind Flight in Theory and Practice*. They argued that one could control an airplane without a proper visual reference through instruments. But blind flight entailed two interlinked problems: one, flying the aircraft without external visual reference, and two, controlling the aircraft without outside vision. Again, these two problems demanded that pilots learn to interpret and trust the instruments. Granted, this was not a simple task; other efforts continued to provide more realistic training. The French Air Service used old Blériot monoplanes with a truncated wingspan and had trainees taxi at 40 miles per hour. They called this use of a reduced-wingspan plane the "penguin system." The American version of this method, using stub-winged Curtiss Jennys, was called the "grasscutter." In both versions, the aircraft didn't take off, so the students could merely get acquainted with the basics of flight dynamics.

While working in the piano factory, Ed Link stayed active at his local airfield, rehearsing his barnstorming skills, including barrel rolls and parachuting stunts. He was ecstatic about Lindbergh's transatlantic achievement in the summer of 1927. He learned about the Sanders Teacher and the penguin system. But it was Ocker's work that compelled Ed Link to initiate his own training system. Ocker had recently blindfolded pilots during training, turned them around on a swivel chair, and asked whether they turned clockwise or counterclockwise. "And that was one of the things that gave me the idea that you could make a whole airplane to train a pilot to do everything," Ed Link recalled for an oral history project five decades later. "He was merely demonstrating just what I repeated: that you couldn't tell where you were going by sight or feel. You had to have an instrument that told you were turning and whether you were flying straight or level and so forth." But with the ground-based trainers Ed Link envisioned, pilot training could happen readily at scale. "They just thought that 'twas silly, but time has proved differently."

As the engineer turned sculptor Alexander Calder—heralded as the painter of movement—once said, "Just as one can compose colors, or forms, so one can compose motions." In late 1927, Ed Link began engineering his motion trainer—part piano and part plane.

Dec. 4, 1934.

E. A. LINK, JR

1,982,960

ILLUMINATED AERIAL DISPLAY

Filed April 27, 1934

12 Sheets—Sheet 1

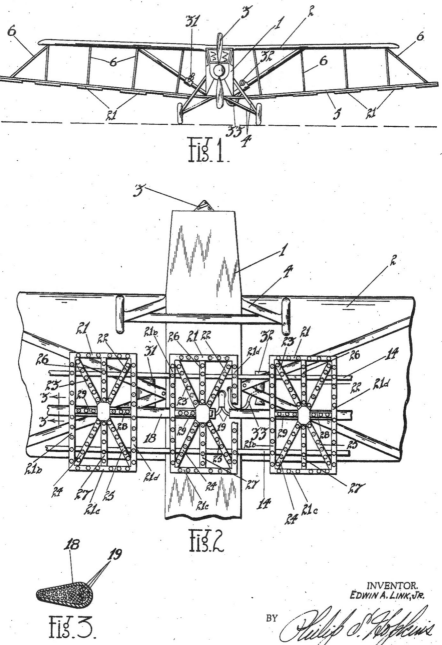

Fig. 1.

Fig. 2

Fig. 3.

INVENTOR.
EDWIN A. LINK, JR.

BY Philip S. Hopkins

ATTORNEY.

Chapter Two

Wicked Vagueness

ON APRIL 22, 1993, IN CANTON, OHIO, KE MING LI DROVE home in her Volvo 850. Her six-year-old daughter, Diana Zhang, was in the right front seat. A boxy Volkswagen Golf abruptly stopped ahead for a turn. Li stopped in a panic and rear-ended the hatchback. The low-speed collision triggered her Volvo's front airbags. The 2,600-pound blow exploded outward in milliseconds, striking her daughter and pinning her to the inside roof. The young Diana died of severe head injuries three days later. Li survived with a chin scrape. She sued Volvo for product malfunction and negligence. The court sided with Volvo for two reasons. First, the company's testing data demonstrated that its product was "safe" if used properly. Second, the daughter was not wearing a seat belt.

Airbags were first fitted in American cars in 1950. Even after their launch, engineers were uncertain about how the airbags worked in different conditions. Typically, the scope of the design process makes engineers operate in "partial ignorance." They don't necessarily have all the insights from the get-go or have the luxury of waiting for all explanations. At some point, constraints force engineers to forego additional tests and simulations for the sake of launching the product. Therefore, final product testing is usually carried out by users of the product after it leaves the factory. Monitoring how others use the technology usually drives further product improvements. This iterative aspect of design has been called a "social experiment," where "ongoing success in engineering depends on gaining new knowledge, as does ongoing success in experimentation."

But engineering experiments in the real world differ from laboratory studies in two ways. First, one can't randomly sample customers buying cars with or without airbags. There isn't a control group of people who don't receive the treatment that would distinguish the effect of the experimental treatment on those who receive it. Second, obtaining informed consent from the public can be tricky, even though, as users of potentially dangerous technology, people should have a say in its development. Through social experimentation, engineers eventually handled the "hard vagueness" of airbags. But as we'll see, achieving an acceptable social solution took many lives and years, requiring a blend of public policy and public relations.

But first, let's consider vagueness. The concept lacks a crisp definition; it's imprecise. Philosophers love to use bald people as examples to describe vagueness. If someone has over half a million hairs, they aren't bald, but with none, they are. But if someone has a couple of thousand hairs, are they bald? Without much precision to mark the boundary, how's one to define who's bald and who isn't? Again, someone with zero hairs is bald. If you apply the "tolerance principle," say, someone with n hairs is bald, then someone with $n + 1$ hairs is also bald. Then is a person with 1 hair still bald, or "less bald"? How about 2 hairs, or 7 hairs, or 13,287 hairs?

In this line of thinking, even though the premise is valid, the conclusion comes out invalid. So vague situations are also susceptible to the sorites, or the heaper paradox. A contemporary of Aristotle, Eubulides of Miletus, asked: What's a heap? Can 1 grain of sand create a heap, or does it take 100,000 grains? If 1 grain isn't a heap, and adding an extra grain doesn't make a heap, then by extension, 100,000 grains aren't a heap either. Akin to sand, our words, too, can pile up vagueness. Where does one system end, and where does the other begin?

Such vagueness permeates our daily lives. Consider the paradox of *mañana*, which relates to an indefinite future that scholar Dorothy Edgington describes as "the unwelcome task which needs to be done, but it's always a matter of indifference whether it's done today or tomorrow." These paradoxes are profoundly relevant in considering how gradual change affects us: increase in prices, pollution, population, and the planet's temperature.

One scholar argued that it's unproductive to think of vague boundaries as "a grid, a net, a system of pigeon-holes, a way of drawing a line, dividing a field." Instead, we should visualize vagueness as "magnetic poles around which some objects cluster more or less closely and from which are more or less repelled." Imprecision can be appealing with concepts like love, justice, beauty, and sustainability. When scrutinized, even our most simple ideas can become vague. H. G. Wells observed this problem with the definition of a chair. Chairs could be armchairs, reading chairs, dining room chairs, kitchen chairs, dentist chairs, thrones, and opera stalls, and could cross over to become sofas, bar stools, dorm sleepers, and park benches. Given the variety, Wells wrote, one could "defeat any definition of chair or chairishness." Ultimately, the philosophical puzzles of hairs and chairs help us formulate better problems. But while philosophers can debate definitions forever, engineers must deliver practical designs, even in uncertainty.

As a version of hard vagueness, uncertainties can mean a lack of knowledge or a lack of trust in knowledge, which can increase the risk of failure. Engineers may know the key variables but not necessarily their values, and those often require probability calculations. In the case of airbags, engineers had to make essential design trade-offs and discovered issues from field use.

In 1972, two decades after airbags were first fitted, an insurance company offered customers discounts for vehicles equipped with them. Then, in 1973, a Chevrolet hit a delivery truck at 35 miles per hour. The exploding balloon killed an unbuckled seven-week-old infant in the car's front seat. Incidents like this raised the vague question: Were airbags harmful or helpful? The resulting debates inflamed emotions and influenced policy for the next two decades.

In the 1980s, legislation to boost passive safety—that limit rather than avoid the damages from an accident—increased airbag use. But in 1983, only 14 percent of American drivers used three-point lap-and-shoulder seat belts, which were shown to forestall fatal injuries. Without the widespread use of seat belts, airbags could only help so much. By the early 1990s, the use of dual-front bags had dramatically expanded, just as the identity of airbags evolved from being merely *passive* restraints to *supplemental* restraints. In 1992, manufacturers introduced bright information labels to warn about potential injuries from airbags. A 1994 National Highway Traffic Safety Administration analysis reported that airbags reduced fatalities from frontal damage by 30 percent. Dual-airbag installation proliferated. In 1995, manufacturers phased out automatic seat belts. Over 70 percent of American vehicles came installed with airbags; a year later, it went up to 90 percent. And in cases of accidental airbag inflation, over 96 percent of the resulting injuries were minor.

Shortly after the death of six-year-old Diana Zhang in 1993, the National Highway Traffic Safety Administration issued a warning that parents should not seat their children in the front passenger side. This warning would prove tragically ineffective. In 1996, a mother in Boise, Idaho, drove her Volkswagen Jetta for Thanksgiving shopping with her one-year-old daughter in the front seat, unsecured. A minor fender bender in the mall parking lot activated the airbag, decapitating the child. The traffic-safety agency was now forced to combat the airbag problem with something far more potent than a warning.

A month later, in a presidential address, Bill Clinton encouraged auto companies to develop a new generation of "smart" airbags with advanced sensors adapted to the passenger's size. Clinton pressed manufacturers to install specialized, gentler airbags for children and higher-risk adults. Car dealers could deactivate airbags at a customer's request. Carmakers derided requests to install switches that turn off airbags; they pleaded the case for simultaneous seat belt use. Many litigations emerged primarily from custom-ers' fears that airbags can be perilous even in minor mishaps. An irony appeared in these struggles. The automobile-safety debate had now regressed to where it originally began, with seat belts. And countries like Australia, Canada, and Germany, with superior seat belt compliance, had reported greater success with airbags.

The design and deployment of airbags, and ultimately their pub-lic acceptance, illustrate "hard vagueness." Scholar Jameson Wet-more observes that airbags show how engineering uncertainties involve more than technical dimensions, even in the case of a thor-oughly tested and diligently developed product. Even with over 40 years of design experience, engineers lacked full knowledge about these devices, and the only path forward was further testing and technical refinements. Engineers improved anthropomorphic test devices and crash protocols and ensured the propellant wasn't too abrasive. They stepped out of their factories, met with lawmakers, and offered testimonies to assure them that airbags offered max-imal protection. Yet, the unnecessary deaths from low-speed col-lisions continued. Reports from early 1997 showed that airbags killed 3.5 children for every 1 saved. And even after a regulation requiring airbags in every car, in 1998, 40 children were killed at speeds under 15 miles per hour.

Engineers knew the range of optimal outcomes of airbags when appropriately deployed. Still, they couldn't quantify their failure prob-abilities in social settings with grave consequences. Redesigning the airbags from scratch or recalling them from millions of cars would not fix public-perception problems, especially when people weren't following the recommended guidelines. Jameson notes that the chief uncertainties surrounding airbags relate to the fact that they are not a plug-in component. Each incident varied depending on the vehicle—sometimes, the airbags deployed too quickly or slowly, pressuring engineers to account for different crash scenarios. They concluded that the safe performance of airbags relied on how *informed* the indi-viduals were about them. Engineers acknowledged that although they were well versed in technical testing, passengers' social and behavioral nuances added vagueness. Their hard problem required more than launching a product and monitoring its performance. It required educating the public about the risks and uncertainties.

Working with insurance companies, media, regulators, elected officials, civil servants, lawyers, and sociologists, the engineers guided a not-so-obvious solution. Their efforts included public relations campaigns reminding people to wear seat belts. Subsequent data collection on airbag performance led to the revised crash- and barrier-testing protocols and advanced simulations of other failure modes. Insurance firms closely tracked the claims of their policyholders for trends and, using those insights, generated public service announcements that reminded them that an automobile is a human-machine system.

These nationwide efforts combined the insights from both the laboratory and the field. They were inspired by the notion that "a social solution must be pursued because a safe and effective technical solution could not be implemented soon enough," as Wetmore observes. Had it not been for this collaborative approach, "it would have taken much longer to both understand and deal with the problem." The engineers, regulators, and advertisers devised an even simpler three-point message clarifying drivers' responsibilities: restrain all passengers properly; seat all children up to age 12 in the back; and maintain a 10-inch distance from the airbags. As this message spread, the National Highway Traffic Safety Administration reported increased use of seat belts, infant restraints, and rear seating. Between 1994 and 2000, fatalities per million airbags plummeted almost 10-fold in the United States, saving tens of thousands of lives each year over the following decade.

No technology is without risk, just as no freeway is free of crashes. Characterizing the hard vagueness in safety-critical products such as airbags, or even roads, often focuses on two interlinked questions: Are the technologies designed to meet current standards as safe as they *can* be? And, are the technologies designed to meet current standards as safe as they *should* be? These questions pose uncertainty because "standards tend to be written not with crashes but with crash surrogates in mind," notes scholar Ezra Hauer. For example, one can use sight distance, vehicle-separation distance, and driver comfort as substitutes for safety measured in crash frequency and severity. However, there is no clear link between established standards and experienced safety.

Engineering under uncertainty is like sometimes being unable to "decide how much salt to put into a soup without an anticipation of its taste," notes Hauer. As a result, technologies are usually not safe, not unsafe, or not appropriately safe. Safety failures aren't either-or but a matter of degree. They aren't like a roof collapsing but like pavement cracking or a beam deflecting. And as with degrees of baldness, how can one draw clear boundaries among these degrees of safety? These insights and standards of safety come from time and experience, reflecting "what a committee of professionals of that time considers to be a good practice." As a result, most transportation technologies have been reactive, with an "unpremeditated level of safety."

As we'll see next in soft and messy vagueness, technology linked to human behavior doesn't always guarantee straightforward progress. However, like airbags, civic engagement can apply lifesaving pressures to such problems.

✳

ZEN KOANS ARE RIDDLES for meditation. They help train one's mind to attain illumination, presenting a path through the paradox. One well-known anecdote relates to an accomplished academic visiting a Japanese master to question Zen philosophy. The monk served the visitor tea and kept pouring it as the cup overflowed. The visitor pointed out that the cup was overfull and couldn't take any more tea. The master smiled and made his point. Like the cup, the visitor's mind was overflowing with beliefs and biases. Emptying is a prerequisite for self-inquiry.

Bill Rouse is the systems engineering version of a Zen master. "Avoid jumping into solutions," he says with his trademark calm and easy smile. "If you think you know the answer immediately, you are probably wrong." Rouse disliked chemistry in college: "I needed more visual feedback than mixing liquids," Rouse said. He chose mechanical engineering. "You can study gears and shafts, but you cannot see electrons," he said. Then began his lifetime pursuit of visualizing abstractions.

For a college internship, Rouse wrote a simulation program for a defense contractor to plan for the availability and maintenance of sonar spare parts on submarines. But he realized that his equations didn't include humans, even though people operated, maintained, and were affected by those engineering products. In the early 1970s, Rouse's interest led him to graduate studies in human-machine systems at the Massachusetts Institute of Technology. He prototyped an air-traffic control simulator for his thesis and learned the essential difference between human behavior and performance: human behavior is about what people do, and human performance is how well they do it. Both factors influence how people make accurate decisions and, for example, in Rouse's study, how people coordinate aircraft landings. Rouse's understanding of human-machine systems evolved as industry practices changed to accommodate more human factors. Sooner or later, experts can no longer relate to being trainees; they stop learning. Rouse vowed not to fall into that trap. He was going to be a perpetual pupil.

Rouse is a tall man in his seventies with, call it, *hair*odynamics. If you compared the professor's helmet of white, wavy hair to Newt Gingrich's, he would fail you and advise that the correct response is Jay Leno. He has cultivated a mind that focuses on what others are saying or trying to say and then assembles their views. As a young postdoc, Rouse imagined ways for technology to enhance collaboration on social problems. He organized citizen focus groups. One, of interfaith clergy, debated income-tax policy recommendations; another group reviewed proposed investments for a government research lab; and another worked on improving library operations. Each participant voted in real time using handheld thumbwheel switches. Rouse tallied their votes in his minicomputer and displayed them on an electronic tote board. "I easily remember hauling around this equipment to meetings in my 1969 VW bug," Rouse said. "There was a lot of wire!"

Rouse's research is not only about collaboration but also about how humans navigate the ambiguity of language. Language, though, is notorious for its "soft vagueness." There is lexical ambiguity—different meanings for one word; and then there's syntactic ambiguity—many meanings in one sentence. Consider the many interpretations of the statement: "He saw the man on the mountain with a telescope." And this: "Time flies like an arrow; fruit flies like a banana." Then there is the ambiguity inherent to hedge words like "roughly," "possibly," "presumably," "usually," "might," and "could."

Rouse's work builds from this vagueness, asking who the stakeholders are, what their interests are, and how they communicate them. He then assembles their assumptions into a model depicting visual-spatial dependencies people didn't realize existed. "If two people are opposed, usually they just argue with each other," Rouse said. "I arm them with our models and then see how they argue." Rouse acts as an orchestra conductor, allowing stakeholders to argue while maintaining balance. The basic idea is to help participants to think across multiple levels of influence: as individual agents, as service providers, as organizations, and as a society. This multilevel-abstraction model provides a richer visual of social and behavioral phenomena.

The polymath Bertrand Russell said, "Who speaks of vagueness should himself be vague." Decades before Rouse's multilevel models, Russell wrote a proof in set theory. In essence, a car could belong to a set of cars, but a set of cars couldn't belong to a set of cars. A set of cars isn't a single car to belong in the set of cars as the first car did. In explaining this "theory of types," Russell showed that a set of cars isn't a member of itself. Russell offered a vision for multilevel thinking: a car, cars, sets of cars, sets of sets of cars, and so forth. We can see that a hundred cars are a parking lot, a thousand cars are traffic, a million are a megacity, and a billion are climate change. At each level, new phenomena appear, requiring us to envision them.

Multilevel visualizations aren't intended to be simply cosmic; these high-level visualizations should promote experimentation and reflection. The Danish engineer Jens Rasmussen proposed that visualization can be more profitably applied in hierarchies of abstraction and aggregation. Each visual can produce a fresh way for the viewer to zone in on the function of interest. For example, an SUV can be decomposed into propulsion, battery, radiator, alternator, axles, steering, suspension, and muffler. The propulsion can be further pieced into motors, controllers, converters, and inverters. Much like the precision-engineered Bach cantatas, these parts can be aggregated at different levels of abstraction, from propulsion to planetary temperature. Building-information modeling in the construction industry allows analyses at different "levels of development." In BIM, color schemes with sliding scales of intensity, saturation, and opacity help engineers view uncertainty and ambiguity. Visualizing these levels of abstraction equips them to resolve issues better.

Some years ago, Rouse presented a complex model of the US health-care system to a group of corporate and policy leaders. He explained the benefits of interactive simulations and visualizations to make sense of the fragmented systems. Rouse's bullet-pointed presentation on "multilevel enterprise modeling" was demanding even for a group well versed in jargon. Quickly, Rouse rebranded his model as a "policy flight simulator." Decision-makers could fly the future before writing the check for it. Now, he had the group's rapt attention.

Following the Affordable Care Act of 2010, which implemented reforms to the health-insurance markets, hospitals confronted a vague environment where they were unclear about threats and opportunities. CEOs in the New York area sought guidance from Rouse and his colleagues about the kinds of technologies that they should invest in and about how to consolidate their negotiation power when mergers and acquisitions were becoming common. Rouse's team set up 8-foot-by-20-foot touch screens with a 180-degree immersive visual experience. They aimed to give CEOs a highly interactive feel of mergers and acquisitions beyond their financial statements.

Rouse's team blended information about 66 hospitals in Manhattan, the Bronx, and eastern Long Island. They presented the CEOs with a complete visual model to show how the hospital markets changed due to the new health-care law based on their financial performance. The model's zoom-in features engaged participants in the complexity of their decisions. Rouse presented scenarios: What would happen if hospitals invested only in their operations? What if they merged with competing hospitals to increase negotiation power with insurance companies? Or what if they focused only on acquiring successful private physician practices with high reimbursements? He ran 30 demos for over 200 hospital executives, insurers, and state government officials to bring them all on the same visual page. Each participant took over the "controls" and surveyed different options for the group to discuss those trajectories. More debates occurred, with business decisions following the model's prediction. The results were resolutions rather than hard solutions. As Rouse noted, this technology enabled a new kind of discussion: "A canned PowerPoint presentation cannot enable such rich interchanges and stakeholder buy-in."

Rouse's multilevel tool encourages questions like: How might the population health of Miami evolve over the next two decades? How can technology drive the quality and costs of medical care in Texas? How can the Gulf region prepare for future hurricanes, the West Coast better manage wildfires, and the United States respond to epidemics more successfully? Rouse learned from these exercises that not everyone should be able to test-drive all futures, and having a few vetted scenarios was helpful for baseline discussions. Creating new conditions "on the fly" may be detrimental and lead to riskier business decisions. Almost uniformly, participants in Rouse's group exercises believed themselves to be strategic and systems thinkers. "Often, they were very myopic in what they were looking at. But you can't say that," he said. "I know that's a no-fly zone. Like engineers, the CEOs were also keen on finding solutions quickly."

But quick and precise decisions are not always the most effective. Models for vagueness should avoid what the investor John Bogle called "worship of hard numbers." In the case of investing, the "momentary *precision*" of stock price may seem more legitimate than the "eternal *imprecision*" of measuring a corporation's value. But this kind of temporary precision can give a false vision of the long term. Indeed, some large-scale models Rouse created can be done without numbers. The group exercises proved that cognitive engagement could be as valuable as computing power. "The exploration should be more like cultural anthropology than mathematical physics," Rouse notes. "It involves observing and listening to stakeholders, and representing what you learn, often in sketches and diagrams. You are looking for an understanding of their assumptions rather than answers."

Numbers can provide pleasurable clarity. Quantifications are seductive and sticky because they make murky goals feel tangible and actionable. Scholar Sally Engle Merry described an "indicator culture" that places a premium on numerical forms of data for decision-making. The indicator itself is a "broad and vague term"; it can be a ratio or a rational order, as Theodore Porter described of the French engineers. These indicators can and do shape our motivations and assumptions, often appearing more precise and potentially pernicious than they are.

Multilevel models may not change the Earth's physics, chemistry, and biology. Even so, they can help change our viewpoints and actions. Like a Japanese Zen garden, such models are equivalently composed of rocks, trees, and water, with winding paths and different perspectives. They are manicured imitations of nature, not nature itself. "We should calmly reconsider our typical modes of operation and explore alternative ways of sense-making," Rouse said. "It sets the stage for renewed effort, mindfulness, and concentration. One can become highly aware of the greater context of one's undertakings and very existence. All the pieces fit together."

✳

SOMETIMES, THE WORLD GETS thrown off its axis, and such disorders defy description. Unlike déjà vu, these situations leave us with no familiar feelings, only foreign frustrations. Scholar Karl Weick labeled such strange moments "cosmology episodes."

Brian Collins's personal cosmology episode began in April 2010, with tea, toast, and BBC breaking news. "The bloody thing just exploded, and I couldn't pronounce it," Collins recalled. "You have to speak from somewhere down there below your larynx." He referred to volcano Eyjafjallajökull (*AY-yah-fyah-lah-YOH-kuul*), a glacier atop the mountain overlooking the Icelandic islands. Collins's pronunciation had a bass-baritone tremor, rhyming like "I forgot the yogurt." *Time* magazine later declared Eyjafjallajökull one of the top 10 buzzwords of 2010, in the company of "austerity," "anchor babies," "bedbugs," and "Bunga Bunga."

When Eyjafjallajökull exploded, Brian Collins was the chief scientific adviser to the UK Department for Transport, responsible for coordinating an emergency response. "It was a bugger because it came down, went round a low, and then came back again," Collins recalled about the ash plume. "It wasn't a density issue; it was an accumulation issue." Despite warnings, the UK government didn't anticipate the consequences. After all, the pilots in the UK were used to cloudy takeoffs and landings.

Public emotions erupted over the subsequent week—a stoppage that lasted twice the duration of the shutdown of American airspace after the 9/11 attacks. Airline executives called the shutdown "scandalous." Among those desperate to travel, a group of corporate lawyers was willing to pay six-figure sums for a private aircraft and crew but found no takers. The *Monty Python* legend John Cleese took a taxi from Oslo to Brussels and recalled an old joke to a reporter: "How do you get God to laugh? Tell him your plans." The Royal Navy and private luxury liners came to the rescue, repatriating civilians stranded in different parts of the world and evoking comparisons to the Miracle of Dunkirk.

In early 2009, the year before its eruption, Eyjafjallajökull moaned with multitudes of mild earthquakes. The ice-topped stratovolcano, about 75 miles east of Reykjavík and 1,000 miles from London, is part of southern Iceland's east-west range of faults and fissures. After the slow start, the summit boomed on March 20, 2010. Eyjafjallajökull blazed against the emerald northern lights with a mixture of strombolian lava bombs and vulcanian cannon fires. The result was a "tourist eruption" or, as one scientist put it, a "volcano Disneyland." Ignoring safety warnings, hundreds of sightseers took to super Jeeps, snowmobiles, and quads. Others witnessed the wheezing warmth of the fire fountains from helicopters. Two visitors died.

The volcano rekindled on April 14 and began expelling magma from beneath the glacier. Meltwater flooded the region, forcing evacuations. Heat-cold reaction roared out plumes of fine ash reaching 32,000 feet, ultimately dispersing over 250 million tons. About 60 percent of the ash was silica; over a quarter of the particulate matter was under 10 microns in diameter, posing significant cardiovascular and respiratory risks. One official impulsively equated the event to the 1986 Chernobyl nuclear disaster, likening its ash to radiation. In his 1922 poem "The Waste Land," T. S. Eliot wasn't predicting Eyjafjallajökull when he wrote, "I will show you fear in a handful of dust." But fear and dust did spread as the 2010 "ashpocalypse" traveled furiously across European airspace. Although its impacts felt surprising, like all volcanoes, Eyjafjallajökull was predictable.

In 1822, when human flight was still a futuristic fancy, Eyjafjallajökull spewed a sporadic smokestack of clouds. Since this last eruption, with increased transcontinental air traffic beginning in the 1970s, volcanic dust has become a more serious threat to aviation. A local disturbance can now bring the global economy to a standstill. In this case, it did. The London Volcanic Ash Advisory Centre, run by the UK's national weather service, put out an advisory. The International Civil Aviation Organization guidance instructed pilots to avoid volcanic ash. The "zero tolerance" for ash inherently buckled the busy hubs of Heathrow, Frankfurt, and 300 more airports, creating the continent's "worst peacetime aviation crisis." In a week, canceling over 100,000 flights stranded 10 million passengers during the Easter holidays, slamming the airline industry with over $1.7 billion in losses. The shutdown also damaged some African economies. For instance, Kenya exports about 500 tons of fresh flowers to Europe daily. With grounded cargo, growers furloughed thousands of workers.

Of course, the idea that volcanic ash can be problematic isn't novel. Between 50 and 70 volcanoes erupt yearly. Scholars Amy Donovan and Clive Oppenheimer point out that there are currently over 700 volcanoes located within 60 miles and over 1,300 volcanoes within 150 miles of international borders, each with effects that can cross borders. On June 8, 1783, 28 years after an eruption of "Angry Sister" Katla that Iceland was still recovering from, hell broke loose again. The Laki craters belched a fire fountain, spewing poisonous ash into the pastoral landscape. Within minutes, everything turned dark; a blizzard of powder-like coal ash fell out of the clouds. For the next 12 days, Laki spilled two Olympic swimming pools of lava each second. Over the next eight months, sulfur-rich effluvium rose into the heavens, creating a haze that Benjamin Franklin feared "was of a permanent nature." The volcanic veil cooled the world and collapsed communities; the death of livestock and crops fueled famines, perhaps even contributing to the French Revolution. The Nile and the Niger shrank; tens of thousands succumbed to breathing problems in Europe, and far more perished over the years from the ensuing drought and hunger stretching from Iceland to India.

But ash's specific threat to aircraft became obvious on a summer night in 1982 over Indonesia. A British Airways 747 to New Zealand flew into ash from the growling Galunggung volcano. Passengers first assumed that the smoke inside the cabin was from cigarettes. Then the first engine flamed out, and the remaining three failed within minutes. The pilot made a heroic announcement: "We are doing our damnedest to get them going again. I trust you are not in too much distress." It became a "masterpiece of understatement." The pilot glided the aircraft from 37,000 feet over a dramatic 23 minutes, managing to restart three engines during the descent. The plane made an emergency landing in Jakarta using a two-inch view on the side window clear of ash. All the 263 passengers and crew landed unhurt.

Within days of this extraordinary event, a Singapore Airlines jumbo to Australia experienced something similar, as did a KLM 747 a year later over Alaska. The Redoubt volcano burped vaporous ash, blanketing nearly 8,000 square miles. The mechanical damage to the aircraft cost over $80 million. And in 1991, the hot and hostile awakening of Mount Pinatubo in the Philippines destroyed a $100 million aircraft. Despite these experiences, the aviation industry had not proactively conducted tests to determine how much ash the turbines could handle before they failed.

For an aircraft operating at 600 miles per hour, volcanic ash is mechanically, electrically, and chemically asphyxiating. The melting temperature of the ash is close to the operating temperatures of large engines, which puts airplanes at risk of an immediate shutdown if they fly through volcanic plumes. The "zero tolerance" policy in 2010 was controversial because the Eyjafjallajökull ash could be sensed only by radar, LIDAR, and infrared satellites. It was invisible to the naked eye. The CAA, the UK's Civil Aviation Authority, wouldn't take the risk of prematurely reopening the airspace, even partially. It now needed to confirm the "safe" ash levels for flight, which required real-time tracking of ash clouds.

The CAA asked the Met Office to run its Numerical Atmospheric-dispersion Modelling Environment code to produce ash-concentration charts. The CAA also tasked airframe and engine manufacturers to determine safe thresholds for flight. Lacking time, data, and insights into the damage patterns, the modelers confronted unavoidable uncertainties. Their vagueness stemmed from inexact knowledge; some had to extrapolate from previous ash encounters of Redoubt and the Galunggung. The industry developed a "Safe-to-Fly" chart with multiple levels of probability modeling.

Back in Westminster, "messy vagueness" was brewing. Brian Collins and the interagency team were vexed with a situation out of solutions and out of control. There were many stakeholders, some already very upset at the financial and social impacts. Getting community members to understand and respond to an invisible threat can mean disagreements. Still, Collins trusted the process of convening and coordination. Starting April 16, 2010, the CAA brought over 100 organizations worldwide into discussion. "We met seven or eight times a day with hundreds of participants to make various bits of this jigsaw come together," Collins said. From jet-engine manufacturers to regulators, insurers, and public officials, the participants had different views, and reconciling them wasn't always easy. But their common goal was to reopen the airspace safely. If dialogues can cause messy vagueness, they can also help dissolve it. Sometimes it's about simply waiting out a temporary crisis; sometimes, it's about using the winds of an emergency to unite people that need to work together for mutual safety; sometimes, it's both, and sometimes, it's neither. "No one talks or writes about the importance of these discussions," Collins said, "because they are not newsworthy."

On April 20, 2010, the industry reported two milligrams per cubic meter of air as an acceptable safety limit for flight operations, and the CAA agreed to that requirement. Later that day, European airspace lifted the travel ban. At the same time, test aircraft monitored Eyjafjallajökull's ash until the ejection rates subsided 10 days later. Passengers arriving at the UK airports were jubilant. One passenger compared the feeling of getting the boarding pass to "winning a golden ticket to Willy Wonka's chocolate factory." *The Guardian* pondered "how liberating it is sometimes to be powerless before nature."

That evening, Collins returned home exhausted and sleep deprived. He sat on his couch with a beverage and flicked on the BBC. Slowly his eyes widened to the channel's jarring split screen. One half showed the continued earthy effervescence of Iceland. The other broke the news of a human-made eruption. The *Deepwater Horizon* marine rig was aflame across the Atlantic. Over the next three months, tireless torrents of toxic crude gushed four million barrels into the Gulf of Mexico.

✳

BRUCE NAUMAN SAID MAKING art is "like going up the stairs in the dark, when you think there's one more step, and there isn't." Good art and engineering can arise from vagueness—incomplete, invisible, and ignored. As we saw, language can be a notorious cause and effect of vagueness. The engineer-philosopher Ludwig Wittgenstein quipped that "philosophical problems arise when language *goes on holiday*." A vague condition, of course, can be frustrating. And one economist estimated a "several-thousand-year efficiency loss" worldwide because of our tolerance to vague language. But "certainty can be fatal," wrote marketer Harry Beckwith, and "often what we believe we see is not really there."

Whether wicked systems contain vagueness is itself a vague thought. In its trio of hard, soft, and messy forms, we should study what vagueness reveals, when it arises, and what we lose without it. Even the word "efficiency" can become vague. Its context defines whether it's pursued as the technique for a target or the target for a technique. It can be a controlled method and a method to achieve control. We can't avoid vagueness; like efficiency, it drives progress.

Engineers have applied fuzziness in product development, recognizing that everything is a matter of degree. Your washing-machine, cruise-control, and air-conditioning units "know" when to guide themselves from one setting to another: still greasy; slow down; cool. At the heart of this thinking is what black-and-white thinking leaves out: the so-called excluded middle. A fuzzy mindset eschews binary logic; it thrives on the imprecise states in between. It's not hot or cold; it's about "just right"—the rest is manipulating the soap, speed, and sensors to get there. Scholar Lotfi Zadeh, who engineered fuzzy logic in the 1960s, called it "computing with words."

Vagueness is an instrument of choice for policy and diplomacy, as we are unclear on what's being said. As a central banker, the former Federal Reserve chair Alan Greenspan said, "I've learned to mumble with great incoherence. If I seem unduly clear to you, you must have misunderstood what I said." Vagueness offers flexibility in thinking. Blurry terms like "middle class," "family values," "jobs," "fair wages," "freedom," and "greatness" can invigorate masses and win elections. "Life, liberty, and pursuit of happiness" and "One for all, all for one" are misty mottoes rather than concrete commitments. Athletes tend to exaggerate their abilities on vague qualities such as "fitness" and "endurance" rather than more precise ones such as running speed and scoring rate. People also tend to favorably rate themselves on a vague attribute such as "defensive driving" over a clear-cut task such as parallel parking.

Consider diet programs, a nearly $80 billion market in the United States. People give up on or switch from a weight-loss regimen if they fail to meet overly precise metrics. When appropriately applied, vagueness can motivate wellness but can also be a source of distortion. Products labeled "low fat" imply they are healthier, even when they can have as many calories as the alternatives. When calories are given as a range in the nutritional information rather than as an exact number, some may use this vagueness to justify their food choices.

Vagueness is also why many of our ethical principles to guide technology development aren't so helpful. And vagueness is also why, when an airbag, volcano, or viral matter bursts open, we find ourselves in an OOTLUF, an Icelandic-sounding acronym for "out-of-the-loop unfamiliarity." As scholar J. L. Austin observed, you "stand a good chance of not being wrong if you make it vague enough," whereas with precision, it's the opposite.

To beneficially engage with the types of vagueness inherent in wicked problems, one needs to zoom in and out of the levels of abstraction. Collaborative planning exercises can employ helpful vagueness to generate broader viewpoints. In one training effort after the 2010 Eyjafjallajökull outbursts, a scenario-planning exercise in Iceland imagined and prepared for the effects of Katla, a volcano with even greater explosivity and emissions. The practice prepared for the degrees of variation and the vagueness we can expect from volcanoes. Some were gas rich, others ash loaded, with each hazard requiring customized responses. The simulations enabled participants to simulate flight paths out of a crisis.

"The governance of integrated systems doesn't necessarily mean the system of governance is integrated," Brian Collins said in his office in University College London's Chadwick Building, named after the famed social reformer. Indeed, social reform involves engineering in core ways. "In those days, the realization was things were all joined up," Collins observed. "You know, there's a lot of buzzwords around collaboration, co-creation, and joint learning these days, and that's because everyone is now getting hold of the fact that they can't do it in isolation." Collins had déjà vu moments during the Covid-19 pandemic, with far more lethal viral aerosols taking the place of volcanic aerosols and killing orders of magnitude more people. The House of Lords organized a virtual hearing on risk planning for the Delta strain of the pathogen, which was outfoxing London's efforts to curb the pandemic's next wave.

Collins testified as vice-chair of the UK National Preparedness Commission. The stately icons of the esteemed Parliament appeared as thumbnail icons on his laptop screen. Collins observed that the UK faced a colossal risk from its inability to respond to disasters. "We are very good at mass heroics," he said, "but some of these events are so dangerous for the well-being of the nation that we cannot depend on that." Collins lamented the disjointed approaches to assessing risks and uncertainty. He argued for "joined-upness," the critical prudence and prevision in planning for civic contingencies. States and nations may isolate themselves until they realize they are all coupled. "Are we a resilient country?" Collins asked in a reflective moment. "It depends on what you mean by a 'country.' "

Refrain

Roll

O RVILLE AND WILBUR WRIGHT BEGAN PURSUING AERO-
nautics as a sport and reluctantly entered the scientific side of
it. Twenty years later, Ed Link experienced the opposite effect with
his flight trainer. He was reluctant to see his engineering become
entertainment, but he wasn't surprised. As his patent claimed, the
Link Trainer provided "great value as means of instruction for student
pilots" and "affords an interesting and unique entertainment device."

Ed Link's generic aviation trainer adapted the suction principles he'd used to great success in player pianos. But instead of
making piano keys go up and down, the electrically activated bellows drove compressed air that mimicked the aileron, elevator, and
rudder functions. Like the piano pedals, the stick movement filled
or emptied the bellows to duplicate pitching and banking motions.
The bombinating device flew and felt like an airplane. The pilot
trainer's initial version featured a hooded cockpit mounted on a
turntable, a turn-and-bank gauge, a compass, and indicators for
airspeed and rate of climb. A universal joint provided the connection between rotating shafts. The machine's overall setup looked
like an eight-foot bumblebee turning in a 15-foot diameter.

In 1930, Ed Link opened a flying school featuring his unique
invention. He charged a flat $85 fee for the course. In it, pilots
safely rehearsed at a fraction of actual flying costs without concern
for losing life or equipment. The intercom allowed the student and
the supervisor to communicate during the session. The instructor's

desk also contained an automatic plotter of the student's course. Driven by a motor and three wheels, the crab-like record keeper allowed the students to review and correct their mistakes in subsequent training sessions. In the ensuing years, the instructors, often not pilots themselves, proved as vital as the instruments.

The flight was defined by "physical movements and sensations, the signals and indications from the cockpit's instrument panel, or the rules and procedures from manuals." Flight training had to combine these factors and accurately replicate physical sensations, mechanical signals, and official procedures. In Ed Link's "virtual flier," simulation and reality were conjoined. Observers often contend that the Link Trainer was a forerunner of virtual reality. Yet, as scholar Frank Cardullo said, "virtual reality is an oxymoron. Geometrically, a virtual image isn't located where it appears to be, while a real image is where it is. The more precise technical term would be 'virtual environment.'" With its first training sessions, the Link Trainer created an unprecedented form of a virtual environment and inaugurated a distinctive conception of flight.

The experiences provided by the Link Trainer may have been simulated, but their economic significance was real. "It was a low-hanging fruit," said Roger Connor, a pilot, and historian at the Smithsonian National Air and Space Museum. The merits of the Link Trainer in successfully unifying technological and psychological training for pilots became evident, and more profoundly so in establishing a new protocol for training in the aerospace sector. Most powerfully, the trainer consistently reinforced the flying fundamentals: position, direction, distance, and time.

Previously, pilots applied a step-by-step "1-2-3 method" to maintain straight and level flight. They adjusted the rudder pedals, control stick, and airspeed to influence how the aircraft's nose and wings moved. Even with this rudimentary method, pilots had to go against their instincts. The Link Trainer reoriented pilots' default dependence on their intuition. It trained them not to monitor their instruments but to trust them. Critically, the Link Trainer didn't merely mimic an aircraft's movement but prepared pilots to adjust instruments in response to varying conditions. The trainer added signals, subtracted distractions, multiplied viewpoints, and divided tasks. "It wasn't so much that the Link Trainer was an ideal instruction platform; it certainly wasn't," Connor said. "It was about knowing where you are now to visualize where you want to go next." The blue box was as much a skill-set trainer as a mindset trainer.

IN 1930, THE SAME year Ed Link opened his flight school, the Link Piano and Organ Company went out of business. His father, Edwin Link Sr., moved on to opportunities in real estate. For Ed Link, much depended on the newly formed Link Aviation. Around this time, the shy, wiry Ed Link met an adventurous journalist, Marion Clayton, who was assigned to profile him for the local newspaper. A romance took off, and Marion Clayton repeated throughout her life that she married her best subject.

Ed Link ran a government-approved airplane-repair shop in an unheated hangar to support his growing family. Between repair jobs, he built trainers by hand and ran his flight-training courses. He faced criticism that his trainer was mere "hangar flying" and could never simulate or supplant actual flight. But Ed Link's rebuttal highlighted his trainer's ability to cut cost, risk, and training time while producing proficient pilots. Case in point: When he trained his brother George on the Link Trainer in 1928, this would have taken 10 hours in actual flight, but the simulator cut training in half. George could safely solo after 4 hours of trainer time and 40-odd minutes airborne.

As the Great Depression ground on, Ed Link's revenues came mainly from air shows. Crowds applauded his skydiving, speed tricks, and balloon-bursting in the Pitcairn autogiro, also known as the "flying windmill." Ed Link tried to use his aircraft exhibitions to promote the trainer. In one instance, he and George towed the blue box on a borrowed Ford Model T to an event in St. Louis. The organizers refused to make it an exhibit, calling it a carnival ride. Coin-operated entertainment, once the province of the Link family's nickelodeon pianos, now included the Link Trainer.

Some circuses purchased blue boxes as amusement attractions. A barker beseeched the patrons in a state fair: "Just a quarter and you pilot a plane yourself." Marion kept the books and sold hamburgers at these shows. During the week, they lived off the leftovers. "Marion made more money selling hotdogs than I did flying airplanes," Ed Link recalled. An ever more desperate Ed Link was tempted to offer half his business to anyone who gave $500. He eventually sold 50 trainers to amusement firms, miniature golf courses, and only 2 to aviation firms, at $450 each. Even these purchases were for showroom demos rather than the actual purpose of training.

Frantic to bolster their finances, in 1933, Ed Link began experimenting with aerial advertising. He designed a forced-air mechanism using a borrowed monoplane to write words in the sky from under the aircraft wings. "I put a venturi tube in the slipstream of the airplane, and that created a vacuum," Ed Link recounted. "It was a player piano in the air!" The lights turned on and off from the ad text punched in piano rolls. He also equipped an aircraft with organ pipes. The flying billboards now had music. "Spaulding's Cakes Are Fresher," "Drink Utica Club Beer," and "Enna Jettick Shoes Are the Most Comfortable." The words beamed out of the night skies from 2,500 feet with an aerial roar.

As Ed Link once again proved his inventiveness over the skies of Binghamton, a political fiasco was ready to explode in Washington that would make the blind flight a national imperative. It started with a suicide club.

IN EARLY 1925, the US Congress passed the Air Mail Act, which required the postal service to privatize mail delivery. In response, bidders pooled their resources and formed the first aviation start-ups. This burgeoning group of contractors took over the operations from the postal pilots in 1927. Some commercial carriers called themselves "airlines" but were far from it; they didn't even care about installing seats on their planes. A historian noted that carrying airmail and freight—which paid three dollars per pound for a 1,000-mile trip—was far more cost-effective than carrying passengers. If a 150-pound passenger were the equivalent weight of airmail, they would have had to pay "a prohibitive $450 per ticket" (or around $8,000 in today's dollars). But cost differentials weren't the only counterintuitive aspect of early aviation start-ups.

In 1928, a letter sent from Boston to Seattle would be routed by one airline to New York, then flown to Chicago by another carrier, then to San Francisco by a different company, and finally to Seattle. Mail navigated a "crazy quilt of routes," notes scholar Erik Conway, resulting in sunk cost, lost time, and wasteful effort from pilots. In 1929, the engineer-turned-US-president Herbert Hoover dismantled this process through his postmaster general. Hoover sought rational efficiency in mail delivery, and "shotgun weddings" of consolidation among the many airline start-ups involved in the mail business soon followed. More logical and lucrative routes caught on.

No matter who delivered the mail, whether the Post Office or the private companies, forced landings in lousy weather were an enduring problem. Over 6,500 forced landings occurred in the nine years the Post Office operated the airmail service. Their competition, after all, were the all-weather railroads. Some years earlier, at a winter air meet in California, pilots wet their fingers in their mouths. And they held them up to see if one side cooled quicker than the other; call it a quick and dirty wind vane. When pilots launched in poor visibility or bad weather, their lack of instrument training made routine flights extremely dangerous. Conway highlights two types of weather-related accidents during the airmail years. The first type of accident occurred when the pilots flew too low in order to keep sight of the ground, often hitting trees, smokestacks, and buildings. Contact flying in poor visibility was lethal. The second type of accident occurred when pilots stayed hazardously high in the clouds. Planes entering the clouds were prone to spin, nose-dive, and crash on the ground.

A year before his celebrated Paris trip, Lindbergh worked as a mail contractor and wrecked his plane twice in dense fog during routine runs. Despite the known dangers of poor-weather flying, Lindbergh's employers still expected him to deliver the gasoline-drenched mail promptly. Even with the 1933 completion of an illuminated network of skyways and beacons, wicked weather buffeted profits and took lives. Conway observes a crucial point: if flight safety is paramount, one should stay on the ground. But safer flying didn't mean *not* flying: "Aircraft that sit on the ground produce losses, not profits."

Concerned about airline profiteering, in February 1934, President Roosevelt canceled the private-airline mail contracts and directed the Army Air Corps to distribute the mail. Previously, in 1918, the army had temporarily delivered mail at a much smaller scale. This time, about 250 army lieutenants delivered mail in Boeing pursuit planes and Curtiss bombers. Before their dispatch to mail delivery, the young pilots had no reason to fly in poor visibility, and few had nighttime flying experience. Because aircraft training remained expensive and lethal, the Army Air Corps sent out a troop of gravely unprepared pilots.

Within the first month of the army's postal service, airplanes spun to the ground in dense rain and snow. In 70-odd days, 60-odd crashes took over a dozen lives. Roosevelt was under fire, and World War I fighter ace Eddie Rickenbacker declared it "legalized murder." An aerial demigod by then, Lindbergh denounced the president's actions as "unwarranted and contrary to American principles." Billy Mitchell, a US Air Corps legend, questioned the true proficiency of the pilots. "If any Army aviator can't fly a mail route in any weather, what would we do in a war?" Soon, the Army Air Corps halted its mail operations. It resumed delivering mail—but only during the daytime and in clear weather.

Two days after Roosevelt canceled flight contracts, but before the Army Air Corps mail service started, Ed Link landed in Newark, New Jersey, in dense fog. His associate Casey Jones had arranged a demonstration of the Link Trainer for the Army Air Corps officers. The army ordered six trainers, at $3,400 each, which quickly bolstered Ed Link's business. With the delivery of the Link Trainers to the army in June 1934, Ed Link kick-started a revolution in aerial training.

In 1935, with government approval, Ed Link sold 10 trainers to Japan. The following year, the 31-year-old Ed Link sailed to Japan with Marion to provide instructions. After reaching Tokyo, he found that the technicians had disassembled one of his trainers, apparently intending to copy its entire design. He refused to help reassemble the machine, citing a lack of essential tools, and returned home. In 1936, more trainer orders bubbled up from the United Kingdom with the condition that the boxes be built in the Commonwealth. The Soviet Union, Italy, Spain, China, and eventually over 30 more countries followed suit.

Now adopted by military powers across the globe, Ed Link's engineering soon found its pivotal place in the lead-up to World War II. With enough faith and persistence, what had once been a circus thriller was finally a proper aviation device.

THE ROMANTIC POET Samuel Taylor Coleridge wrote that "shadows of imagination" required a "willing suspension of disbelief, which constitutes poetic faith." Blind flight is a matter of blind faith in instruments instead of rhyme. The Link Trainer was engineered to encourage pilots into a willing suspension of disbelief; it regimented real-time editing of sensory impressions gathered in flight. Pilots interpreted sensory stimuli, even when they understood that they were simulations. One aviator described the experience: "You start with the box that frames the problem and into the box you put everything that contributes to trainee involvement—the controls, indicators, etc. You omit from the box all that is not fundamental to the training situation. At least you do so to the extent that absence will not have a detrimental effect upon the trainee. Outside the box, you develop the instructor control . . . and then you connect the two areas through an appropriate interface. Finally, you insert the student into the box and close the lid." The key was to shut off external distractions so the trainees could be absorbed in the virtual environment.

Some resisted learning to fly in a box. "They didn't like the idea," Ed Link said. "If they just made them think it was an airplane, they enjoyed it better. So it was purely for psychology. It was the psychology of somebody getting in something that looked [more] like an airplane than like a box." Once "airborne" in the trainer, the pilots often thought the gauges were erratic and provided false information on their turns. "The turns did happen, of course," notes Conway; "that was the point." This new cognitive approach required pilots to "substitute their own psychological perceptions for the 'truths' created by their instruments," observes historian Timothy Schultz. But to better understand how the Link Trainer succeeded in rescuing pilots from depending on intuitions, let's look at a moment that initiated the instrument-flight era some years before Ed Link's fog landing in Newark.

VISION IS PERHAPS THE most reliable of our senses. Unlike with touch and hearing, without sight, our grasp of balance and direction is muddled. If you walk blindfolded, your sense of balance goes awry. Without sight to orient us, our inner ear can send mixed messages about where we are in space, resulting in a conflict of sensations. The pilots of the pre-instrument era flew dizzily in clouds. Lacking a visible horizon, they couldn't disregard their senses; they became disoriented and often spun out of control. Weather and vertigo were the archenemies of aviation.

As we saw earlier, William Ocker and Carl Crane's experiments in the 1920s provided initial insights into what happens to balance when lacking visual reference. But subsequent blind-landing experiments themselves lacked vision. In one approach, aviators lined airfields with tethered balloons to guide landing; another, led by the British and French air forces, laid an electric cable around the landing field for signaling. This attempt also failed, lacking precision transmitters and receivers to communicate with the aircraft. Other navigators attached long tail skids and weights to the planes that scraped the runway before touchdown to give pilots a physical signal. Eventually, a breakthrough emerged. In 1926, the mining magnate Daniel Guggenheim and his son Harry established a philanthropic fund to promote aeronautics. Their goals were clear: to ensure safe and reliable flights in all weather. They sought to identify ways to fly through the fog and address the general problem of blind flying. It involved a short, stocky, bald test pilot with a crooked nose.

At age 13, Jimmy Doolittle was adventurous. He built a wood-framed glider from the instructions in a 1910 issue of *Popular Mechanics*. Doolittle tied the contraption to a friend's car and successfully launched himself into the air, gliding briefly before he crashed. Doolittle also built an engine-powered glider with the extra cash made from boxing matches. It also flew successfully before a thunderstorm took it down. In 1925, at age 28, Doolittle earned the nation's first doctorate in aeronautical engineering from the Massachusetts Institute of Technology for studying how wind velocity affected flying characteristics. Later that year, he bagged another first: the Schneider Trophy, awarded annually to the winner of a race for seaplanes and flying boats. Doolittle topped the speed record in a Curtiss float biplane at 232 miles per hour. The next day he outraced himself at 245 miles per hour, more than quadrupling Glenn Curtiss's record from a dozen years earlier. Then followed an existential crisis. At the time of his win, Doolittle was still a first lieutenant in the Army Air Corps. With two kids and elder-care duties, becoming a captain was a distant dream. What next? Where was he going? Should he quit flying? Doolittle's phone rang. Harry Guggenheim borrowed him from the army to join their "fog flying" program.

The Guggenheim fund focused on five linked problems: first, dissipating the fog; second, effectively locating landing fields from the air; third, developing instruments to accurately measure how high the planes were; fourth, improving tools for flying in the fog; and finally, penetrating fog by light rays. The fund encouraged concepts to disperse the fog. Ideas included heating the atmosphere to above dew point; penetrating the fog with moisture-absorbing charged particles; spraying a drying agent over the fog to aid condensation; and, more mechanically, breaking up the clouds by propellers.

Doolittle and colleagues focused on improving the aircraft panels. They upgraded a Consolidated NY-2 biplane with an array of instruments. On September 24, 1929, Doolittle flew into the thick morning fog swirling over Mitchel Field in Long Island. Technicians pushed the biplane out of the hangar and prepared the radio system and landing localizers. Doolittle roared the 220-horsepower radial-piston engine. He taxied and took off into the fog, reaching 500 feet. He landed 10 minutes later, relying entirely on instruments.

A jubilant Harry Guggenheim encouraged Doolittle to repeat the feat, but this time with the cockpit covered by canvas sheet to achieve a fully blind flight. For safety, copilot Benjamin Kelsey sat in the open front cockpit with a full view. Doolittle lined up the aircraft for the liftoff from under the hood in the rear cockpit. The plane climbed a thousand feet and leveled. Flying some more miles east, Doolittle turned back to Mitchel Field, steadily descending and eventually landing. Kelsey sat still and relaxed, with his hands behind his head. "However, despite all my previous practice, the approach and landing were sloppy. The whole flight lasted only 15 minutes," Doolittle wrote. "This was just ten months and three weeks from the first test flight of the NY-2."

Doolittle's blind landing ended the seat-of-the-pants flying era. The Guggenheims shut down their fund's operation that year, declaring their mission successful. The age of instrument flight initiated by the Guggenheims then evolved from a stunt to a survival strategy, thanks to Ed Link. The Link Trainer provided calculated torture to those trained in it, teaching them to fight their instincts and trust their instruments. Doolittle and Ed Link's innovations worked in tandem: Doolittle's composite flight panel, outfitted with the latest instruments, enabled safe blind landings, and Ed Link's flight trainer, engineered to recreate all the sensations and technology of a cockpit, successfully trained a composite flier. But it took another deadly mishap—a political sandstorm—to elevate Ed Link's engineering to standard practice.

ON MAY 6, 1935, shortly after 9:00 p.m., Bronson Murray Cutting, a 46-year-old Republican senator from New Mexico, boarded a red-eye to Washington to debate a veterans' bill. Cutting flew aboard the Transcontinental and Western Air Flight 6—also called the Sky Chief service—featuring the twin-motored Douglas DC-2 and the roomiest cabin among passenger planes. The 14-seater had an 80-foot wingspan and cruised at 170 miles per hour. But some called the Sky Chief "too big to fly." Among the 10 passengers were a young woman with an infant daughter and a six-member Hollywood crew. When fueling was complete, the 28-year-old pilot and his 24-year-old copilot had no checklist duties. This practice wouldn't arrive until later that year, starting with the Boeing bomber B-17 Flying Fortress. As the wheels went up, Cutting went to sleep.

The route was routine. The weather, clear; the flight, smooth. But after crossing Wichita, a bad-weather alert squawked through the captain's staticky radio. Headwinds slowed the plane. The pilots prepared to fly on instruments, with a dense-fog warning issued for Kansas City. As the conditions worsened, Kansas City became zero-zero: no ceiling, no visibility. As the descent began, a TWA dispatcher directed the pilot to a government landing field in Kirksville, Missouri, 120 miles east of Kansas City. The conditions in Kirksville were a 1,200-foot ceiling and four-mile visibility. The flight heading toward the new destination was fog-shrouded, with misty conditions in even the few clear spots. The pilots attempted contact flying, with plans to track a concrete freeway. They dropped to 100-odd feet above the ground. At 160 miles per hour and 15 miles from the destination, the plane's left wing slammed into the top of a tree. The pilots lost control, and the aircraft rolled. At around 3:30 a.m., the Sky Chief crashed in flames on the edge of a pasture. Senator Cutting and four others died instantly.

As the congressional inquiries into air safety launched, so did the finger-pointing. Some briskly blamed the inaccurate weather reports, and others cited defective radio apparatus on the plane. Another party blamed the pilots; they were ill-qualified to fly, the hissing went. Even the beacons, located north of the crash area, were condemned. The Department of Commerce had lowered the beacon wattage to cut costs following the budget pressures of the New Deal. This cost-saving measure resulted in the "mushover effect"; the pilots couldn't pick up the low-intensity beam through the fog.

Royal Copeland, a Republican mayor in Michigan and later a Democratic senator from New York, led a sweeping congressional investigation. In May 1935, the Copeland committee held six days of hearings, eventually assembling 900 pages of testimony from the Post Office, the Weather Bureau, the Bureau of Air Commerce, the army, the navy, the airport-construction authorities, and the airline pilots' association. Within a month of the accident, the investigation concluded that the probable cause was the aircraft's "unintentional collision with the ground" due to poor visibility. Other contributing factors included the Weather Bureau's failure to predict hazardous weather, the aviation authorities' choice to grant takeoff clearance in Albuquerque, the pilot's judgment errors, and the Kansas City TWA ground crew's delayed decision to redirect the plane. TWA was penalized, and efforts commenced to structure and standardize the country's commercial air regulations.

The Copeland probe precipitated the 1938 Civil Aeronautics Act signed by Roosevelt. The act centralized aviation activities and diverted the Department of Commerce's responsibilities into a new, independent Civil Aeronautics Authority, which quickly divided itself further into two new groups. The first group, the Civil Aeronautics Administration, oversaw air-traffic control, safety programs, and airway development. The second, the Civil Aeronautics Board, centered on rulemaking, including pilot training and certifications. Although it unfortunately took a tragedy to compel government intervention, these new regulations for commercial air travel ushered in a new, safer age of flight. Meanwhile, as technological advances forged a new kind of cockpit, the pilot's role rapidly changed from throttle jockey to analyst.

✈

WHEN ED LINK TOOK his first flying lesson in the 1920s, his instruments—the stick, pedals, dials, switches, compass, and rudder controls—were tools from aviation's earliest days. Soon, he'd learn to use the artificial horizon and the altitude indicator. In World War I, pilots benefited from the addition of the airspeed indicator. By World War II, the number of instruments grew, featuring a magnetic compass and devices for airspeed and turn-and-bank indication. By the 1950s, the cockpit panel became sophisticated enough to prevent stalls and crashes, crisscrossing a maze of pneumatic pipes and electrical controls. In the military arena, by the 1960s, a fighter plane alone had over 40 instruments. An aircraft's instrument panel could be adapted and configured for high-altitude fighting, reconnaissance, or ground-strike missions.

With the "instrument explosion" during this era of aviation, pilots took on an additional role: consulting analysts monitoring the systems under various conditions and constraints. "The day of the throttle jockey is past," noted a former head of the FAA, pointing out that the new breed of pilot was "becoming a true professional, a manager of complex weapons systems." In the more provocative phrasing of a retired air force colonel: "A modern pilot is like the corpse at a funeral: his presence is traditionally expected, but he doesn't do much." Pilots "rarely 'hand-fly' the aircraft as they comply with company directives to surrender control to the autopilot except for takeoff and landing. Sometimes even the landings are machine-flown." Test pilots complained that the barrage of information mentally drained them.

To realistically reflect the instrument explosion pilots would face in real cockpits, Ed Link's flight trainer, once equipped with rudimentary indicators, added directional and radio gauges. By the early 1940s, the trainer could simulate wind conditions with an analog computer. One 1945 news article declared that the trainer flew more like an airplane than an airplane. "Don't make the mistake of thinking this 'dummy' plane is easy to manage," aviator Assen Jordanoff wrote. "In fact, the trainer is far less stable and much more sensitive to control than a regular plane, for the simple reason that you can get much better training in a hypersensitive plane than you can in a slower and more stable machine."

Partly because the trainers could accurately reproduce the newest cockpit instruments, Ed Link's company tripled in size within a decade. Near the end of World War II, Link Trainer production hit its zenith: one blue box left the Binghamton assembly line every 45 minutes. While engineering transformed the pilot's role, societal progress and a pioneering vanguard of female and Black aviators redefined *who* could become a pilot.

ANY SYMBOL OF PROGRESS conceals its regressions, and aviation was no exception. Through the mid-20th century, women's opportunities in aviation reflected the roles women typically played in society. They started out as cooks and clerks, and recruiters told women that the aviation apparatus was similar to kitchen appliances. Women were hired for crucial support roles that would release men for military duties that women weren't permitted to perform. In public perception, women diminished their femininity by taking on the aviation roles of men. Jacqueline Cochran, who later broke the sound barrier, wouldn't exit the plane she piloted without touching up her makeup. That prompted a comment: "Though she flies like a man she hasn't become trouser-minded."

In the early 1920s, Amelia Earhart worked odd extra jobs as a stenographer and truck driver to fund her pilot's license. Her first airplane ride was 10 minutes long and cost her $10. Her lessons from aviator Mary Snook, the first woman to own a commercial airfield, cost her $500 for 12 hours. Earhart encouraged younger women to take up flying, even though this pursuit was unaffordable in an era before the Link Trainer. She argued, "They are simply thoroughly normal girls and women who happen to have taken up flying rather than golf, swimming, or steeplechasing."

In June 1928, a year after Lindbergh's flight, Earhart became the first woman to cross the Atlantic, as a passenger in the Fokker Trimotor *Friendship* piloted by Wilmer Stultz and Louis Gordon, from Newfoundland to South Wales. A year later, Earhart finished third after Louise Thaden took the top honors in the first transcontinental air contest for women; it was pejoratively called the Powder Puff Derby. Then, in May 1932, Earhart soloed across the Atlantic in her Lockheed Vega from Newfoundland to Northern Ireland, with winds stopping her from repeating the Lindbergh feat of five years before. Later that summer, she flew "her lovely red Vega" nonstop from Los Angeles to Newark in 19 hours, setting the record as the first woman to make this cross-country flight.

In 1930, United Airlines hired Ellen Church, a nurse trained as a pilot, as the first flight attendant. At the time, airlines wouldn't hire female pilots, and women were primarily employed in aircraft sales. While men could defray the costs of expensive flight lessons by working on airfields, women had to pay cash up front. Even for licensing, women were subject to a stricter standard than men; they needed to take an exam in addition to the requisite flight hours. Despite the additional barriers to entry that they had to navigate, female pilots were often the subject of tasteless jokes and crude comments about their competence. In one distressing instance, as historian Joseph Corn points out, a federal investigator rejected the possibility of engine malfunction in an accident and attributed the failure to neglect by the woman pilot.

For much of the 1930s and the 1940s, women's wartime participation transformed their role in the nation. Over six million women entered the workforce during World War II, taking advantage of remarkable opportunities previously unavailable to them, including as Link Trainer instructors. Their letters conveyed patriotism and passion, as well as newfound independence. "You are now the husband of a career woman—just call me your little Ship Yard Babe!" wrote one Indiana woman, adding that opening a checking account for the first time was a "grand and glorious feeling." A 19-year-old working at Douglas Aircraft Company as a blueprint manager wrote to her Marine Corps fiancé: "Imagine, [me], little Betty, the youngest in her department with seventeen people older than her . . . under her. . . . I am the big fish in my own little pond—and I love it." Another woman, from Cleveland, Ohio, told her husband in a letter: "Sweetie, I want to make sure I make myself clear about how I've changed. I want you to know now that you are not married to a girl that's interested solely in a home—I shall definitely have to work all my life. I get emotional satisfaction out of working and I don't doubt that many a night you will cook the supper while I'm at a meeting. Also dearest—I shall never wash and iron—there are laundries for that! Do you think you'll be able to bear living with me?"

In the summer of 1942, the US Navy formed the Women Accepted for Voluntary Emergency Services, a reserves program. The agency picked 74 women from the first class of WAVES to serve as Link Trainer instructors. In 1943, Ed Link himself was present to congratulate the WAVES graduates. Over the next three years, WAVES recruited 84,000 young women, many of whom were assigned to naval aviators training. The white, middle-class women were better educated than most of the seamen. In late 1944, the program admitted Black women only after pressure from activist organizations and approval from Franklin Roosevelt, just a month before his reelection to an unprecedented fourth term of presidency. The WAVES provided Link instrument training to novice naval gunners and aviators, including future US president George H. W. Bush, and recurrent training to experienced pilots.

The quasi-military Women Airforce Service Pilots program, or WASP, was formed in August 1943 from the merger of the separately organized Women's Auxiliary Ferrying Squadron and the Women's Flying Training Detachment. The WASP program selected 1,830 healthy 21-to-35-year-old women with a pilot's license from over 25,000 applications. About 1,100 women ultimately fought and 38 died on duty. Marion Stegeman Hodgson served as a WASP and was among the first women to become military pilots. She wrote to her mother: "The gods must envy me! This is just too, *too* good to be true. . . . I'm far too happy. . . . Honestly, Mother, you haven't lived until you get way *up* there—all alone—just you and that big, beautiful plane humming under your control." The WASPs flew every aircraft and received instrument-flight training, with some serving as Link Trainer instructors. In total, over 400,000 brave women from the various programs served the army, the navy, the marines, the Coast Guard, the Army Nurse Corps, and the American Red Cross.

Before and during World War II, the Link Trainer was also critical in expanding training for Black pilots. The US military had existed as a racially segregated institution since its inception. But Eugene Jacques Bullard dreamed of piloting machines in the air just as young white people did. Born in 1895 to a formerly enslaved family, Bullard ran away from his home in Georgia. In 1912, he boarded a freighter to Scotland, where he boxed and later joined a Black entertainment troupe in London. After a trip for a welterweight prizefight in 1913, he settled in Paris, joining the French Foreign Legion, toiling to receive his wings in 1917. As America entered World War I that year, Bullard tried to join the US Air Service and received a rapid rejection. But Bullard then proved his value in the French combat flight assignments he performed until 1919. After espionage entanglements during World War II and postwar stints as a security guard and a civil rights activist in New York, in 1959 the French government made him a Chevalier of the Legion of Honor.

Another pioneering aviator, Bessie Coleman, persevered in the face of racist and sexist US policy. In 1915, 23-year-old Coleman was working in a Chicago barbershop when a newsreel inspired her to take up aviation. All the US flying schools rejected the young woman because of her race and gender. So, she faked her birth date in her passport and set sail to Paris, determined that she could do anything French women could and better. After completing a 10-month course in 7 months, she received a pilot's license in 1921. Racial barriers proved insurmountable even as she excelled in aerial spins and swivels. Coleman's dreams to establish the first Black flying school in the United States crashed, and she pivoted to make her living off air shows. She refused to perform if Blacks and whites couldn't enter the grounds through the same entrance to the field. Her popularity won her enduring fame as "Queen Bess." Tragically, her life was cut short when her plane flipped in midair and threw her to her death in 1926.

Given the barriers to joining the military during this period, many Black aviators like Queen Bess made their living and honed their craft through air shows. One famous daredevil of the era was the Trinidad-born Hubert Fauntleroy Julian, nicknamed the "Black Eagle." In 1923, he parachuted into Harlem and received a ticket for causing a traffic jam. Julian bought a war-surplus seaplane and christened it *Ethiopia* after the kingdom he'd later defend. Unfortunately, his plans for a transatlantic solo flight failed due to his lack of financial backing.

In the 1930s, pilot William Jenifer Powell from Kentucky was considering not only how to join the field of aviation for himself but also how to open up the industry for all Black Americans. Powell enlisted in the segregated armed services during World War I. A toxic gas exposure in Europe ended his military career. Like Bessie Coleman, Powell then faced rejection from flying schools. With his engineering background, he knew aviation was an instrument of world commerce. Powell envisioned Black people taking roles across the industry. With two business associates, he founded the Bessie Coleman Aero Club. Its members included James Banning, who in 1932 was the first Black pilot to fly coast to coast. Banning's mechanic, Thomas Allen, fixed a rickety plane with surplus parts. Reacting to the old Curtiss motor, Banning said, "While originally it developed 100 horse-power, I have reason to believe that some of the horses are dead." Banning and Allen called themselves "Flying Hoboes," completing their trip from Los Angeles to Long Island in 41 hours and 27 minutes over three weeks. Landing "with the gracefulness of a bird, rolling only a few feet before stopping. I had made it!" wrote Banning, "and, immediately, presto-change, a great transformation occurred. My head erect, eyes to the front, shoulders squared, I was a different man."

After decades of pioneering feats by Black pilots, in 1938, the US military developed its civilian pilot-training program across colleges and extended training to Black cadets at historically Black colleges and universities. Pilot Charles Alfred Anderson was hired to head the Howard University program in Washington, DC. The following year, at an isolated campus in Tuskegee, Alabama, founded initially by Booker T. Washington, the US military created a new training program with Anderson as the lead instructor. In January 1941, the War Department started the all-Black 99th Pursuit Squadron, which was to be trained at Tuskegee. Later in May, Eleanor Roosevelt visited the campus. She asked to take a flight with one of the pilots, saying, "I always heard that colored people couldn't fly airplanes." Anderson flew the First Lady on a Piper Cub even as Secret Service remained uneasy about the ride. This highly publicized visit ensured the White House's support for Black pilots, sending the nation a strong message about their proficiencies and patriotism.

Some war-propaganda posters now featured Black pilots, in one instance with the message "Keep us flying! Buy War Bonds." Mrs.

Roosevelt maintained long-term correspondence with some Tuskegee pilots. In a summer 1942 letter, one airman wrote to the First Lady, alluding to the Link Trainer: "Your letters and gifts have been very inspiring and have prompted me to try to be a better soldier. My work is very interesting. . . . Soon this short radio course will be over and I'll be of some service to Uncle Sam." "Chief" Anderson directly instructed hundreds of cadets at the Tuskegee Institute. And over time, just under 1,000 Black cadets trained on Link Trainers. In 1943 and 1944, the 99th conducted multiple attacks on enemy aircraft, resulting in some of the most successful days of the European campaign. By the end of World War II, Tuskegee Airmen earned decorations for their magnificent record of nearly 1,600 combat missions.

While these developments were underway, a group of army pilots was scaling heaven in Binghamton, thanks to Ed Link's newest and stellar creation.

IN 1939, ED LINK began developing a Celestial Navigation Trainer to help aircrews simulate a bombing mission and improve their accuracy in night raids. The critical element of celestial navigation was a device called a sextant. Ed Link developed a training sextant by modifying an instrument used for maritime navigation and applying insights from his close friend (and decorated naval officer) Philip Van Horn Weems. The sextant displayed the angular distance between the stars and the horizon without electricity or the need for stillness. The finely numbered metal arc formed one-sixth of a circle, which gave the device its name. A minute, one-sixtieth of a degree, measured a nautical mile.

Although he came from the conservative naval tradition, Weems was an iconoclast with big ideas about applying maritime navigation principles to flight. Weems met Lindbergh and introduced him to the maritime sextant for the 1927 flight to Paris. Lindbergh plotted his Great Circle course with piloting, guided by fixed ground reference points during daylight. And at night, he used dead reckoning to deduce his present location based on already flown distance and course. Weems thought Lindbergh could have traveled at higher altitudes, reaching the destination earlier using less fuel.

In his influential book *Air Navigation*, Weems wrote that understanding the fundamentals of navigation could have saved countless lives in the early days of aviation. For instance, in the Dole Derby of 1927, the winner, *Woolaroc*, used celestial and radio navigation to reach Honolulu. *Aloha*, the second-prize winner, used a marine sextant. Those who perished in the *Golden Eagle* flew past the northern part of the Hawaiian Islands. They lacked navigation skills and eventually ran out of fuel on the breathtaking blue. A decade later, the 39-year-old Amelia Earhart and her navigator, Fred Noonan, disappeared in their silver Lockheed Electra over the South Pacific. For their pioneering 29,000-mile journey, the pair traveled without navigation or Morse-code communications. Eventually, they lost visibility, fuel, and their lives. Weems devised an almanac convenient for celestial sightings, which proved helpful for Ed Link's next invention.

Ed Link once equated his trainer's benefit to the value a globe had in presenting geography. But the Celestial Navigation Trainer was an artificial paradise. The 45-foot-high silo looked like an abandoned granary, but its chief feature was a star-studded synthetic sky, like a planetarium. A chicken-wired cupola projected over 400 electric stars overhead, including the principal constellations of the Northern Hemisphere. Motor and gears precisely rotated the pretend celestial sphere 360 degrees in 24 hours.

An expanded Link Trainer at the center of the silo accommodated the bomber crew. An instructor on the ground floor set the flight conditions and plotted the crew's performance using the tracker on a chart. The navigators supplied the flight course using the view of the constellations and radio aids. They pinpointed their "plane's" location against the moving stars produced by a collimated effect. The Celestial Navigation Trainer projected 250,000 square miles of photographed terrains as a mosaic on a silkscreen, varying altitudes from 3,500 feet to 35,000 feet. The British Air Force used board models with an elaborate jumble of hand-painted color prints in England. These celestial "fixes" oriented the trainees, who set the targets on a checkerboard of images the trainer "flew" over. Once the trigger was pulled, a glow on the screen indicated whether the mission was accomplished or not.

Over time, the Link Trainer and its descendants enabled the lifesaving technological and psychological advancement of learning to fly without leaving the ground. Nonetheless, questions about the training procedure remained: What behaviors must be trained, and to what criteria? What is the optimal mix of theory, practice, and reflection? How does one evaluate the effectiveness of training and make the best use of costly assets? Building on these considerations, systems engineer Valerie Gawron offers generic training criteria that she has developed over the years, calling them the "ten C's": completeness, clarity, conciseness, consistency, compactness, communications, competence, correctness, constructing from previous knowledge, and staying current. Arranging these Cs becomes critical for trainees to effectively transfer what they have learned in the trainer to a real scenario.

Advanced versions of the Link Trainer mimicked artificial storms, fog, icing conditions, rough air, glare, and the zero-zero condition that led to Senator Cutting's death and many other unfortunate deaths of passengers and pilots. The trainers became sophisticated with so-called upset-recovery training, in which pilots reacted and adapted under extreme circumstances. And with newborn computing power, flight trainers were on their way to becoming flight simulators.

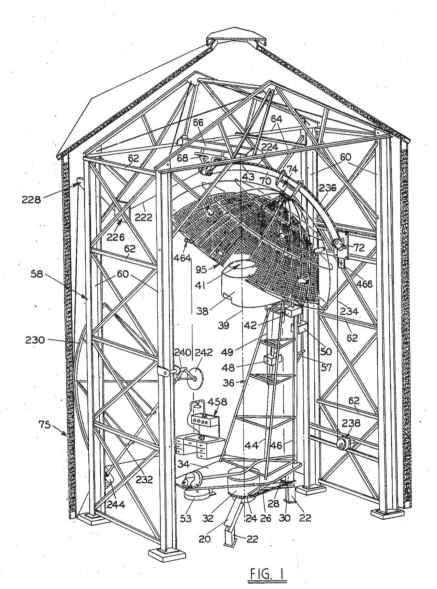

FIG. 1

EDWIN A. LINK
INVENTOR.

BY *Donald T. Hillier*

ATTORNEY.

Chapter Three

Wicked Vulnerability

A ROUND 2:00 A.M. IN JULY 1918, ISAAC GONZALEZ, A Puerto Rican electrician, stepped out of his clammy Back Bay tenement in Boston. The heat was unbearable, and his fever dreams were intolerable. He looked around and ran into the darkness as if drawn by the menacing powers of Mount Doom.

Gonzalez lived in a boisterous Italian neighborhood with cramped dwellings. Like many others in the area, he worked for Purity Distilling Company, which converted molasses from the Caribbean into industrial alcohol. During the distilling process, the first boiling of sugar cane yielded syrup used for cooking. The second boiling produced less sweet but premium molasses for commercial-grade edibles. The third created the puckery black-strap variety, fermented and distilled to make industrial alcohol, mainly for munitions. Purity owned the coal-gray tank on Commercial Street, about 200 feet from the port. The container could hold 2.5 million gallons of crude molasses.

Ever since the slave ship *Desire* sailed out of Salem in 1638, molasses has played a significant role in American history, notes historian Stephen Puleo. It wasn't a mere by-product of sugar cane; John Adams stated it was an "essential ingredient in American independence." As early as the 1760s, sailors smuggled almost half a million gallons of molasses into New England. The rum and sugar trade furthered the young nation's participation in the transatlantic slave trade. Some sailors received their wages in rum and could exchange them for enslaved people in West Africa. In the 1790s, whiskey manufacturers used one-half of the American molasses supply. The remainder went to colonial delights such as gingerbread, baked beans, shoofly pie, and apple-and-spice pandowdy. The Quaker leader William Penn endorsed molasses beer as a "very tolerable drink," as it became a popular substitute for drinking water.

In April 1915, Purity began building a molasses tank. Workers riveted seven overlapping vertical steel plates ranging in thickness between one-half and five-eighths of an inch. The assembly, one of the loftiest in New England, was 50 feet tall and 90 feet in diameter. Builders expedited construction when the United States entered World War I. Manufacturers faced increased demand to convert more molasses into alcohol for ammunition. Pressure loomed to get the tank in operation quickly.

Management didn't have time or interest to conduct thorough leak tests. The tank barely survived 50-mile-per-hour gusts during construction but was declared "sturdy, sound, and ready to use." When a 35-year-old man slipped and fell off a broken staging plank into the tank and died, the manager was upset about losing half a day of work. The goal was to have the tank ready for a 700,000-gallon delivery of molasses by that New Year's Eve.

The steel container gave a creak like the hungry cry of an animal. Molasses seeped from the seams and gelled around the rivets. Gonzalez dreaded the tank's collapse, a fear that compelled many of his nocturnal nightmare-driven runs in the summer of 1918. These were times in which anarchists posed violent threats, protesting unfair wages and unsafe working conditions. Radicals attacked companies such as Bethlehem Steel Works, DuPont, and Westinghouse, injuring and killing many, and even blew up a room in the US Capitol in 1915. When a pipe bomb intended to blow up the molasses tank failed due to a faulty fuse, Purity arranged for 24-hour security. Yet, Gonzalez's fears remained.

On January 12, 1919, Boston was a frigid 2 degrees Fahrenheit.

Around 11:00 a.m., the 400-foot-long *Miliero* steered into the harbor. The workers pumped 600,000 gallons of molasses from the steel-hulled steamer into the tank. The container was near capacity, totaling 2.3 million gallons. As the new batch mixed with the cold leftovers, the tank groaned. The workers ignored it because they were familiar with the noise. Molasses started seeping between the seams. Because of the temperature changes, the mixture fermented and produced gases that spiked the pressure against the steel walls. The air temperature had risen to 16 degrees when the pumping finished a day later.

On January 15, the temperature rose to an unseasonable 40 degrees. People in the neighborhood set aside their peacoats to enjoy the weather. Around lunchtime, they talked about the flu pandemic and armistice. Some remembered Theodore Roosevelt, who had died two weeks earlier. Others gossiped about Babe Ruth's threat to resign from the Boston Red Sox, and Charlie Chaplin's latest hit, *Shoulder Arms*, the silent comedy costarring his brother Sydney. Firefighters were playing cards. A pair of railroad workers had gotten into a tiff.

All seemed normal when a deep rumble rippled across Commercial Street. Steel bolts snipped, and fastener rivets snapped from the colossal container, flying in all directions like shrapnel. With a roar, the molasses tank split open and set the tsunami of syrup loose. The surge—30 feet high, moving at 35 miles per hour— engulfed everyone and everything in all directions. The gluey liquid poured into every opening, swirling into cellars and shoving houses off their foundations. The sweet lava, thick and slick, splintered furniture, tossed farm equipment, smashed windows, and upended vehicles and wagons, making debris of barrels and boxes. The elevated train trestle buckled from its steel supports and tumbled. An oncoming train careened and skidded to a stop. Its passengers froze in fright as the people below skated and fell. Schoolchildren returning from their morning classes were rolled like fudge candies.

The brown wave suffocated 21 people. Some bodies were found

immediately, and others were pulled out from the pier months later. The sticky ripples injured 150 people, and many animals perished. "Molasses, molasses everywhere," boomed the front page of the *Boston Globe* the next day. Rescue workers set up limelights for round-the-clock cleanup of the viscous sludge, freeing people trapped "like dinosaurs in a tar pit." Fireboats sprayed saltwater on the streets and buildings. The molasses was still gummy, but efforts continued to siphon the slop—from bathrooms to basements and butcher shops. A sugary smell haunted Boston's downtown for decades to come.

The "hard vulnerability" of the tank was known and ignored. Molasses, weighing nearly 12 pounds per gallon, stressed the steel plates at over 31,200 pounds per square inch and nearly doubled the pressures on the rivets, which were rated for only 10,000 pounds per square inch. Studies of tensile failure along the row of holes, shear failure associated with rivet pressures, and the bearing failure linked to the damage between the fasteners and their holes confirmed that the stresses far exceeded the allowable limit. Based on the technical manuals of the time, the Purity tank's safety factor should have been at least four times the strength chosen. The tank's thin walls, selected as a cost-savings measure, proliferated a rapid fracture. Since the 1850s, when steel production ramped up, low-temperature brittleness has been a problem. Microscopic analyses from the Boston Navy Yard showed that the tank's fractures were herringbone, a zigzag damage pattern. The breakdown of riveted constructions remained common well into the mid-20th century; during the winters of 1943 and 1944, World War II Liberty ships literally broke in half, embrittled. Another unrecognized insight in 1919 was the complex effect of manganese on steel fracture.

After this peculiar and preventable plight, the parent company, United States Industrial Alcohol, faced 125 lawsuits. They set in rotation a circular blame game that lasted five years. Court proceedings had over 3,000 witnesses and 45,000 pages of conflicting testimony. As groups conducted explosion tests with large tanks and scale models, Harvard and MIT professors disagreed on why the tank burst. The company claimed that bomb-tossing anarchists destroyed the tank. In the end, despite the material defects of the tank, its poor design and fabrication, and the improper mixing of batches that topped the tank pressures, the court didn't indict the company. It ordered a $630,000 fine (about $11 million now), with families of victims receiving $6,000.

Engineers can solve hard vulnerabilities like those in the molasses disaster through improved protection regimes, testing, and certification requirements. Since the incident, higher technical standards have governed the design, fabrication, and installation of storage tanks. Advanced safety and maintenance measures have led to new materials that can withstand unusual stressors. Well-regulated construction codes and corrective actions for pressure vessels and piping have prevented collapses worldwide. Yet, a 40-year review of 242 storage tank accidents from the 1960s onward noted that fire and explosion accounted for a disproportionate number. A third of those were "caused by human errors including poor operations and maintenance." Sounder engineering would have forestalled these accidents.

"There's no disaster that can't become a blessing," the novelist Richard Bach wrote, and "no blessing that can't become a disaster." The molasses disaster remains a prominent case study of vulnerable design and negligence. It is also a cautionary tale in constructing below safety margins and placing profit over public safety. While regulations tend to fix these issues, any hard vulnerability is bounded by clear beginnings and endings. As we will see, that's not necessarily the case with soft and messy vulnerabilities. People often judge based on the risks they are personally willing to accept, much like the soft efficiencies they are willing to pay for. The resulting vulnerabilities change with changes in properties, like how the failure mode of some metals transitions from ductile to brittle under certain conditions.

One can see soft vulnerability lurking in the molasses disaster regarding how much (lack of) safety was acceptable to Purity. And in a messy proportion, the superstructure of molasses and rum production was built on human exploitation and endangerment, including slavery. While hard and soft problems are often nested in messy problems, the molasses flood shows that even seemingly hard problems have soft and messy moments. A material vulnerability can be simultaneously mechanical and moral. Just because something has a hard vulnerability doesn't mean it's amenable to a quick technical fix. Safety programs that produce hard preventive solutions do require the softer processes of training and quality management. As solutions are part of resolutions, so can resolutions eventually become solutions.

Around 2:30 a.m. in late August 1918, Gonzalez ran again toward the tank in the ill-lit streets with his same tormented dash. His neighbors grew suspicious, and his wife became distraught. He looked at the area from Copps Hill. The night was dead still, and Commercial Street calm, with some seagulls cawing. The molasses tank stood tall as if invulnerable. His frequent warnings about the tank were unheeded. In October 1918, just three months before the sticky terror was unleashed, the exasperated Gonzalez resigned. There were no more bittersweet dreams to haunt him.

✳

IN *HERE IS NEW YORK*, E. B. White wrote that the natives give the city its solidity, settlers give it passion, and commuters give it its tidal restlessness. Guy Tozzoli was one such restless commuter, and in the 1970s, he helped New York reach further heights. As the Port Authority director, Tozzoli oversaw the design, construction, and eventual leasing and maintenance of the iconic World Trade Center.

Around 7:00 a.m. on that severely clear, 70-degree Tuesday morning in September 2001, the 79-year-old Tozzoli departed Hoboken for lower Manhattan in his black sedan. The seasoned engineer hosted conference calls with his colleagues during the hour-long commute in his suit and tie and gravelly voice with an Italian flair. Tozzoli was a disciple of Robert Moses, a domineering personality who charged and polarized urban development. At the end of his 40-year career with the Port Authority of New York, in 1987, Tozzoli had been named president of the multinational World Trade Centers Association. From his 77th-floor office in the North Tower, he could see the Empire State Building, the Art Deco titan and runner-up in New York's race of heights.

Tozzoli worked for Moses to plan the grounds of the 1964 World's Fair in New York, but the Twin Towers, at 110 floors each, were no cookie-cutter constructions. Rising from two giant ditches on the tip of the Hudson, the skyward cities were sturdy, slender, and sapphire. With a population of 40,000 workers and four times as many visitors daily, the towers came with their own zip codes. Steel: 200,000 tons assembled just in time. Concrete: 425,000 cubic yards. Aluminum: 2.2 million square feet. Marble: 6 acres. Glass: 43,600 windows with crawling automatic washers. Wires: Electrical cables stretching the round-trip distance between Acadia and Yosemite National Parks. These weren't just specifications; they were political statements.

On September 11, Tozzoli ran late for his 9:00 a.m. meeting, and a bus accident set him back. Just as he was entering the Holland Tunnel, at 8:46 a.m., a Boeing 767 struck the North Tower over a dozen floors above his office. The building swallowed the 200-ton aircraft arriving at over 450 miles per hour. Thousands of gallons of jet fuel dispersed on impact, splashing over ducts and shafts. The smoky orange fire billowed on television screens worldwide.

"It's going to take us a long time to fix that," Tozzoli muttered. His mental reference was to the 1993 terrorist bombing with a truckload of nitrourea that killed 6 and injured about 1,000 people. He knew all the exit routes from the buildings. At 9:03 a.m., a second plane crashed into the South Tower at over 550 miles per hour, striking near the 80th floor. The estimated force at impact was 25 million pounds. Tozzoli tried to rush to the scene, telling a cop that he had built the place. "I don't care if you're the pope" was the response denying his entry.

The South Tower collapsed at 9:59 a.m., and the North Tower at 10:28 a.m. Nearby seismographs picked up a magnitude of 2.4, and debris damaged as many as 400 buildings in the neighborhood. Two fireboats pumped water from the Hudson at over 300 gallons per second. The ensuing cascade collapsed critical utilities, halted power distribution, burned internet servers, blocked wireless calls, leaked natural gas, crashed the New York Stock Exchange, and damaged the telecommunications systems of Verizon in the Financial District—a network with the capacity equivalent of Austria, Denmark, and Egypt combined. Phone networks were overwhelmed with over a billion long-distance calls that day. AT&T mobilized large semitrailers with backup mobile command centers and brought them online in crucial staging areas. Dust and debris filled every physical space, just as fear and grief started filling the minds and hearts of a nation under attack.

Unlike the molasses that sugarcoated the blocks of Boston, the effects of 9/11 mutated beyond the streets of Manhattan. The tragedy has functioned as both a metaphor and a metonym for disasters. Developing a single model to understand all the vulnerabilities that fueled the failure was unfeasible. It requires an appreciation of complex adaptive systems, distinguishing the two very different meanings of this term. The first meaning is a complex system that's adaptive as a whole system, as in a multicellular organism. And the second is a complex system comprising elements following their own locally adaptive strategies, as in traffic. The second type of complex system doesn't necessarily organize itself to operate as an integrated whole, as in the first. Indeed, the second type could lead to many maladaptations witnessed in vulnerabilities of 9/11: people working at cross purposes, failing to communicate.

A key reason why these two types of complex systems aren't distinguished well enough is that both are interdependent and can show unpredictable behavior. That's why, compared to the decades of papers published on "complex adaptive systems," there are hardly any publications on "complex maladaptive systems." Every social vulnerability is marred by maladaptation and can reach a level of complexity that crushes modeling tools. Still, one must consider their whole-system impacts.

An individual model is inevitably weaker in decoding "soft vulnerabilities" of maladaptation. In contrast, ensemble models can bolster a more complete explanation. For example, why were there differences in how the fire burned and persisted in each tower? Why did the second tower collapse first, even though the first tower was hit first by a similar aircraft? What were the smoke-flow and ventilation patterns across the floors? What led to intense fires in specific locations of the towers? What role did aviation fuel play in the fire spread? In an elaborate question-behind-the-question exercise, the National Institute of Standards and Technology (NIST) built linked ensemble models to investigate why the Twin Towers collapsed. Between 2002 and 2005, engineers and scientific analysts blended information from physical model testing, burn experiments, lab studies, statistical processing, and visual evidence. The investigation circumstances were daunting, and simulated tests couldn't reproduce the original failure conditions. Instead, the investigators composed a daisy chain of models and explanations, each separable for testing. In eight volumes, over 10,000 pages, the resulting analyses reported on aircraft impact, building and fire codes, structural steel failure, fire-protection systems, heat-release patterns, emergency responses, and people's behavior.

The 9/11 destruction was one of the most photographed events in human history, but therein lay the challenge. The investigators analyzed 7,000 photos sifted from 10,000 in total. They reviewed 75 hours of video clips from a raw count of 300 hours. How could they piece together all these images? How to verify a precise time sequence for the collapses? How to predict the fire's behavior and damage course in retrospect? Thousands of photos had no time stamps. Even when they had time stamps, in some cases, camera settings mistimed the photos. In other instances, the broadcasts had time lags. Like making a pointillistic portrait, the investigators meticulously constructed a window-by-window profile for the four faces of each tower. From airplane assaults to building breakdowns, investigators assembled a consistent and cohesive timeline by piecing together the fragmented information.

The investigators also scrupulously studied the properties of metal in the Twin Towers. Without knowing their original location on the towers, the team cataloged 236 structural steel elements, later determined to be from exterior panels, core columns, or floor trusses. The independence among models was crucial to constructing the "most probable collapse sequence": the model of aircraft impact and damage, the model of fire and heat conditions creating a "thermal history of each floor," and the model of the structural deformation and collapse. One added complication was that the steel came from multiple suppliers, including one in Japan. While a standard construction project has two or three different steel grades, the World Trade Center had 14 grades, with yield strengths from 36 to 100 kilopounds per square inch.

While these modeling processes multiplied questions, they also created the capacity to answer them. The results fed into the other modules of the overall investigation. They contributed to a global simulation that unraveled the mystery of how the towers' core structure weakened, bowed inward, and sagged the floors, leading to the collapse. And over the years, the 9/11-failure analyses showed ways to improve fire-resistance rating for frame structures. They also enabled a fireproofed object to remain protected longer, shaped exit stairways for rapid evacuation, and advanced radio communications for emergency operations.

The towers were anywhere between one-third and one-half occupied when the aircraft struck. Had they been fully occupied, with 20,000 occupants each, the NIST report estimated that 14,000 people could have perished. The evacuation time could have been three hours or longer. Modern builders still ask: Can skyscrapers be made fully resistant to terrorist attacks? And if so, can they survive them? Society's appetite for accepting soft vulnerabilities and making relevant trade-offs is at the core of such questions. Even the sturdiest walls of the Pentagon couldn't deter a plane from penetrating the building. Concrete bunkers may survive explosions, but they aren't our preferred designs for daily life.

One way to cope with soft vulnerabilities is to engineer more nimbleness into product or service functions. New York City operations had previous experience with short-term responses. Because operations dealt with up to 600 water-main breaks yearly, the city had a greater capacity to handle water breaks on September 11. The redundancies engineered within the transit system to work around train failures enabled rapid recovery. Similarly, the energy company Con Edison's ability to tackle frequent underground fires gave it the fitness to cope with the catastrophe. And when advanced technologies failed, the New York City Fire Department adjusted its response using old radios and manual overrides.

Another means by which engineers reduce soft vulnerabilities is by designing multiple protective features. Our bodies have evolved many-layered protections to guard against illness. The first stage of protection begins to break down when we are young. Small tumor-like growths may develop but don't spread as cancer because protections limit tumor growth. Then as we age, the next set of protection fails, and another compensating mechanism generates. A polyp may appear and disappear because the body, correcting for errors, can quash cancer growth. And then, as another safeguard develops, the tumor may spot a vulnerability and start to spread.

Suppose the body gains a new protective feature in this long evolutionary relay race. The robustness can confer an advantage, while the older system layered under the new degrades. More new protections arise, layered on top of the existing protections, which begin to decay. Older defenses become more variable and less reliable compared to more recent protections, potentially allowing mutations and maladaptations. Scholar Steve Frank calls this increasing system performance with decaying system components the "paradox of robustness," highlighting biological vulnerabilities in disease progression. As we'll see in "messy vulnerabilities," an engineering version of this evolutionary paradox relates to upgrading levees for hurricane protection over an existing system that can sometimes worsen a disaster. Policy problems arising when newer fixes are introduced on top of deteriorating ones are common but not well theorized. The destruction of the World Trade Center and its aftermath showed that resolving soft vulnerabilities involves addressing technical, legal, behavioral, and emotional events. In messy vulnerabilities, we need to be prepared for more surprises.

One of Tozzoli's joys of working at the World Trade Center was watching visitors from many parts of the world express their amazement at the engineering. With a twinkle in his eye, he told an interviewer, "I'd ride the elevator to the 107th floor just for the hell of it because I wanted to hear what people were talking about, getting to floor 100." On September 11, as the Twin Towers disappeared, a new emotional weight set in, particularly for Tozzoli. He listened to the radio on his drive home in the hours-long procession. As a citizen, he felt helpless. The towers erected with clocklike precision were now clouds of smoke and dust; 12 million square feet crushed to 1.2 million tons of debris. As an engineer, he knew the towers' legacy was more than what they were in stone. As in Pericles's sentiment of the Funeral Oration, the "unwritten record of the mind" will grant them life in a new form. That night, Tozzoli's eyes were wide open in tears.

✳

HURRICANE IVAN GAINED CATEGORY 5 strength as it passed Cuba and twirled toward the Gulf Coast. New Orleans–based scholar Shirley Laska was worried that it could be the "big one." On September 13, 2004, Ivan landed near the Alabama-Florida border as a Category 3, with 120-mile-per-hour winds that wrought $26 billion in damage.

"What If Hurricane Ivan Had Not Missed New Orleans?" is how Laska titled one of her articles, writing that the possible ruin could have cost over $100 billion. A 17-foot storm surge would have been pushed into Lake Pontchartrain, overtopping the levees and barriers designed to protect New Orleans to the south. The city's 20-foot flood would have taken 9 days to pump out, and the evacuation of 600,000 people would have taken over 10 days. Per the American Red Cross, the consequences of Ivan could have been far worse than a powerful Californian earthquake.

On August 29, 2005, Laska's anxiety came to life. With an energy equivalent of 100,000 atomic bombs, Hurricane Katrina barreled over the Gulf and made its second landfall at 125 miles per hour, dumping 20 inches of rain. Over 1,800 people died, and 1.2 million evacuated the area—a migration four times the displacement from the 1930s Dust Bowl. The 1970 Bhola cyclone in Bangladesh killed over 500,000 people. The 2004 Indian Ocean tsunami took 227,000 lives and displaced seven times more people. Yet, Katrina was a disaster of disasters provoked by "messy vulnerabilities." Total losses were around $200 billion. Katrina damaged about one-half of the 71 pumping stations, creating 50 major breaches in protective structures. Unmaintained pumps aggravated the floods. The temporary pumps cleared the water in 53 days.

"Messy vulnerabilities" are often or always a mishap of many hands. They can arise from geological as much as geopolitical susceptibilities. New Orleans was established in 1717 as the deepwater port between the Mississippi River and the waterway Bayou St. John leading up to Lake Pontchartrain. The French Crown developed it as a petite 14-block settlement, with a draining ditch on each block. The first levee was erected in 1719 by Jean-Baptiste Le Moyne de Bienville, the French Canadian Father of New Orleans. In 1722 the Great Hurricane pummeled the coastal land with a 10-foot storm surge. For economic reasons, the French decided the city would continue functioning where it was. In 1734, the Mississippi broke the levees and swamped the settlement, forcing the city to fortify its natural levees. The Louisiana Purchase in 1803 prompted more settlers to drain the swamp and build walls to protect their plantations. Over the next two decades, the walls stretched sporadically from New Orleans to the estuary of the Ohio River.

The steamboat era of the early 1800s gave New Orleans a year-round advantage for commercial shipping, ranking the city second to New York in commerce. Between 1833 and 1878, the city built over 35 miles of canals to dewater the lower parts that housed the city's poor people, many of whom succumbed to yellow-fever outbreaks. Yet an 1844 flood wrecked the farms of Arkansas, Mississippi, and Louisiana. In 1849 floods ravaged the Lower Mississippi Valley, and the Swamp Land Act of 1850 turned the political focus to taming rivers. The levee system was built on the backs of enslaved people, and their repeat ruins seemed like slavery's "curse to the soil."

In 1861, after long episodes of nervous breakdowns and just months before the Civil War, Captain Andrew Humphreys delivered his long-overdue *Report upon the Physics and Hydraulics of the Mississippi River*. In it, the future chief of the US Army Corps of Engineers and a founder of the National Academy of Sciences wrote, with a young lieutenant, Henry Abbott, that "the first object of the settler has always been to secure himself from inundation during the high stages of the river." Humphreys and Abbott believed in the sole use of levees to control floods, giving their deeply technical report a peppy political vision. The army engineers liked what they heard, and so did the politicians. Congress approved funding, and the "levees only" approach profoundly shaped how America managed its water resources for many decades.

Then the flaws of this policy began to float up. The rivers were now swollen, potentially more destructive, forcing engineers to make the levees taller. Other proposals were dismissed. Pump stations began to appear around 1895, but Humphreys and Abbott's logic of lengthening levees remained the standard for the next three decades. Then the Great Mississippi Flood of 1927 breached levees

and buckled communities, leaving 700,000 people homeless, surpassing a billion dollars in damages, about a third of the federal
budget. Following the $325 million federal Flood Control Act of
1928 (totaling about $6 billion today), the Army Corps of Engineers
led the Mississippi River and Tributaries Project between 1931 and
1972. New Orleans's population started to boom in the 1960s with
new schools, suburbs, and the Superdome. Then, beginning in
the late 1980s, engineers began constructing concrete floodwalls.
They based their assumptions on Hurricanes Betsy and Camille,
which battered Louisiana in 1965 and 1969. But the new barriers were not tested with worst-case scenarios of the new reality,
and worse, their construction wasn't complete because of conflicts
between state and federal officials on the details and funding when
Katrina struck. And days later, Hurricane Rita brought devastat-
ing rainfalls, reflooding parts of New Orleans. As evident in the
levee-failure investigations, engineers are never off the hook. They
are held accountable for explaining the shortcomings of creations
that precede them and are responsible for preventing future losses
responsibly and effectively.

Following Katrina, an Interagency Performance Evaluation
Taskforce (IPET) studied 284 miles of levees and floodwalls, including the 169 miles that were severely damaged. IPET concluded that
the breaches in the London Avenue Canal and the 17th Street Canal
had led to over 80 percent of the flooding in downtown New Orleans and over one-half of the fatalities. "The system did not perform
as a system," the IPET report said. "The promise or perception of
a system when it does not exist is perhaps more dangerous than no
system at all." Like the 9/11 modelers, IPET had its own version
of the question-behind-the-question exercise. What conditions did
Katrina create where the levees and floodwalls needed to operate?
How did the performance of the hurricane-protection system compare to its design intent? What can be learned from the damaged
and undamaged levees to plan their repairs and improvements?
What role did or could pumping and drainage play? And what were

the broader social consequences of the flooding, and how to plan for future risks?

Systems engineer Zach Pirtle has studied the different models IPET used to analyze the levee failures. IPET pursued three branches of modeling, each of which led to a commonly shared conclusion. One approach used finite element modeling. This high-resolution method deconstructed the levees into over 3,000 cells. Then, the analysts computed stresses and strains on the material deformation. From this, they arrived at the desired safety factor to reach a stronger capacity than the system can support. They ran slow-motion reconstructions of the levees fractions of seconds before they failed.

A second approach, called limit equilibrium assessment, analyzed the preconditions of failure. The water in the Gulf, the lake, and other canals overtopped some levees and eroded them. In some cases, the levees failed before the water reached the top, creating critical vulnerabilities for deadly breaks and gaps. Some of these levees were designed for lesser storms; in other cases, they couldn't withstand the water levels they were intended to withstand. While many levees had a "hard vulnerability" of suboptimal safety factors—like the molasses tank—there were several other critical vulnerabilities. These vulnerabilities included the unanticipated way the levees settled underground, leaving gaps between the structure and soil.

The third approach that IPET took, a centrifuge model, used miniature prototypes of the levee system. Models cannot replicate the original conditions of the disaster, nor can they replace active operations for destructive testing. Engineers refer to the scale model approach as "similitude," which can guide strength tests with "toys that save millions." Practical full-scale models of history have far preceded the scientific theories of structures, perhaps why the Great Living Chola Temples in India and elderly European cathedrals are still admired, for strength and soul alike. The unbreached analogs for Katrina nearly matched those of the levees, so the stability and erosion testing were comparable. Put to the test, the centrifuge model agreed with the other models IPET reviewed. Their damage assessments converged on a key finding: the failure of I-walls across the flood-protection system of New Orleans, with particularly catastrophic consequences from the 17th Street Canal breach with weak underground clay. The result was a "hydraulic short circuit" that worsened seepage and erosion.

Pirtle and his colleagues observed that these independent models provided confidence about the failure progression, each with different operating principles. Moreover, of the three models, one was physical, and two were numerical. Their analyses of levee breach were fundamentally different, as were their scales and several abstractions, which, as we saw previously, are helpful tools to manage vagueness. Pluralism encourages a collection of models, with each model's incompleteness being compensated by another. The models highlight the trade-offs in generalizing at the expense of realism and precision. Then there are occasions where gain in realism is because of a sacrifice in generalization and precision. And sometimes, precision is acquired without the strength of generalization and realism. Within this mix of compromises—where multiple models with multiple methods run on multiple realities— lies robustness, the democratic essence of models. Or, to put it in the helpful phrasing of scholar Richard Levins, "Our truth is the intersection of independent lies."

Two decades before Levins, writer Jorge Luis Borges reflected on the obsession with excessive exactitude. In his profound paragraph-length story, expert cartographers made a map with point-to-point precision of their country. The map became as large as the landscape itself. The perfect creation was deemed worthless, and the land became a habitat for animals and beggars. Maps can unexpectedly become the territory, and the image, the reality. The result is classic confusion between the model and the system it claims to represent. As orderly as the London Tube map may seem, for a character in a hypnotic Neil Gaiman novel, it's "a handy fiction that made life easier but bore no resemblance to the reality of the shape of the city above." But what Levins and Borges allude to is the hallmark of the democracy of models. A model's confidence comes from combining many perspectives.

Dissolving messy vulnerabilities requires more than pure truth-seeking; it's about trust-building. The Katrina models were built from disorganized knowledge, even the ideal of "organized igno-rance." If the soft vulnerabilities of 9/11 were resolved into global security measures, then the messy vulnerabilities of Katrina point toward engineering better policy capabilities. Technical failures have been part and parcel of New Orleans's natural history. But Katrina reminds us that design considerations such as safety factors shouldn't simply be based on past practices as if it's a hard problem. The messy issues converged when even well-intentioned emergency-response and reconstruction efforts were dysfunctional. These are inevitable surprises that merely depart from expectations.

✳

AS LONG AS VULNERABILITIES have existed, so have our desires for immunity against them. Leonardo da Vinci proposed undam-ageable bridges in one letter to the Duke of Milan during the Italian Renaissance. He believed his structures could resist fire, withstand combat, and yield to no army. Yet, vulnerability can be a shadow concept, visible only when the light is cast at a slant. The Romans applied it to describe how a wounded soldier on the battlefield is prone to further injury. A vulnerability has a time dimension: a prior wound may curtail the capacity to respond effectively, evi-dent in the soft and messy problems of 9/11 and Katrina. Vulnera-ble people, even countries, find it challenging to reconstruct after a significant event. But many vulnerabilities aren't necessarily per-manent. Vulnerabilities shift as one goes through the many stages of life, like changing jobs or homes or immigrating to a country. These may well be " 'situations' which people move into and out of over time."

Like vagueness, vulnerability can be imprecise. As in looking at the impacts of carbon emissions on communities, how do they affect climate targets, and what best options exist for adaptation? Answers to these questions can be dollar values or morality debates. However, vulnerability offers us a greater perspective as a starting point. Who is vulnerable to climate change, and how can one address their specific needs as a priority? If we frame vulnerability as an end point, that assumes that adaptive capacity determines vulnerability, writes scholar Karen O'Brien. In contrast, "viewing vulnerability as a starting point says that vulnerability determines adaptive capacity and hence adaptations." Like efficiency, making vulnerability a starting point offers more pathways for intervening in a wicked system.

Further, vulnerability is often confused with risk. But swapping risk and vulnerability is like interchanging weight and strength. Akin to the molasses tank, many organizations have weight but not necessarily strength. The flexibility and endurance that characterize strength are also features of vulnerability. To the point of disaster vulnerability, historian of the Dust Bowl Donald Worster outlined the crucial difference between flood control and irrigation. He compared flood control to holding an umbrella when it rains and irrigation to directing the rain elsewhere. If flood control is about actively managing risks, then irrigation is about actively engaging with vulnerabilities.

Many see vulnerability as something to be eradicated. But the absence of vulnerabilities isn't possible, and even if it were, it would not improve the world. Minimal vulnerabilities can make us adaptive. The Dutch scholar Wiebe Bijker, who has studied the social construction of disasters such as Katrina, observes that the inner workings of vulnerability depend on the external environment. The United States and the Netherlands prioritize different coastal-defense methods because of their political cultures, even if the goal is the same: keep the water out. A person "clad in many layers of wool, can be invulnerable up in the Himalayas but vulnerable in

Hyderabad, India," Bijker writes. His point is to appreciate that people in Louisiana, Bangladesh, and the Netherlands face the chronic risk of flooding but are acutely vulnerable. While risk is readily quantifiable, vulnerability is based on perception. Vulnerability reduction carries a narrative power of lost lives, families, and livelihoods unrelatable by risk ratios. Thus, in wicked systems, vulnerability is more relevant than risk.

One can also manage risks without successfully reducing vulnerabilities. Before 9/11, the World Trade Center was vulnerable to terrorist attacks by airliners. Engineers anticipated this vulnerability when designing the Twin Towers in the 1960s by learning from a July 1945 incident. A B-25 bomber supposed to land at Newark airport crashed into Manhattan's then-tallest tower, the Empire State Building, which was hidden in the fog. In similitude testing, engineers considered the resilience of the Twin Towers against a Boeing 707 crash, an aircraft similar to the Boeing 767s that ultimately toppled the towers. The tests concluded that the structures would not collapse and would be safe except to suffer local damage.

Every day, engineers conduct design improvements to restrain risks and refine performance. Consider the difference in capabilities between a fourth-generation combat aircraft and a fifth-generation striker. The design improvements can be efficient and vulnerable: millions more lines of code, twice the software-controlled functions, over five times as expensive, ten times more subsystems, and a hundred times more interfaces. In a world focused solely on risk—or efficiency—a simple malfunction or a pathogen from a wet market can give the world a bitter taste of complexity.

Chauncey Starr, a developer of modern risk analysis, was well known for his elegant analogies. In one probable scenario, he explored the public perception of risk if tigers escaped a zoo. Standard risk-management approaches might argue for hard solutions, such as eliminating all tigers from zoos, defanging the tigers, or securing the tiger cage with cameras and alarms. But Starr argued that none of these risk-based approaches influence the public acceptability of zoos, even though wild animals sometimes flee their enclosures. Most people consider zoos safe to visit with their children. Here, trust and reliability become vital proxies of vulnerability, a form of confidence that no numerical risk analysis can provide.

Engineer Roger McCarthy, who has led high-profile failure investigations, notes that even with decades of experience, engineers struggle to differentiate between the reasonable and the not-so-reasonable risks. "Now, if I know how to measure risk, I'll measure the goddamn thing," McCarthy said. "But if I show that your unadjusted lifetime risk of dying from the mere fact of you possessing a driver's license is 1 in 200, your reaction is 'Nah! It can't be remotely that high.' We presume that driving risk is almost entirely under our control. Yet, we all have known personally somebody who's died in an automobile accident."

Risk quantification is common in engineering models, even if disregarding human behavior is convenient. "Assume no people are in the equation" engineers like to half joke when modeling complex risks. Once, an automaker asked McCarthy to compute vehicle parameters for a new pickup design with less rollover. He characterized the performance of Ford, Chrysler, GM, and other brands. He boiled down his analysis to 200 measured vehicle parameters. McCarthy fed them into a multivariate regression model, a statistical technique that finds patterns among many variables simultaneously. He then solved for variables that saliently predicted rollover and found that he could virtually predict the rollover rate of every automobile model on the road with high accuracy. How? "I had so many variables that the machine could arbitrarily choose coefficients on the most abstruse variables and make them fit," McCarthy said. "Was this engineering? Was I learning any design? No, but I was doing marvelous mathematical curve fitting." Ultimately, he could match any rollover rate and vehicle parameters with his algorithm.

Unlike risks, vulnerabilities aren't amenable to numbers because they often arise from human perceptions of outcomes. The immediate effects of hard, soft, and messy vulnerabilities make sense only in their contexts, even if their results were ultimately far-reaching. As a composite of hard, soft, and messy conditions, wicked vulnerabilities tell us what might have existed and what could exist. One can see a version of the molasses tank incident in the 9/11 attack and some aspects of both incidents within the Katrina disaster. In wicked systems, systemic vulnerabilities are not always attention-grabbing; they can be "boring apocalypses" that slowly become irreversible. We shouldn't presume that future vulnerabilities will look like the past. Overcoming such faults will require us to overcome our defaults.

Refrain

Yaw

TAKEOFFS ARE VOLUNTARY, AND LANDINGS ARE MAN-
datory. Pilots James Morton and Maurice Watson were well
trained in this maxim.

At half past midnight on March 30, 1967, Morton, a trainee captain, guided Delta Flight 9877 out of the gate at New Orleans airport. His instructor, Watson, accompanied him as captain of this
routine proficiency check. Their airplane, the Douglas DC-8 aircraft, had completed a regular passenger trip from Chicago earlier in the day. Three Delta employees and an inspector from the
Federal Aviation Administration were also aboard this short practice ride. The training agenda was straightforward: takeoff, circle,
and touch down with a simulated two-engine out on one side of
the aircraft.

Nearly equal in training, Morton had completed over 17,000
flying hours, and his instructor, Watson, was approaching 20,000
flight hours. The cockpit atmosphere was relaxed as the flight
began to taxi at 12:40 a.m. At 12:43 a.m., the aircraft reached its
critical engine speed. When it was too late to reject the takeoff,
Morton and Watson simulated the first engine failure by powering
it down. A minute later, they simulated the second engine failure
during the climb. At 12:46 a.m., with both "failed" engines idling at
1,200 feet and 200 knots, Watson informed Morton that the rudder
power was lost. With a warning light pulsing on the dash, Morton
quickly dropped the nose of the plane to steady it, leveling at 900

feet. Morton reduced the speed to 180 knots and flaps on the wings to 25 degrees, too much and too early. The aircraft lurched to 1,100 feet. Captain Watson reminded Morton of what to do.

12:48 a.m.
Watson: Don't let that thing get below a hundred and sixty [knots].
Watson: Ball in the middle, Jim.
   Whatever it takes, put 'er in there now.
Morton: Get my landing gear for me.

Under the jet-black skies and a nearly full moon, New Orleans sparkled. Morton quickly turned the aircraft toward the illuminated runway two and a half miles away. Fellow crew members encouraged Morton, "Okay, Bud, looks good," "How 'bout that," and "Now we're straightened out." Watson took over the landing checklist and lowered the landing flaps.

12:49 a.m.
Watson: Wing flaps, landing flaps . . .
Morton: Call my airspeed for me.
Watson: One forty.
[Sound of engines beginning slight spool up]
Watson: One thirty-five.
Watson: See, you're letting her get. . . . 'ut the rudder in there. . . . You're getting your speed down now; you're not going to be able to get it.
Morton: Uh uh.

They had five-mile visibility; the landing plan was a 2.5-degree glide path, a regular touchdown. But the airplane's angle was 3 degrees, and Morton didn't compensate for the increased drag from the flaps by adding speed for the descent. Watson didn't correct the trainee captain. Morton decreased the descent by raising the nose angle instead of increasing the power. He lost control.

12:50 a.m.
Morton: CAN'T HOLD IT BUD.
Watson: Naw, DON'T, let it up, let it up . . . let it up!
[End of recording]

The aircraft hit a large tree about 2,300 feet short of the runway, then slashed through two more. As it hovered out of control, the plane sliced through houses, damaged vehicles, and pulled down power lines. The D-8 aircraft scattered its tail and engine parts, sparking fires and gouging a 30-foot crater. Next, it slammed a railway embankment with a metallic screech and skidded toward a Hilton hotel, where it exploded, destroying about 50 rooms. The sea of fuel and the wreckage killed 9 high school girls, one of whom was blown out of the building. The girls were among 23 students who had financed their trip to New Orleans from bake sales, car washes, and babysitting. The total death toll was 19, including those on the plane.

The accident report concluded: "It is obvious from the total evidence that the causal area lies in the human element." It identified the probable cause as improper supervision by Captain Watson and Morton and Watson's inappropriate use of flight and power controls. The report reasoned the possibility that "because of the near equal status of the two pilots, the instructor was more hesitant to take control of the aircraft," which he should have done earlier in the flight. The report also noted that Watson's intense workload and minimal rest likely contributed to his inappropriate supervision. These compounded factors resulted in an avoidable and "nonsurvivable accident." The fact that a simulated proficiency test could have spared 19 lives didn't go unnoticed. Forced to confront their complicity in this tragedy, Delta and other airlines turned to flight trainers as the exclusive training method. But airline companies could have reached this conclusion three years before Watson and Morton's mortal mistake.

On another New Orleans–based flight in February 1964, Eastern Air Lines Flight 304 took off toward Washington Dulles. The crew aboard the Douglas DC-8—the same aircraft flown by Watson and Morton—lost control over Lake Pontchartrain. Unable to yoke up the plane from a full nose-down position, the aircraft plunged into the lake. All 58 passengers died, including a recently wed 21-year-old stewardess who had successfully fought the airline's policy of not employing married flight attendants. After the accident, investigators recovered the flight recorder from under 20 feet of Lake Pontchartrain silt. The unit was too damaged to yield a cockpit transcript. The investigators studied the plane's maintenance records and other functioning and failed DC-8s for the cause. They concluded that the accident resulted from a faulty pitch trim compensator destabilizing the aircraft. In a cruel irony, the part that caused the crash was due for service the next day. Notorious for their mysterious and menacing part failures, the Douglas DC-8s earned the monicker "death cruisers." But mechanical errors weren't the only contributing factor. Investigators also concluded that the pilots' lack of emergency preparedness contributed to this preventable disaster.

As aviation tragedies go, the 1964 and 1967 New Orleans crashes were darkly telegenic. These dramatic and seemingly avoidable disasters sparked public outcry and influenced the FAA to change its flight-training requirements. Since then, commercial pilots have trained exclusively on simulators to practice emergency protocols.

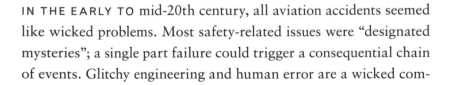

IN THE EARLY TO mid-20th century, all aviation accidents seemed like wicked problems. Most safety-related issues were "designated mysteries"; a single part failure could trigger a consequential chain of events. Glitchy engineering and human error are a wicked combination, and it seemed impossible—even politically risky—to assign blame for accidents after the fact.

For most of aviation history, flying made enormous demands on a pilot's cognition. In the late 1930s, one physician reported a condition called "aeroneurosis," a nervous disorder in which the stress of flying, mainly over oceans, fueled fatigue. In the years and decades that followed, airlines introduced policies to curb pilots' exhaustion, including backup pilots and crew replacements. Engineering enhancements also helped alleviate the demands on a pilot's attention. The jetliners of the 1950s boasted additional safety features and automated meters that made them more manageable than the older piston-engine aircraft. Even with these advancements, a pilot's work remained exhausting and complex, with life-and-death consequences.

Since their development in the early 1940s, flight-data recorders have been considered controversial. The Air Line Pilots Association derided the "black box" recorders (which are, in fact, bright orange) as "nothing but a mechanical spy." Indeed, the black box intends to record, if not spy on, a pilot's actions in the cockpit. Engineered by David Warren, the son of an Anglican missionary who was killed in an unresolved flight crash, the black box could track a plane's motions. The data allowed investigators to reconstruct what happened in the event of an accident. With essential work by engineer James "Crash" Ryan, the American flight recorders were developed by Lockheed Aircraft Services and, oddly, Waste King, the garbage-disposal company, and General Mills, celebrated for cereal brands. These early models recorded and transmitted the plane's speed, altitude, vertical acceleration, and compass data. The black boxes came in a heat-resistant, crash-protective, battery-powered case.

The recording units were popularized in the 1950s after the "Comet crashes," in which three new De Havilland planes were all catastrophically damaged in the air. Soon after, the recorders became an unassailable feature of the aircraft. In the debates preceding their deployment, developers decided on the best place for the equipment: the aircraft's tail, usually the last part damaged in a crash. As one professional said, they "never saw an airplane back into a mountain."

The black box is sometimes called the "indestructible machine." As scholar Greg Siegel points out, the black box was protected "for every conceivable calamity, its mission was to go to hell and back." The institutionalization of black boxes, Siegel notes, coincided with Cold War surveillance, interception, and decryption. In 1966, a year before the Morton and Watson accident, the FAA regulated minimum performance requirements for the black box. They determined that an aircraft's flight recorder should withstand impact shocks up to 1,000 g's for five milliseconds. For perspective, at zero g's, we feel weightless; at 1 g, we can firmly stand on the ground due to the Earth's gravitational field. But at 100 g's, a human body is crushed. The FAA also mandated that recorders endure 5,000 pounds of force and an impact penetration of 500 pounds dropped from 10 feet. Black boxes had to withstand corrosive fluids such as oil and fuel for 24 hours and be able to stay under seawater for 30 days.

The flight recorders weren't that different in concept from Ed Link's motorized recorder on the instructor's desk to track the fictional course of flight training from no place to nowhere. As the black boxes were untangling the mechanical mysteries of plane crashes, the blue boxes unraveled another aspect: psychological training for disaster preparedness.

✈

IN THEATER LEGEND Sanford Meisner's technique, an actor strives to "live truthfully under imaginary circumstances." Within the Link Trainer, pilots were asked to achieve a similar goal. In the 1950s, analog processors sped up simulators' capacity for aerodynamic calculations. Increased computing power enabled an elaborate Link training system for an all-weather fighter aircraft. It involved "8,300 engineering drawings (using 180,000 square feet of various kinds of paper), 8,150 pounds of structural steel, two and one-half tons of aluminum, 5,000 pounds of electrical cable, enough power to heat, comfortably, two 6-room houses, enough air-conditioning to cool two 6-room houses, enough electronic tubes to build 82 television sets or 282 radios, 250,000 feet (about 44 miles) of wiring (equivalent to the amount required for 125 well-lighted homes), 352 transformers of various types, 10,636 resistors, 93 servo motor generator sets, 85 simulated aircraft instruments and 225 parts from the actual F-89 aircraft." Once completed, the simulator was 10 feet high, 24 feet long, and 23 feet wide. The simulator accurately reproduced the experience of locating and destroying adversaries with 1,000-pound bombs at the peak speed of 530 miles per hour. Within the Link Trainer, fighter pilots learned to live truthfully under imaginary circumstances. They gained skills that might one day prove invaluable.

While the devices of the 1940s and 1950s emphasized *flight training*, the digital computers of the 1960s and onward gave rise to the more popular and durable term "flight simulation." While all squares are rectangles, not all rectangles are squares. Similarly, all trainers are simulators, but not all simulators are trainers. The Variable Anamorphic Motion Picture, developed by Link engineers in the 1960s, used distortion lenses and mirrors to display widescreens from 70 mm color projectors. The VAMP showed videos of actual takeoffs and landings filmed at the Chicago O'Hare airport and later programmed them into pilot exercises. The films were so convincing that the movie adaptation of the best-selling 1968 novel *Airport* by Arthur Haley used Link's footage, echoing the Link Trainer's origins in entertainment during the Great Depression. In 1977, the first six-dimensional motion system was launched—capable of pitch, roll, yaw, surge, sway, and heave—a hexapod with hydraulic jacks and limbs. The "synergistic six" swiveled to mimic, however imperfectly, the scenes and sensations of actual flight.

In the 1990s and 2000s, "simulation syndrome" set in with the development of a slew of specialized simulators. Soon, the tail gunners, radar operators, and command posts each had their own simulators. These advanced systems were equipped with precise aerodynamic effects and facsimiles of the instruments that trained operators in different, specialized scenarios. These digital simulators did not merely train pilots how to fly. Instead, they taught pilots complex techniques, including managing an aircraft's systems or exercising aerial combat maneuvers, such as dogfighting. The machines produced repeatable and reliable actions in the pilots by faithfully reproducing human sensations and machine controls in code.

Today's simulations have transcended Ed Link's idea of using only the necessary ingredients to produce the essential experience. The general goal—even an FAA requirement—of modern simulators is to have as many elements identical to the actual flight conditions as possible. Ideally, trainees should feel no difference between flying a real aircraft and "flying" a simulator. The more convincingly features are reproduced, the better the training transfers from the simulator to the cockpit. Or, as the scholar Sherry Turkle has written, "In the culture of simulation, if it works for you, it has all the reality it needs."

✈

THE GENERATIONS OF PILOTS trained at FIT Aviation, Florida Tech's division for flight instruction, have offered living proof that simulated training produces real-life success. Originally built in the 1940s to train navy and marine aviators, FIT Aviation still occupies part of the beachside airfield southeast of Orlando, once a flat cow pasture surrounded by a salt marsh. Nearby West NASA Boulevard teems with aerospace and defense contractors' buildings. On a periwinkle February day, professor Donna Wilt waited at the Center for Aeronautics and Innovation entrance, ready to take me on a flightless flight. "My dad was an engineer in maintenance and trained on the blue box. So I just had that knack," Wilt said, walking to a hall of simulators.

"This is an infinite runway," she said of a visual emanating from a wraparound projector from a specialized simulator. It was a thrill ride from the nearby Disney World. "Normally, you only have a few seconds to land because the runway runs out. But this one goes on forever." This simulator allows beginners to hone the essential motor skills for hand, eye, and foot coordination through repetitive practice. Another exercise involves landing with different cross-winds. Trainee pilots can compare their results, like video gamers vying for the highest scores. Wilt described those landing contests as "hangar parties" where the simulators were used for entertainment. "It's fun to have a party with a bunch of flight instructors."

A century ago, Ed Link focused on delivering an affordable training platform. These days, simulators offer what Wilt described as "over-the-counter coaching." Many products strictly compete on price now. "So basically, we get a Microsoft Flight Simulator with a really cool box around it," Wilt said, referring to the software series older than the company's most famous product, Windows. "Even the instruments are not real. They are simulated." The trick with modern flight simulators is determining their mission requirements and how their fidelity is tailored to the syllabus. "If you are training for knob switching, this is perfect. If you are teaching them procedures and checklists, it's ideal. And basic instrument skills? You can't beat it." Wilt described another simulator, noting how procedure trainers and full mission trainers for the same vehicle can coexist in the same course and place. As Ed Link understood, a simulator's realism affects a student's ability to adapt the training to fly. That's the idea behind the "transfer of training." A positive transfer means the students can effectively carry the knowledge, skills, and abilities from the virtual environment to the actual flight. A negative transfer occurs when habits, emotions, and prior experiences inhibit that crossover.

Even the low-cost, mass-market simulators that Wilt demonstrated rendered a blade of grass in fine detail. "That's a high fidelity, sure, but it's also a gimmick for the look and feel and to increase sales," Wilt said. The question of real proficiency gain for students remains an open question among flight instructors. Putting someone in a swanky simulator doesn't mean that person will be a competent pilot. How real should a simulator be? And how much detail is enough? There are many forms of fidelity. *Physical fidelity* is about how a trainer looks, feels, and sounds. *Cognitive fidelity* relates to the trainer's reproduction of the physical and psychological workload in piloting. And *motivational fidelity* reflects whether the device is friendly and engaging in a way that encourages students to adhere to training.

A vivid visual experience is engaging, but it may not necessarily teach pilots the best techniques for handling, for example. That kind of know-how comes from practicing the yoke or rudder controls to achieve a natural effect, such as a coordinated flight turn. Despite the effectiveness of simulators, some practitioners believe that no simulation is complete without the experience gained on a real plane. In recent years, the FAA has increasingly valued simulator training over actual flight experience, which remains an expensive and risky alternative. The crosswind landing trainer Wilt demonstrated at FIT Aviation was purely for personal practice, not FAA training credit. However valuable for process training, it was an altogether stationary experience, lacking motion cues that Ed Link thought were indispensable for flight trainers.

In fact, the debate over whether motion cues are necessary for flight training has been a long-standing one. During the post–World War II slump and eagerness to sell more units while saving costs, the flight-training industry hatched a plan to do away with motion, the defining feature of the Link Trainer. Newer builders argued that motion trainers didn't provide g-forces as experienced during flight. Ed Link disagreed. In 1947, he wrote that the "total lack of vertigo" in stationary trainers was a detriment. Ed Link believed that "because vertigo is definitely experienced in the air and part of one's training is to overcome this sensation," if these physical cues weren't present, instrument training would not be satisfactory.

Not every engineer agreed with Ed Link's convictions about the necessity of motion. Five years later, the B-47 jet bomber simulator replicated the six-engined Boeing Stratojet. But the simulator dubbed the "sitting duck" was stationary and had no motion or acceleration effects. One company executive said, "The pilots become so preoccupied with the realism of their flight duties in the simulator that they don't appear to miss the sensations of the motion." But Ed Link never changed his mind about the inferiority of motionless simulators, and the US military seemed to agree. In the 1960s and 1970s, the military was reluctant to accept simulators that didn't result in the "pucker factor," slang connoting the stress, specifically rectal contraction, affecting pilots in full alert and fear. Motion fidelity is linked to the adrenaline response in pilots—palpably, a whole-body response that could affect their ability to fly. One Australian wing commander stressed: "If there's no pucker factor, it's no bloody good, and it doesn't fly like the real aeroplane."

The motion-no-motion debate occupied center stage during the 2010 merger of United Airlines and Continental Airlines. Continental trained its crew on fixed-base simulators, requiring no motion. In contrast, United regarded full-motion training as highly realistic and necessary to prepare its crew for aircraft stalls, upsets, engine malfunctions, and landing problems. Continental made an economic argument to the FAA, noting that no-motion systems were cost-effective in developing crew proficiency. In contrast, United strongly opposed that idea, as Continental tried to expand the practice to United. Mergers, just like marriages, are as much about dealing with incompatibility as they are about cherishing compatibility.

At FIT Aviation, I was ready for my first ground lesson on an FAA-approved advanced aviation training device. Ron Mock, a retiree from the air force turned training and simulation manager, was setting up the instructor console. It was a single-engine, four-seater utility plane, Piper Archer, ready to take off from the Melbourne airport.

"The master switch is on the lower left over there," Mock said. "Yep, that one."

I turned on a switch for the engine. "All right! Who wants to fly?" I said.

"Wait, the engine stopped running," Mock said, pausing. "The key must be off."

"I thought we were already in the air," I said.

"Not yet," Wilt said. "You turn the left knob there."

(Relief.) Everything now seemed under control as I steered the aircraft for liftoff. It was going to be a smooth flight, I told myself.

"Now . . . what's that ticking noise?" Wilt asked.

"What ticking noise?" I asked.

(Silence.)

"It's the hydraulics," Mock answered.

"Right," I mumbled. I pulled the stick to lift the nose. As I gained altitude, we heard a screaky beep. "What did I do?"

"I have no idea. That's not a sound you hear in a Piper," Wilt said. "Oh . . . so I see the problem. This says we're about to crash, but we obviously aren't."

(Confidence boost.) The sound was gone a moment later.

"Off we go into the wild blue yonder, climbing high into the sun."

The top view of the terrain was clear—some scattered clouds. The world looked different from the flight deck. Roads, trees, houses, and shores. The scenery was generic but realistic. There was no pressure of flying a real aircraft, and no lives at stake, including mine.

"We don't know our angle of attack directly," Wilt said, "but things like airspeed are critical for estimating it. I'm watching that. With airspeed and the power setting, I have information about the plane's drag and thrust, which will determine our performance and descent rate," Wilt added, checking off all the readings. "That looks good, and that one's good too."

"I feel like I'm on a blind flight," I said. "I don't know what's happening, but it looks like something's happening."

"You're banking a little, so level," Wilt said. "Look out that wing and see how much ground is above. See where the horizon is? Now, look at this side. And if they're the same, then you're level."

"Yeah. I think I am."

We settled at 3,000 feet as the plane whirred.

(A few minutes later.) "Let's lose some altitude," she said. "See where the horizon is above the nose? Let's level this."

I brought it down to 2,300 feet.

"I see the runway," I said, adjusting myself like a Top Gun correcting for a carrier landing.

"No. Don't land there," Wilt instantly corrected me. "That's the shuttle landing strip. Kennedy Space Center."

I thought it was an excellent time to pull in a barrel roll like a barnstormer, but just then, Wilt said, "Oh, it's raining."

"I inserted a weather effect," Mock said with a grin.

I giddily steered the stick, all set for a thunderous roar on the runway.

The touchdown was anything but that.

I landed the Piper on a small puddle.

"That was a nice touch."

✈

AS THE FLIGHT-SIMULATOR INDUSTRY has grown, so have the tendencies to "gamify" the experience. As early as 1944, Ed Link received a suggestion for a pure game based on the blue box and the Celestial Navigation Trainer. "Not that I wish to throw any cold water, but I have had some experience in using technical subjects and gadgets to amuse people, and found that the public is very fickle and sometimes unreceptive to amusements that might even improve their own good," Ed Link wrote in a letter. "In other words, it is a fickle and unpredictable business which we, as technically minded people, get enthusiastic over, and sometimes misjudge human nature."

Today's flight instructors can attest to the perhaps uneasy

union of entertainment and technicality in simulators. Donna Wilt recalled a recent experience trying out a new full-motion simulator. "I flew it right into the Grand Canyon," she said. "The whole time I was down there, it was going 'Terrain. Pull up! Terrain. Pull up!' whereas I was having fun." Then she offered the controls to her copilot, a retired airline pilot. " 'Let's get the hell out of there,' he said. He didn't like it at all." Whether a mega-franchise movie or a cheap trainer with off-the-shelf simulator software, on-screen effectiveness is a function of realism. Being highly realistic doesn't necessarily mean the simulator is effective.

Visual effects are "euphemistically called 'hybrid' or 'empirical' techniques, and more candidly called 'grotesque hacks,' " Alex Seiden, a technical director for seven Oscar-winning films, has written. "The dinosaurs of *Jurassic Park* were not 'simulated' any more than pre-World War II Los Angeles was 'simulated' for *Chinatown*." Since the mid-1990s, when Seiden wrote this, gaming has dramatically influenced Hollywood and aviation, with people accustomed to much richer, more realistic imagery.

But how does one balance purpose and presentation? This is a perennial question. Seiden discussed an aspect of this problem related to designing dinosaurs for the screen. "Often, a precise simulation would not only be more complicated but would also be aesthetically undesirable; for example, the scale of dinosaurs in *Jurassic Park* changes dramatically from shot to shot and sequence to sequence," as he put it. Seiden's point applies to modeling and simulating flight terrains. Mountains cannot resemble pyramids, and the greenery must be blended with gross imperfections to convince the eye. Such visuals should combine with the pilot's senses to provide a stable reference and understanding of the operating environment.

Like Seiden, former Link simulation engineer David Gdovin understood, perhaps better than anyone, the balance between purpose and aesthetics when designing simulated environments. In 1995, Gdovin and fellow engineer David Peters cofounded Diamond Visionics to produce "virtual worlds in real-time." "Every single building in this view is procedurally generated from data," Gdovin said, explaining their latest software. Their simulator displayed the city of Seattle with 14,000-foot Mount Rainier meditating in the background. Their software engine ingests data and satellite imagery and runs them in real time to produce the simulated vision of Seattle. "The program gives us the shape of the footprint. It gives us the location of the footprint. It gives us the orientation of the footprint. It gives us the height of that footprint," Gdovin said. "And bingo! We take all that data and render Seattle 60 times a second from the raw source data we're given." This means that, true to life, the Diamond Visionics simulator never displays exactly the same terrain twice. Graphical programming is generative—a tight dance of polygons and pixels, allowing three-dimensional scenes to be built and modified on the fly, now a common feature in many applications.

Indeed, some of Diamond Visionics' earliest work happened in parallel with simulation for laparoscopy, a minimally invasive surgery in which a thin telescope tube with a camera is inserted to see inside the body without needing a large incision. The simulation to train surgeons in laparoscopy required special effects of lighting, wetness, and reflection from the internal organs, as well as realistic images of blood, tissues, and surgical instruments, precisely what a surgeon would see. Gdovin's team adapted the idea for military simulations: replacing the surgeon with a warfighter, changing blood to soil terrains, and all the rest.

"So what's our destination, San Francisco or Honolulu?" Gdovin asked Jason Petrick, one of his lab engineers.

"Let's go to Honolulu," Petrick said, and he launched a descent scene.

"Here's the out-the-window day scene," Petrick said. A few more mouse clicks. "Here's the night scene. And here's the infrared."

Looking at the demo was like riding a magic carpet over Oahu. The reflections of the sky on the ocean and waves appeared Pixar-quality.

"Do you notice what's happening to the runway here?" Petrick asked Gdovin.

"Put runway edge lights every 30 feet," Gdovin said, asking for a replay.

"It just takes a few moments to roll in that new data," Gdovin said. "Boom! We got edge-lit runways, and it's as simple as that."

My attention turned to the waves in the ocean.

"We're working on shadows of the waves, just to make it more realistic," Petrick said.

Jonathan Richards, a quality-assurance engineer who tests for glitches and bugs, was sitting across from Petrick. "Yeah, everything should be realistic," Richards said, zooming in and out of midtown Manhattan on his monitors. Richards scrutinized the rendered images for lag and hardware interruptions that could make the experience unpleasant. The result, "barfogenesis," is a motion sickness caused by visual media.

Back at Petrick's monitor, the descent scene in Honolulu was advancing. The distant flight strip came into view. As the aircraft banked, the vistas varied dazzlingly with the vantage point. With a slight turn, the reflective textures of the ocean changed. Buildings, billboards, and their shadows swept past, conveying the essence of what philosopher Paul Virilio dubbed a "dromoscopic vision," a vision born from speed, as if on a race course. "The world flown over is a world produced by speed."

Gdovin asked for some weather effects.

"How about some rain on this paradise?" Petrick asked.

Gdovin had a twinkle in his eye.

"How about a blizzard?"

The power and ability of Diamond Visionics' simulations are literally dizzying. "Think of any red brick building, and I can render that red brick building," Gdovin said. But the quality of the visuals isn't the only meaningful component of their simulations. "Now, the pilots don't care whether it's a red brick building. They only care about not hitting it," Gdovin said. "The stakes are very high in this kind of simulation. Lives depend on it."

To determine whether the produced visuals are meaningful, one should consider three focal levels of perception. The first level of perception is image-referenced and linked to the simple identification of things, say, a tree. The second level is world-referenced, which requires a contextual understanding of the visuals; for example, the tree is on a mountain. The third level is knowledge-referenced, involving the importance of the context; say, the tree on the hill is approaching closer and faster, which may mean the pilot's losing altitude. All this is to say, specifying a simulator's level of visual detail is as much an engineering challenge as it is a computing requirement. How, then, does one establish the minimum acceptable detail for a simulator?

The high-performance geospatial demos and immersive experiences available today may have been unthinkable to Ed Link and his contemporaries. But, the challenge of specifying a simulator's level of detail would have been quite familiar. In the early 1940s, he grumbled about the requirements set by the government for the trainers. Ed Link felt that the regulators had "been adding one thing after another which makes the Link Trainer considerably more costly and more complicated, which takes more equipment and men to produce, and considerably complicates the production problem." He saw no value in producing a complicated trainer for novice training and intended to simplify the instrument trainer further. "The cockpit will be square like a box, will have no wings or tail group. There will be complicated and hard to produce wind drift. There will be no useless gadgets such as a leveling device," he wrote. "I am cutting out everything that is not being used on primary instruction." While you can find highly complex demos like Gdovin's, many of the simulators that Donna Wilt demonstrated at Florida Tech were designed with Ed Link's original objective of simplification.

Flight-simulation professionals often argue that "it's not how much you have but how you use it." As scholar Eduardo Salas observes, this idea has created a standstill in aviation because advances in simulation and simulators have overshadowed the training itself. Learning develops through an array of influences, including how the learning environment is designed and how able the instructor is. Training is ultimately a cognitive and behavioral activity, and we "must abandon the notion that simulation equals training and the simplistic view that higher fidelity means better training."

WORLD WAR II CREATED business stability for Link Aviation Devices in the 1950s. But at his business's apex, Ed Link began losing interest in flight simulation. Ed Link complained that the ingenuity in aviation was not what it used to be in his earlier days. "I sort of lost interest in the space age," he said, "when they computerized everything, although it was a very natural and important step. But I was more interested in originating than looking at a computer."

In 1953, Ed Link stepped down as president and became chairman of his company's board. A year later, Ed and his brother George sold the company. In the years to come, Link's future holding companies, including the Singer Corporation, produced a striking array of simulators designed for everything from commercial flights to weapon systems. They included specialized tasks for everything from vertical takeoff and landing to cockpit procedures, radar landmass simulators, submarine trainers, power plant simulators, and industrial simulators, with far more advanced instructor consoles. Now there are generic simulators for a suite of tasks beyond aviation, including for cars, trucks, ships, locomotives, surgeries, and bomb defusions—with lifelike sounds and experience.

"I started as a grease monkey with dirt under my fingernails," Ed Link told a reporter, "and now after making my fortune as a swivel chair executive I am back as an experimenter."

Ed Link had moved on to a different medium.

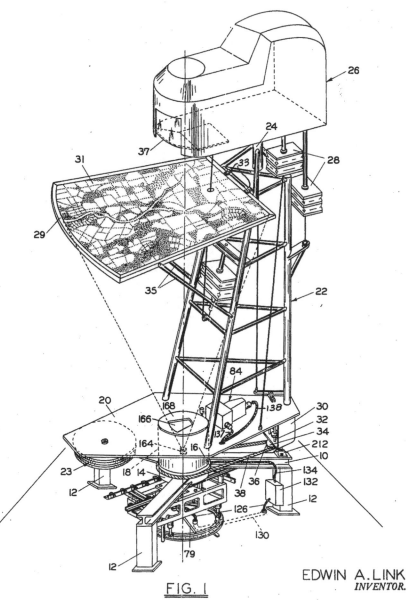

FIG. 1

EDWIN A. LINK
INVENTOR.

BY *Donald T. Hillier*
Philip S. Hopkins
ATTORNEYS.

Chapter Four

Wicked Safety

HARRY AND BESS HOUDINI GAZED AT THE GRAVESTONE in Piershill Cemetery of Edinburgh. "He fooled them in life and he fooled them in death," the magician told his wife. "I envy him." The epitaph read, "Sacred to the Memory of the Great Lafayette."

The vaudeville headliner the Great Lafayette was a German-born American émigré, originally named Sigmund Neuberger. He started his career imitating a Chinese conjurer, performing sharpshooting tricks for two dollars a week. In the 1890s, his performances became extravagant: flashy costume changes and spellbinding stunts, waterfalls, statues coming to life, and midair bird catching. The Great Lafayette played to packed audiences in New York, New Orleans, and Nashville. His sets were glamorous with stage design and celebrity endorsements. Lucrative offers, luxurious houses, and lush coach services followed. By 1900, the 29-year-old master illusionist became a top-selling entertainer.

At his career high, Lafayette unveiled his signature illusion, "The Lion's Bride," an enchanting 25-minute coda whose 1910 Christmas performance at the London Coliseum sold out months in advance. In the act, a rogue sultan kidnaps a beautiful princess who survived a shipwreck. Her lover, the Great Lafayette, goes to rescue her. Given a choice between marrying the sultan or being fed to a lion, the princess chooses death by feline. As a lion prowls around irritably on stage, the princess enters the cage, and the majestic cat pounces on her. Through special effects, Lafayette appears, slashes the lion, and rescues the princess to thunderous applause.

On May 9, 1911, Lafayette's troupe performed "The Lion's Bride" for an audience of 3,000 at the Edinburgh Empire. The legendary theater was a successor to three short-lived venues on that site that had all burned down in the previous decades. The show was underway with the usual cast of acrobats, fire-eaters, and horses. Lafayette was ready for the critical lion scene when a stage lantern caught fire. The audience was awestruck and stuck to their seats. Viewers assumed that the inferno was part of the razzmatazz and did not rise to escape. After the iron fire-protection curtain fell, the conductor ordered the orchestra to play "God Save the King" to signal the show's end and rouse the people to exit. Luckily, the audience vacated the venue unharmed, with minimal complications from smoke. Legend suggests that everyone evacuated in the two and a half minutes it took the orchestra to play the British national anthem.

This theatrical tragedy of 1911 unintentionally created a design standard—the presumed exit time for most built environments. Engineers subsequently assumed that in two and a half minutes, people should be able to reach an exit door or a place of relative safety. This standard was essential: between 1787 and 1897, theater fires killed nearly 10,000 people. Even if engineers claimed the structures were "fire-resistant," the theaters themselves were often ill-planned, with reforms added on to create safer evacuations. Upgrades included obstruction-free stairs from the balconies and foyers, updated lobby designs, mandated seating distances, fire alarms, lighting, and marked exit pathways. In 1899, an observer wrote that emptying the 1,500-seat Richard Wagner's Bayreuth Theater took two minutes. New York's Madison Square Garden, which had a capacity of 17,000, took four and a half minutes to evacuate.

Fire-safety codes specify maximum travel distances, evacuation times, stair widths, and fire hydrant locations, among many prescriptions, some of which were (and are) established arbitrarily. Engineer Margaret Law dubbed such rigid requirements "magic numbers" that can create oversimplified solutions at the expense of high standards of proof. "Hard safety" problems can occasionally be solved by superstition, just as the duration of the British national anthem became the basis for international evacuation rules. There's always a place for prescriptive regulations. However, Law argued that pretending these precise standards will yield the best safety solution is itself an illusion. Magic numbers offer convenient confidence that a building is reasonably safe. But what about the cases of a lesser distance to the door, a different number of evacuation stairs, or another condition? Special circumstances demand that engineers optimally customize hard safety solutions. Even if one complies with all the prescribed codes, it doesn't make a condo complex automatically safe. Indeed, even if we adhere to the magic numbers, absolute safety doesn't exist, and safety judgments are, thus, always less objective. Law observed that asking endless what-ifs about what could, would, or might have happened under different conditions could make even hard-edged safety issues into unbounded ones, as we'll see with the soft problems of nuclear safety. That is, how safe is safe enough?

Most fire-safety concepts of the Western world are linked primarily to 17th-century London, a city confronting multiple apocalypses. In the long dry summer following the Great Plague of 1665, which took about 100,000 lives, London was ablaze with its second Great Fire, which, perhaps ironically, had a cleansing effect on the congested, disease-ridden capital. It started in a narrow one-way street near London Bridge, where a bakery fire instigated the inferno that lasted four days. It caused very few casualties but left hundreds of thousands homeless. The Great Fire of 1666 gutted most public buildings and tens of thousands of dwellings, largely Tudor-style homes with thick roofs.

Following the devastation, polymath engineer Christopher Wren reconstructed the wrecked churches. He developed the new St. Paul's Cathedral, the crowning glory of London. The city's rise from the ashes led to new safety systems and regulations on everything from road dimensions to material ratings, from furnace tests to two-brick-thick party walls. One new rule dictated that laneways must have a 20-foot turnaround point for fire trucks. In its original context, the 20-foot turning radius was deduced from the distance one can humanely coax and turn around a horse-drawn fire engine. This magic number has proved highly influential in the design of urban spaces.

"Books on fire safety are littered with such regulations," says engineer Luke Bisby. "It's easy to see how the two-and-a-half-minute rule for egress time is crazy. It's certainly absurd." Indeed, the Edinburgh Empire experience was so atypical that it was almost unreal. The audience could almost certainly have remained in their seats for longer than two and a half minutes without realizing the danger, and the fire trucks could have failed to show up. Applying the two-and-a-half-minute blanket rule to all buildings is clearly questionable, yet that's become the norm. Hard safety measures preserved in precise-looking numbers stay in place, perhaps because they are assumed to be correct and seemingly haven't caused harmful consequences. "Even the most celebrated building design ends up mindlessly adhering to some regulation with dubious origins," Bisby said, pulling into Old Town Edinburgh, not too far from the Festival Theatre, where the Empire once stood. "Have you designed your building for fire? Yes. What have you done? I've been code compliant. What is the code based on? I don't know . . . some fire in Edinburgh."

A zippy and cheerful Canadian, Bisby has been called a "flame thrower." When he was a graduate student, 9/11 ignited his research interest in fire safety. His current job title, Chair of Fire and Structures, at the University of Edinburgh, seems straight out of Tolkien. Bisby's friends joke that he never discovered his hobbies; now, he and his students burn couches and Christmas trees to study their combustion properties. They conduct dynamic simulations to analyze flame spread and reconstruct vigorous fire events. Edinburgh provides them with rich case studies for research. After all, weeks before Edinburgh's own Great Fire of 1824, the city created its municipal fire service, the United Kingdom's first.

In his lectures, Bisby likes to point out things in a room dictated by fire safety: windows, latches, doors, locks, carpets, wall lining, fire alarms, sprinklers, seat foams, paint, fire doors, and exit signs. Fire-safety engineering draws on numerous fields of knowledge and includes the tedium of paperwork. The practice comes with "code speak," the technical language of fire-safety regulation. Even with a common language, engineers worry that prescriptive one-size-fits-all guidance is overly limiting. But Bisby argues that deregulating fire-safety design practices isn't always a good idea; more prudent self-regulation is needed.

Hard safety has evolved into a performance-based design to unleash engineering innovation in recent decades. The idea is to achieve specific performance levels by any means rather than their being dictated by strict compliance to criteria. In principle, this flexibility can mean lesser costs and more significant safety margins, benefits that have stirred similar discussions in drug development, aviation, and food and fuel production. Still, central questions remain about whether the buildings are sufficiently safe and whether performance-based designs meet society's expectations for safety or create new risks.

"Don't learn safety by accident," goes a workers' motto. People may crave hard, definitive solutions no matter how soft and perceptual the life-threatening circumstances are. A hard engineering solution can get everyone safely out of a building during an accident, including emergency services personnel. Unlike such hard safety, though, soft problems can look like bats turning the global economy upside down, and, as we'll discuss, the tragic trio of an earthquake, a tsunami, and a nuclear accident remixing risks among systems large and small.

On that gruesome day in Edinburgh, the fire trapped Great Lafayette's onstage crew between the safety curtain and the backstage exit. The distrustful Lafayette had locked the rear stage door to block others from stealing his tricks. Ten people lost their lives, along with Lafayette, his horse, and his lion. In a bizarre turn of events, rescue workers recovered a scorched body in Lafayette's costume the following day. They prepared it for a high-profile funeral.

Two days later, the workers were astonished to find an overlooked body below the stage. It was that of the real Lafayette. The earlier body they recovered was Lafayette's body double, whom no one even knew existed. Crowds thronged in the tens of thousands for Lafayette's final journey. The Man of Mystery's ashes were now in a triple-decker, velvet-wrapped oak urn, buried next to the recent grave of his puppy, Beauty, a gift from Houdini.

✳

AS A SYSTEMS ENGINEER, Bill Ostendorff knows the workings of cause and effect. On Friday, March 11, 2011, in suburban Washington, DC, he was boiling water for his coffee when he heard the news. The Great East Japan Earthquake had struck, a magnitude 9.0, rattling over a million buildings and shifting the country eight feet eastward. Pacific waves over 400 miles per hour pierced the northeast coastline and flooded 200 square miles.

Fukushima, the "fortunate island" 150 miles northeast of Tokyo, was elevated to an epoch-defining emergency. Fukushima's nuclear power plant employed six boiling-water reactors and 6,000 workers. Emergency protocols were activated to shut down the functioning units. Fukushima's reactors were designed with earthquakes in mind but were vulnerable to tsunamis. Ultimately, a 50-foot surge inundated them. After a power loss, the cooling systems failed, which ratcheted up the three reactors' core temperatures, pressures, and explosive gas concentrations. Fukushima evacuated over 100,000 residents. While there were no deaths from the radiation exposure, around 20,000 people died from the tsunami. The disaster was "Made in Japan," declared the Fukushima independent investigation commission. The official report noted that the Japanese mindset of reflexive obedience to authority caused the negligence that left Fukushima unprepared. "Had other Japanese been in the shoes of those who bear responsibility for this accident, the result may well have been the same."

Ostendorff was a leader of the American response to Fukushima as a member of the Nuclear Regulatory Commission. As is common in emergencies, he and his team worked in two time frames. "In the initial stages, we lacked information on key safety issues, and we had some homework to do," Ostendorff said. Confident about its safety regime, the United States chose not to shut down its nuclear plants. Over the subsequent, slower stages, Ostendorff and his colleagues performed safety reviews of the hundred-odd nuclear reactors in the United States. Each plant operator assessed their responses to earthquakes, hurricanes, tornadoes, tsunamis, and floods based on the concept of defense-in-depth. If the plant's initial barrier failed to combat risk, the second and the third layers remained intact. Successive safeguards provided backups to backups. Alarms and automation can give such range and redundancy as regular reviews, drills, and checklists.

The industry proposed a suite of "flex" features to safeguard the nuclear plants against hazards. Those features would dependably maintain the core and spent-fuel cooling with auxiliary power. As is typical in safety-critical systems engineering, these "strategies were not risk-based but risk-informed," Ostendorff said. Perfect safety isn't always attainable in a nuclear plant, but progressively lowering risks can ensure its integrity. For instance, riskier options are replaced by safer ones, as in replacing the older fireproofing material with better versions to control combustion temperatures. Performance metrics of sudden reactor shutdowns (called "scrams") and the monitoring of exposure levels in workers have significantly improved. While many of these design upgrades can be carried out by hard technical remedies, "soft safety" questions of determining whether the processes in place are adequate can be contextual. They aren't always amenable to the rigid dictates of code.

The Nuclear Regulatory Commission's work is tied to the Atomic Energy Act by the principle of ensuring "adequate protection." This safety threshold doesn't nullify all risks, however. The law prescribes which risks and what safety standards are judged to be acceptable. Under these circumstances, an engineer's role in the commission is akin to that of an appellate judge—the many unquantifiable elements of "soft safety" come down to a judgment call over how safe is safe enough.

Soft safety problems, like the ones in Fukushima, can compound complexity when a system's efficiency can become a vulnerability in some circumstances and create vagueness in another moment. For example, consider how a quirky geographic split hampered Japan's recovery following the Fukushima accident. The battered northeast regions of the country operated on 50 Hz electricity, the legacy of an initial investment in German generators. However, the unscathed southwest relied on 60 Hz American equipment. This electrical divide stemmed from decisions made in the late 1800s that might have seemed sensible but posed a significant setback later. Factories rated at 50 Hz couldn't draw power from the 60 Hz power stations. As blackouts extended, global supply chains metastasized the harm and affected American automobile production. US manufacturers, lacking essential microchips and specialty paint pigments from Japan, hit the brakes on their orders. Some American factories were briefly shut down.

Just as with Eyjafjallajökull a year before and with Covid-19 a decade later, the disaster revealed fragilities in global supply networks that desperately needed safety-critical enhancements. Corporations can reap the rewards from the hard efficiencies of tightly connected supply chains. While this has allowed stunning successes in stable conditions, companies can be vulnerable to disruptions that aren't always clear-cut, like hard problems. Even in regular times, highly coordinated systems can face uncoordinated risks. Around 2007, in the initial stages of the 787 Dreamliner manufacturing, Boeing didn't know its suppliers beyond the first tier and could not manage the fragmented production delays. Airbus also grappled with glitches when its electrical cables were shorter than specified, built by suppliers lacking integrated design and using incompatible software. Viral outbursts, volcanic outbursts, and nuclear outbursts are exceptional events, and one can prepare for them.

Soft safety can take on different identities. In one version, global seafood consumption from Japan declined, given the concerns about radioactive contamination. American seafood purchases dropped by a third. Even four years later, some Korean groups sought to ban Japanese imports. Food and product safety requirements take different forms; medical devices have benchmarks like transportation and consumer products. Hard safety turns into soft safety when it has a diversity of interpretations, some nebulous.

A few weeks after the Fukushima accident, Ostendorff spoke at a conference about increasing US resilience to a nuclear catastrophe. He briefed the attendees about the NRC's response to the crisis and the future of emergency responses. "How many of you have ever fought a fire?" he asked, eagerly scanning the room of hundreds for an answer. A few hands went up. "Talking about fighting a fire and fighting a real fire are very, very different experiences," Ostendorff said. He reflected on his emergency-response training in the Philadelphia Naval Shipyard from the 1970s. He remembered the hair singed off his eyebrows and the burns on the backs of his hands. "I was darn glad to have had that training," he said.

Enterprises like aviation that involve many strangers working together have developed strong traditions of safety training and standardized operating procedures to maximize predictability. The actions of a pilot and copilot working together for the first time are generally predictable because routinization makes them so. Similarly, aircraft, internet servers, petrochemical equipment, and nuclear reactors are "trained" with extensive testing to be reliable. However, when they fail with catastrophic consequences, even the most reasoned arguments about their past reliability can fade and be unconvincing. Therefore, an ideal safety success for mission-critical systems is if they fail, their effects are less severe. Here's an example.

About 97 percent of aircraft accidents in recent years have been from birds, including geese, gulls, ducks, cormorants, and hawks, with over 14,000 bird strikes annually in recent years. Birds are far from the only natural threat to aircraft. Bats, coyotes, lizards, rodents, rabbits, turtles, armadillos, opossums, deer, and alligators have also triggered tarmac tragedies. Over the past two decades, over 600 bird species, 50 mammal species, and several dozen bat and reptile species have struck aircraft, some more destructively than others. Airports employ avian radars and sonic scarecrows (noise generators) to keep fauna away from turbines. Natural selection hasn't endowed birds with an aversion to airports, observes scholar John Downer in "When the Chick Hits the Fan," a paper containing an analysis of testing regimens. With most birds now accustomed to intimidation tactics, he writes, the "onus, therefore, is on the engine: it must be bird-resilient."

A typical proof test in aviation is simulated bird strikes. Real birds are preferred to pseudo ones; freshly killed ones are better than frozen-and-thawed ones. To meet engine ingestion standards, the FAA requires various tests with volleys of large, medium, and small birds. Then, using the "critical impact parameter," engineers characterize the engine failure modes. A challenge inherent in these tests, as with simulations and testing of airbags, is how to convincingly portray natural hazards, including bird migrations at night. Deliberately damaging aircraft wings with birds or volcanic ash is expensive. So, manufacturers often resort to a satisficing argument showing that their engines are reliable enough for the purpose that regulators can't simply reject by expecting unachievably high reliability in testing phases. The future reliability of infantry rifles can be extrapolated from past failure patterns; such data may or may not be generalizable. However, failure isn't a luxury available to nuclear reactors.

In some proof tests, engineers estimate the mean time-to-failure, or the expected time between two failures, for a billion hours, a run time of 114,000 calendar years! Under such circumstances, the base models and tests must be reasonably flexible and faultless. Achieving ultrahigh reliability "is more akin to a search for 'truth' than a search for 'utility,'" Downer writes, "where even very marginal ambiguities become meaningful." Occasionally, design errors and insufficient planning can lead to calamities such as Fukushima, even if hundreds of past reactors had operated error-free.

Reliability is a nonevent, an indicator of a failure that didn't happen. Unlike most other phenomena, reliability works because of its negative property, notes Downer. Companies may brand and market their products as innovative, disruptive, and transformative. But, engineers achieve safety-critical performance by being the Tories of the technological world. Their conservatism, Downer notes, comes from their belief in verifiable progress "but only by consecrating traditions and building on the hard-earned wisdom of their predecessors."

Ostendorff grew up in Louisiana and is an "extremely devoted Dallas Cowboys Fan." Following a 26-year career in the navy, 16 of those years at sea, including command of a nuclear attack submarine, Ostendorff retired as a captain. He also earned a law degree to supplement his engineering experience. He worked as a congressional staffer for the US House Committee on Armed Services and as a chief operating officer of the National Nuclear Security Administration. After retiring from the Nuclear Regulatory Commission, Ostendorff taught political science and national security at the US Naval Academy.

During a noontime visit to Annapolis, Ostendorff drove past gleaming green track fields, pointing at Rickover Hall, which housed the engineering programs. Admiral Rickover changed the philosophy and precision of the nuclear navy by emphasizing high technical competence, standards, and personal accountability. Rickover brought his czar-like clout to the Manhattan Project, controlling every detail and challenging his superiors. Both visionary and vinegary, Rickover was mystical to some and maniacal to others, even after he was forced to retire at age 82. "I never thought I was smart. I thought the people I dealt with were dumb, including you," the admiral torpedoed Diane Sawyer in a *60 Minutes* broadcast. Previously during an interview, he had rebuked the eminent Edward R. Murrow: "You're looking for easy solutions. The trouble with you is you want easy answers, but you don't know the proper questions."

Rickover's obsession resulted in an uncompromising safety culture of technical excellence. Every process on a nuclear submarine required thorough training, from the reactor start-up to trash handling. Ostendorff recalled a qualifying exam that tested his working knowledge of an attack submarine. "I spent months committing to memory all—I mean *all*—the primary valves and pipes in the reactor plant," he said with a thin smile. "We should never apologize for having high standards for technical competency."

However, the codes and processes that work for hard safety don't precisely transfer to soft safety problems, depending on how people perceive the different risks. The traditional engineering approach of producing "fail-safe" solutions focuses on risk management: What can go wrong, how likely is it, and what are the consequences? In comparison, the concept of a "safe-to-fail" system aims to enable multifunctional backups rather than, for example, reducing uncontrollable risks to infrastructure from extreme weather. A fail-safe system depends on error avoidance; a safe-to-fail system aims to reduce the damage with repair and recovery. In fail-safe systems, a car's safety features can shield its drivers. In a safe-to-fail system, drivers have insurance coverage even if an accident is inevitable. As we'll see next in a messy scenario, agreeing on safety goals and addressing them with vigilance can be difficult.

After an aircraft carrier collided with the destroyer USS *Hobson* in 1952, killing 176 crew members, the *Wall Street Journal* published a much-quoted editorial. "To err is not only human; it absolves responsibility. Everywhere, that is, except on the sea. On the sea there is a tradition older even than the traditions of the country itself and wiser in its age than this new custom. It is the tradition that with responsibility goes authority, and with them goes accountability. This accountability is not for the intentions but for the deed." Rickover, too, extolled the uniqueness of absolute responsibility. "You may share it with others, but your portion is not diminished. You may delegate it, but it is still with you," the admiral said. For three years and over 100,000 miles, Ostendorff commanded a nearly billion-dollar, nuclear-powered attack sub operating under the Arctic waters. He knew the grievous consequences of a potential station blackout. "I made darn sure that our batteries were fully charged before we went under the ice."

In March 2016, five years post-Fukushima, Ostendorff gave his farewell address at the annual Regulatory Information Conference. He reflected on the transparency and accountability demanded in the US response following Fukushima. "Nuclear regulation is the public's business, and it must be conducted openly and candidly," he said. The commission had organized 300 public hearings on safety with an open voting process. Ostendorff cast 25 separate votes on post-Fukushima regulatory reform, all in the public record as his congressional testimonies. "That's how you can hold me accountable for my actions."

✳

IN THE EYES OF philosopher Paul Virilio, each accident is an "inverted miracle." "When you invent the ship, you also invent the shipwreck," he said in *Politics of the Very Worst*. Mishaps—such as the molasses disaster, oil spills, or cyberattacks—often generate tighter rules and safeguards to prevent future blowouts. However, corporations can become complacent with their safety records. An "atrophy of vigilance" sets in as organizations overlook ominous signs, becoming victims of "normalization of deviance," unknowingly or knowingly doing the wrong thing. As scholar Barry Turner observed, "We provoke the hazards whose risks we then have to learn to cope with."

Many organizations have the unenviable task of operating high-risk technologies where mistakes can cost lives, dollars, or both. On an ink-blue Pennsylvania dawn in March 1979, a mélange of malfunctions melted a commercial nuclear power plant at Three Mile Island. Design flaws, equipment frustrations, and worker faults aligned like holes in a stack of Swiss cheese, fostering a free flow of failures. Similarly, risks ratcheted up in the Bhopal gas tragedy of

1984 and the destruction of the space shuttle *Challenger* in 1986, followed by the Chernobyl nuclear accident weeks later. These failures exemplify what the scholar Charles Perrow termed "normal accidents." He meant that the technologies and the systems operating them were such that catastrophic accidents should not be seen as anomalies but as predictable outcomes.

Two key features distinguish technologies prone to normal accidents: they are highly complex and tightly coupled. High complexity means the system can interact in ways its designers and operators didn't anticipate. Tight coupling means a small component failure can spread its consequences through the entire system, triggering a catastrophic collapse. Perrow argued that making such systems risk-free is impossible; catastrophic failures may never be eliminated, but their risks can be diminished. A simple oversight or moment of distraction in the air-traffic control system can lead to an in-air collision between hundred-million-dollar planes carrying hundreds of passengers. Racing cars and roller coasters are further examples. Yet another is the control room of a utility grid, which is as complex as a nuclear reactor. None of these cases comes close to the wicked complexities of ecological traumas, epidemics, financial meltdowns, and climate maladies.

Over the years, some organizations—particularly those that are flexible, non-hierarchical, and highly attuned to risk—have developed nearly error-free approaches to reduce catastrophes, if not eliminate them. In a "high-reliability organization," operations are hierarchical, with a chain of command. But in high-pressure situations, the organizational structure undergoes a dramatic shift; it becomes much more horizontal. Individuals must recognize potential problems and deal with them immediately without approval from higher-ups. The effect is a continuous-learning orientation, what scholars Karl Weick and Kathleen Sutcliffe described as "mindful organizing." Such a mentality strives for "a minimum of jolt, a maximum of continuity." The hallmark of a high-reliability organization "is not that it is error-free but that errors don't disable it." One can pick up a potential problem or error and respond robustly when a work structure is designed to encourage this mindfulness.

The "normal accident" and "high-reliability organization" theories are valuable because they are simple. They advocate for redundancy, the idea of deliberately duplicating critical components of a system to control risk and increase reliability. But more reliability doesn't necessarily mean more safety. Reliability and safety are often confused, so it's essential to differentiate them.

In engineering, safety is generally about acceptable risks and avoidable harms; reliability is satisfying a particular requirement under given conditions over time. One property can exist without the other, but both can coexist and compromise each other. A system can be reliable and unsafe if one disregards the interactions among components, even when they fully meet the requirements. A system can be safe but unreliable if one doesn't dependably follow the rules and procedures. Reliability is neither sufficient nor necessary for safety. Inspecting a pipeline valve can tell much about the valve's reliability but not whether a refinery is safe. Safety is a function of how that valve interacts with other refinery units. That's why an organization, even with a "culture of safety," may be unable to prevent accidents if it's fixated on reliability. In highlighting these differences, systems engineer Nancy Leveson reminds us that safety is an emergent property of a system, not a component property.

Consider this example from Leveson. The coastal parts of Washington State in the Pacific Northwest feature deep fjords and glacier-carved bays, and car ferries connect the numerous islands to move people around. But an unusual complication brought the ferry system to a standstill: rental cars couldn't be driven off the ferries after reaching the port. A local rental car company had installed a security device to prevent theft by disabling their cars if the car moved when the engine was turned off. When the ferries moved, the rental cars weren't running, and all the vehicles disabled themselves. The ferry services were suspended until tow trucks could be found to release the stuck rental vehicles. Was this a simple component failure or a system failure?

In another example, Leveson points to a problem peculiar to some flight simulators whose pedals required frequent replacement. If you perceive it as a component problem, it's easy to blame the parts and invest in better ones or to blame the pilot who damaged them. But you can also see it as an interaction problem among the pilot, the hardware, and the software. It turned out that when pilots braked in the simulator, the software didn't show the minor movements when the aircraft stopped, like the creep you typically see when parking a car. Without this feedback, the trainees braked longer and harder in the simulator than in a real airplane, where the software problem didn't exist. Such issues most likely come not from individual component failures but from processes resulting in flawed requirements and system designs.

"Messy safety" problems occur when it's impossible to exhaustively test the range of system behaviors. Such complexities force design changes that must be consciously considered so that other errors aren't introduced. "Productivity and safety go hand-in-hand for hazardous industries and products. They are not conflicting except in the short term," writes Leveson. In a messy problem, productivity and safety may be contrary approaches for an organization, or some may see them as the same. Whether it's productivity and safety, or affordability and profit, and whether they are reconcilable, safety needs to be understood using systems theory, not reliability theory. Accidents and losses should be seen as a dynamic control problem rather than a component failure problem, Leveson argues, which requires considering the entire social system, not just the technical part. Human error is a symptom, not the reason for most accidents, and such behavior is affected by the context in which it occurs. This view should inform safety from the outset before a detailed design even exists.

Normal-accident theory was part forethought, part fatalism, arguing that avoiding complex hazards is impossible despite planning.

High-reliability organizations are about reducing errors while treating safety and reliability alike. Leveson's systems theory approach suggests safety control, not in a military sense, but through "policies, procedures, shared values, and other aspects of organizational culture" that help dissolve messy problems. Messy safety issues are about lessening the frequency of things that go wrong and increasing the frequency of things that go right. Messy safety should also reveal the biting vulnerabilities often buried in various processes and presentations. Processes are essential for operations but can also be misused and flawed. A process isn't a hard solution to safety or a substitute for leadership. Processes can even worsen failure when mindlessly used. As one observer put it: "In engineering, like flying, you need to follow the checklist, but you also need to know how to engineer."

Similarly, engineers use presentations to communicate design concepts and results without always knowing that their slide decks can invoke "death by PowerPoint." The ubiquitous use of slides for technical presentations is itself a safety hazard. Tragically, PowerPoint has been implicated in NASA's technical discussions in 2003 about the reentry of space shuttle *Columbia* that soon after exploded, killing seven crew members. The slides compressed information on significant risks into poorly formatted bullet points, which was highlighted in the Columbia Accident Investigation Board's review: "The Board views the endemic use of PowerPoint briefing slides instead of technical papers as an illustration of the problematic methods of technical communication at NASA."

PowerPoint is "a social instrument, turning middle managers into bullet-point dandies," as a writer in *The New Yorker* described its "private, interior influence" in editing ideas. "Power-Point is strangely adept at disguising the fragile foundations of a proposal, the emptiness of a business plan; usually, the audience is respectfully still . . . , and, with the visual distraction of a dancing pie chart, a speaker can quickly move past the laughable flaw in his argument. If anyone notices, it's too late—the narrative presses on."

Indeed, the Columbia Accident Investigation Board critiqued

NASA's process and presentation philosophy: "The foam debris hit was not the single cause of the *Columbia* accident, just as the failure of the joint seal that permitted O-ring erosion was not the single cause of *Challenger.* Both *Columbia* and *Challenger* were lost also because of the failure of NASA's organizational system." NASA's culture and structure fueled the flawed decision-making, ultimately affecting the space shuttle program.

Dissolving messy safety problems requires constantly comparing current conditions with emerging requirements and failure modes. The aviation industry wouldn't have thrived without dramatically reducing accident rates. Thanks to their safety and maintenance protocols, public confidence in the aviation industry has steadily increased since the 1960s, when only one in five Americans were willing to fly. While human-induced errors are quantifiable, the behavior that leads to them is not; hence, flight training remains significant in crew preparation. And finally, with messy safety problems, it can be tempting to blame others, but "blame is the enemy of safety," Leveson notes. The goal should be to design systems that are error resistant, not operator dependent.

✳

THE LATE JAPANESE DIRECTOR Akira Kurosawa's *Dreams* is a masterly mosaic of moods, musings, memories, and melancholy. The eight vignettes in this 1990 anthology feature *Mount Fuji in Red*, a stark, musicless seven-minute ballad. In the film, an explosion of nuclear reactors creates an eerie evacuation. A young man rushes into the frenzy of people and their abandoned belongings. He encounters a remorseful scientist. "Where did all those people go?" the young man asks the scientist on the cliff's edge. "To the bottom of the sea," the scientist says. "The dolphins, even they're leaving. Lucky dolphins. They can swim away. It won't help. The radioactivity will get them." The scientist adds that because radiation is invisible, the humans colored their clouds. "But that only lets you know which kind kills you. Death's calling card."

Similarly, after Chernobyl, everything felt dosed—sunsets, forests, and fields. "There's nowhere to hide. Not underground, not underwater, not in the air." In the Nobel-winning Svetlana Alexievich's polyphonic book, an engineer lamented: "It's become the meaning of our lives. The meaning of our suffering. Like a war . . . We're its victims, but also its priests. I'm afraid to say it, but there it is." Kurosawa and Alexievich patently remind us that trust is the nucleus of engineering. So it is with safety.

As we have seen in this chapter, public safety frequently comes down to how safe is safe enough. But how *fair* is safe enough? And how often do engineers make decisions that impose risk on others, and to what extent do they reflect on those acts? Substituting "fair for safe" isn't some linguistic quibble, note scholars Steve Rayner and Robin Cantor. The question of the fairness of safety is about debating the effects of technology based more firmly on trust and social equity than on what's more measurable. This also means that safety-critical engineering, like the ideal of freedom, requires eternal vigilance. And because engineers recognize that absolute safety is impossible, their decisions must be mindful of the risks imposed on others. Engineering design almost always involves trade-offs between safety and cost, feasibility, and practicality considerations. Such decisions create ethical challenges that rarely appear on a PowerPoint slide but are critical to alleviating wicked problems. Trust and fairness are intuitive elements that easy definitions cannot encapsulate. However tempting it is, as the Nobel-winning Toni Morrison wrote in *Beloved*, definitions belong to the definers—not the defined.

The engineer and advocate Jerome Lederer—whose career involved maintenance roles in the earliest US Air Mail Service flights in the 1920s, the preflight inspection of Charles Lindbergh's *Spirit of St. Louis*, and eventually leading FAA and NASA safety programs—knew about the varieties of experience in safety. Lederer was called the "Father of Aviation Safety," and he imagined a "psychedelic" dreamworld made of collaborative, trusting professionals. In it, plaintiff lawyers participated in aircraft design before the accident, not afterward; every safety suggestion resulted in greater passenger comfort; engineers invited others to contribute to design from scratch and were friends with test pilots, not foes; engineers eagerly received new ideas from everywhere: the "NIH" factor "Not invented here" would become "Now I hear." Such manners, Lederer observed, were essential for a working safety culture.

The polymath British judge Lord John Fletcher Moulton, whom Lederer drew upon, said manners include all things "from duty to good taste." In his 1912 article "Law and Manners," the jurist presented "three great domains of human action." One was the Positive Law, where an individual's actions followed binding rules. The second, Free Choice, defines an individual's preferences and autonomy. Nestled between these two domains is the third, the Obedience to the Unenforceable, which was neither force nor freedom. This self-imposed domain was about how a nation trusted its citizens and how those citizens upheld the trust. Obedience to the Unenforceable involves negligence, complacency, decreased oversight, and deviating from responsible practices. Lederer observed that, like doctors and lawyers, engineers have a code of conduct. "Engineers, however, have more problems in putting theirs into effect because they function more often as part of an organization, subject to organizational pressures."

Another facet relevant to wicked problems is that the success of a safety program depends on the context. Technologies often outpace our ability to conduct a safety program or even grasp its complexity. And ironically, as we saw with Steve Frank's paradox of robustness, continuous failure reduction may increase vulnerabilities. In an analysis of "ultrasafe" systems, where the "safety record reaches the mythical barrier of one disastrous accident per 10 million events," scholar René Amalberti observes that an optimization trap can perversely lower system safety "using only the linear extrapolation of known solutions." The design of these ultrasafe systems becomes a political act rather than a technical one, with a preference for short-term yields over more durable measures. Amalberti offers that a safe organization shouldn't be viewed as a sum of its defenses but as a result of its dynamic properties. Learning those would create a "natural or ecological safety," which aims not to suppress all errors but to control their fallout within an acceptable margin.

Add to this the diversity of perspectives that influence how a system produces or prevents errors, deadly in some instances. In fire safety, for example, many different forms of expertise can be claimed in the tightly contested professional domain, each legitimately. Scientists can declare their authority just as firefighters can claim theirs, as can lawyers and regulators. This conflict of views collided in investigating a June 2017 fire disaster in London, the deadliest UK fire in over three decades. Grenfell Tower, a 24-story residential building in London's North Kensington, caught fire, killing over 70 people and injuring dozens more. A refrigerator malfunction in one of the apartments rapidly spread fire in all directions, sped by the highly combustible external facade.

The British prime minister launched an official public inquiry. The commission confronted a grieving and justifiably angry public. Media coverage focused equally on fire-resistant claddings and national fire-safety governance policy. The coverage also emphasized trust, fairness, and expertise and how they underpinned the practice and perception of safety. Luke Bisby was one of the expert witnesses for the inquiry. He wrote six technical reports spanning several hundred pages that were the subject of multiple public hearings. Bisby saw firsthand how a hard safety problem could become a wicked safety problem. The nexus of regulation, innovation, education, professionalism, competence, ethics, corporate interests, social and environmental policy, and politics created a complex mess of interests, power, and, alas, victims, primarily immigrants and people of color.

As an engineer, Bisby was frustrated that there was no coherent voice on this subject and that the opinions came from all directions. The experience, however, made him more reflective. To truly understand what had occurred at Grenfell, Bisby spent five years tracing decades of parallel narratives in government policy, regulation, fire-safety design oversight and approval, and professional responsibility. The effort tested his belief in the primacy of science and rationality and, ultimately, his faith in engineering.

A year after the Grenfell Tower fire, Bisby spoke at an international meeting of fire-safety experts. He lectured about his work on sociological issues in fire-safety regulation. The audience asked him about his expertise in performance-based design. While Bisby is both a proponent of performance-based engineering design and a critic of the often thoughtless rule-following that many building codes enable, he told the audience, "I believe the idea of expertise is a myth." He crossed his hands over his black suit and conference badge. "My definition of an expert is someone who is profoundly aware of their incompetencies." The audience was stunned by what he said. Bisby explained how people are usually generous to themselves by overestimating, not underestimating, their competencies. Bisby tells his students regularly that their engineering degree will make them deeply aware of how little they know. His idea is to instill, early on, a mentality of "chronic unease," an intellectual humility required for promoting public safety. Paradoxically, only by becoming more competent can people recognize their incompetence. Specialized expertise is not a pitfall, but inattention to one's fallibility is. "We all must constantly ask ourselves if we have any idea what we're talking about," Bisby said bluntly. "If we did that, the world would improve."

Refrain

Surge

O N NOVEMBER 21, 1962, PRESIDENT JOHN KENNEDY WAS edgy. He called his science adviser Jerome Wiesner and NASA leaders James Webb, Robert Seamans, and Hugh Dryden into the White House Cabinet Room. He told Webb, the NASA chief, that the lunar mission should be NASA's prime focus. "Jim, I think it is the top priority. I think we ought to have that very clear."

Webb, a former attorney, expressed concerns about the unknowns and was pessimistic about the project's feasibility. As an argument brewed between Kennedy and Webb, Wiesner, an engineer, came down on Webb's side: "We don't know a damn thing about the surface of the moon, and we're making the wildest guesses about how we're going to land on the moon." Seamans, also an engineer, agreed.

Kennedy remained undeterred. The "Soviet Union has made this a test of the systems. So that's why we're doing it. So I think we've got to take the view that this is the key program," he said. "And the rest of it, God . . . there's a lot of things we want to find out about. Cancer and everything else." Webb tried to reason with the president that securing scientific leadership with the space mission would be a better bet than pure political symbolism in the simmering war of ideas between capitalist and communist countries.

"But you can't," Kennedy fired back, "by God, we've been telling everybody we're preeminent in space for five years, and nobody believes us."

"All right, sir, but let me say this," Webb replied. "If I go out and say that this is the number one priority and everything else must give way to it, I'm going to lose an important element of support for your program and for your administration."

"By whom? Who?" Kennedy asked.

"By a large number of people," Webb said.

"Who? Who?"

"Well, particularly the brainy people in industry and in the universities who are looking at a solid base," Webb said. "I'd like to have more time to talk about that because there's a wide public sentiment coming along in this country for preeminence in space."

"Yeah, but see, you got to prove you're preeminent," Kennedy retorted. "Unless this is the way to prove you're preeminent."

ELECTRIC TEAKETTLES CAN BE finicky, even for someone who helped put humans on the moon.

"Let's give it two more minutes, the water's not hot yet," Gerald Gene Abbey said, jiggling the switch up and down. He picked a bag of organic black tea from the tea chest and said: "Yeah, we don't want anything *in*organic."

Abbey was born in 1928. The son of a construction worker, he grew up in Flint, Michigan. "I was never much of a scholar," he said with a coy smile. "I had too much farm work to do." After serving in the Marine Corps, he studied engineering at Lawrence Tech. He failed college algebra but eventually got the hang of it. "It's easier the second time around," he said, adjusting his fatigued blue T-shirt, its divided pocket holding his reading glasses and a pen. After college, Abbey got a job in the General Motors truck and coach division. His first project was to develop fluorescent lights for a bus operator in New York City, the Fifth Avenue Coach Company. Then he was assigned to the multispeed-windshield-wiper project. "Gee, maybe we could make them go slower when it wasn't raining so hard," Abbey said.

In 1956, Abbey was recruited by Link Aviation, then under the ownership of General Precision Equipment. Abbey's first impression of Ed Link came from Tom Watson Jr., the second president of IBM. "Ed Link is a guy I really admire," Watson told Abbey. "If he wants to do something, he just does it." Abbey worked for the Link corporation for over 50 years under different ownerships. He started as a field engineer for the B-52 Stratofortress simulator. Then he led the prototyping of the simulator for the Lockheed Electra, a turboprop airliner from the late 1950s that is still used today.

In October 1957, the year after Abbey began his work with Link Aviation, the Soviet Union launched *Sputnik*. The 185-pound artificial satellite reached the Earth's orbit, earning the USSR a vital win in the space race and stirring American anxiety. Soon after, the newly created NASA launched Project Mercury, aimed to launch a crewed spacecraft into Earth's orbit and return it safely, ideally before the Soviet Union. After countless sessions on Link's Mercury simulator, on May 5, 1961, Alan Shepard piloted the first crewed Mercury flight aboard the *Freedom 7*. From 1961 to 1966, Project Gemini, a mission that extended spaceflight capabilities to be used in the Apollo program, relied on even more advanced space simulators. Abbey was the program manager for the Gemini simulators at Link. "We had evolved as an organization to take on challenges like that," Abbey said. "Newton showed us the physics, and Link did the engineering."

Kennedy's national goal of landing humans on the moon and returning them safely to Earth necessitated increasingly realistic simulations. For Link's Apollo simulator program, Abbey was the associate program manager for systems integration and testing. A key challenge was the complexity of managing the spacecraft configuration, which was compounded by the absence of vehicle-performance data. For aircraft simulation, one builds on test data; when needed, one could test out a real aircraft. With the Apollo simulator, data had to be predicted, and it had to be correct, for lives were at stake. In other words, the "flight" part of the flight test had to be entirely simulated.

A crucial hurdle for the simulation was computing power. In the late 1950s, the first programmable machine ENIAC—Electronic Numerical Integrator and Computer—was a grouping of over 1,500 relays, 6,000 toggle switches, 7,200 crystal diodes, 10,000 capacitors, 19,000 vacuum tubes, 70,000 resistors, and 5 million hand-soldered joints. It weighed over 30 tons and occupied 1,800 square feet. ENIAC consumed 150 kilowatts, "spurring rumors that every time it was turned on, lights dimmed in Philadelphia," one observer wrote. With the dawn of the digital computing era in the early 1960s, IBM machines with speeds reasonable for flight simulation cost around $2 million each. "This was an intolerable cost in view of simulator economics at that time," noted engineer John Hunt, who pioneered the development and use of digital computers at Link. Even if the visual presentation was a matter of basic arithmetic, rapid regeneration of those pictures to simulate motion and control loading cues required unobtainable speed and storage.

In the mid-1960s, Abbey and his team worked with 16-bit microprocessors that could process 25,000 instructions per second. Realistic simulations were the only way to prepare for the moon mission. Sending a flight crew to the moon in order to train for going to the moon wasn't an option. "We got faster through the course of the project, though," Abbey said. "Essentially, we evolved with the available tools." Link rented a spacious hangar at the Binghamton airport for the simulator assembly. The region's chronic cloud cover provided a natural setting for specific tests requiring subdued light. But when the northeast blackout of 1965 disrupted their work, the simulator settings had to be redone from scratch. As designers at NASA changed their minds about the contingencies of the mission, Abbey's team had to respond swiftly to the ever-changing details of the vehicle configuration. To address these alterations, a Link simulator draftsman reconciled spacecraft changes from NASA contractors to keep the reams of production-floor drawings current. "That was our version of rapid prototyping!" Abbey said.

Abbey's learning curve was steep as Link's computing software migrated from machine language to programming languages, changing the protocols for systems integration and testing. "It was just the daily grind of improving, testing, and installing the math models," Abbey said. Then, in January 1967, an electrical fire on the Cape Canaveral launchpad destroyed the command module, killing Gus Grissom, Ed White, and Roger Chaffee, the Apollo 1 astronauts. This tragedy cast further doubt on the already-delayed Apollo mission, and investigators questioned the readiness and safety of the project. As NASA made the necessary changes to the launchpad and spacecraft, the simulators had to be updated again.

Undaunted by the setbacks and the stress-packed environment, the Apollo mission brought together renegades whose work would define engineering and human history. Gene Abbey recalled a bus ride through the large campus of an Apollo contractor. An unassuming engineer sat next to him. The man was blond, crew cut, fit, and two years younger than Abbey. They spoke about airplanes and their hometowns. The civilian test pilot was from Wapakoneta, Ohio. In the early 1950s, he had flown 78 combat missions as a naval aviator in the Korean War, where Abbey had served as a marine. At the end of the ride, Abbey received a scribble on a piece of paper from that person.

His name was Neil Alden Armstrong.

✈

HISTORIAN STANLEY GOLDSTEIN HAS cataloged the human impulse to transcend Earth. Ancient Greeks imagined a powerful water jet that could propel humans to the moon. One 17th-century idea involved a set of rockets propelling an aerial chariot; another, an iron wagon pulled toward the moon by supermagnets. Other dreamers envisioned filling two huge vases with smoke, sealing them airtight, strapping them under the arms, and allowing the smoke to propel one toward the moon. Writer Jules Verne imagined man's journey to the moon in his 1865 novel, *From the Earth to the Moon: A Direct Route in 97 Hours, 20 Minutes*: a columbiad space gun firing a passengered projectile into outer space. Although written a century before the Apollo mission, Verne's satirical vision of an international space race and a crewed rocket to the moon may be the most prophetic vision of space travel in its time.

Many individuals imagined embarking on the half-million-mile round trip to the moon. Still, for those tasked to realize it, training and teamwork were the two most essential skills. Both were paramount in building the mental and physical proficiency to respond to life-threatening situations. During Project Mercury, designed to determine whether humans could function in space, the astronauts spent over 700 hours—over half of their training time—in Link simulators. For Gemini, the project further exploring human capabilities in space and the long-term effects of space travel, astronauts spent 67 percent of the total training time in Link simulators. And in the Apollo program, the combined training time in simulators for command and lunar modules was about 25,000 hours. This was in addition to the dramatic sessions on Boeing Stratotanker, dubbed the "vomit comet." The cargo plane made 40 to 50 steep climbs and dives in a two-to-three-hour period. Each parabolic arc simulated weightlessness by floating the astronauts for 20 to 30 seconds. "The final simulation before a mission was much like a graduation ceremony, except that instead of going out into the world to get a job, we had the task of landing an American on the Moon," wrote engineer Gene Kranz, NASA's chief flight director during the Gemini and Apollo programs.

If Apollo's astronauts were fearless, their simulator had to be peerless. But the simulators hardly looked heroic. The oddly angular boxes housed a jumble of electronics, mechanics, and optics: nearly 700 switches, over 400 circuit breakers, 64 kilometers of wire, and 26 subsystems. Astronaut John Young nicknamed it the "great train wreck." While the original Link Trainers for airplanes did not simulate the external environment, the Apollo simulators required a sophisticated visual experience. Stellar visuals in the lunar module were created by a luminous ball, a miniature descendant of

Ed Link's Celestial Navigation Trainer. The two-and-a-half-foot spherical mirror made a specular reflection. With over a thousand illuminated ball bearings on its surface, each miniature version of an actual star created a perception of faraway celestial fields.

While debates about the merits of motion or motionless simulators remained in the world of commercial flight, for the Apollo moon landing, engineers determined that astronauts had to know what motions, sounds, and forces of acceleration to expect on their mission. Link's engineers achieved force loading on the ground using motors, springs, and sliding shafts to give the "real feel" of the rocket. Precise sound cues indicated a drop in cabin pressure or thrusters firing in countless conditions. Instructors could freeze, fast-forward (by 30 times), and fine-tune practice conditions for astronauts training in the simulators. Of course, these operations required immense computing power. It took four computers to run a simulation—three to drive the equations of motion, the physical variables described as a function of time, and one for the visual system. The host computers were 24-bit machines, each with 350 kilobytes of memory. The 40-ton, 30-foot-tall Link Apollo simulator was a fragile contraption, with 10 tons devoted to lenses, mirrors, and film alone. Engineers had to employ a turboprop airlifter, Douglas C-113 Cargomaster, and 14 semitrailers to transport the simulator.

As one historian aptly writes, "Apollo gave new meaning to the word 'astronomical' in terms of cost and accomplishment." But it also enabled a "world of precreated experience" where reality was "just like the sim." Ultimately, the steep cost and brainpower required to create the simulator were worth it. The Apollo 11 trio put in 14-hour days, practicing at least 50 landing scenarios over three months, making several round trips to the moon. These landing scenarios were akin to practice symphonies of systems that mix photography, dialogue, sound effects, and score into a coherent motion-picture track. All the training was a prelude to an extraordinary cinematic event that premiered on a balmy July day in 1969.

Neil Armstrong, Buzz Aldrin, and Michael Collins reached the moon after a 109-hour journey—12 hours longer than Jules Verne's uncanny estimate from 1865. When the *Eagle* landed, the moon's temperature was 200 degrees Fahrenheit. Armstrong descended from the lander to walk on the Sea of Tranquility. From Houston, astronaut Bruce McCandless, the mission-control capsule communicator, said, "Okay. Neil, we can see you coming down the ladder now." Armstrong adjusted his water-cooled space suit for a "pretty good little jump" onto the fine-grained lunar soil. He stepped off the lunar module, proclaiming an 11-word cosmic soundbite: "That's one small step for man, one giant leap for mankind." Fifty seconds later, while walking on ground previously untouched by humankind, Armstrong's words commended the Link engineers. "It's even perhaps easier than the simulations at one-sixth g that we performed in the various simulations on the ground," he said. "It's actually no trouble to walk around. . . . Okay, Buzz, we ready to bring down the camera?"

✈

APOLLO 11'S SUCCESSES RESULTED from tenacious testing and rigorous refinements over many years. The late engineer and philosopher Walter Vincenti called this iterative practice "normal design." Engineers work to understand how the system operates, what considerations would form a desired solution, and the feasibility of accomplishing it. After years of education and practice, design work becomes intuitive. Engineers can comfortably probe "how the device works" and "what it looks like," a symbiosis of function and form. Through "normal" engineering, the airplane's design has stabilized. That is, cockpit is usually up front and tail in the rear. The design of an airplane has rarely changed, so it's easy to differentiate an airplane from a helicopter if you look at their operating principles: how they work and how they look. This understanding creates an accepted practice. If an engineer veers from those established operations, the result can be a "radical design." The design is more radical if the technology's working principle is fundamentally altered compared to how it looks. Such unique changes produce engineering know-how that may or may not depend on science.

An excellent example of radical design is a weight-loss project at NASA. In early 1965, the 14,850-kilogram lunar lander prompted worries that it was too heavy to do its job. NASA immediately ordered a design diet and formed the Weight Control Board. Grumman, the company that designed and built the lunar module, launched a war against the lander's excess weight. One approach, the "Scrape," searched for opportunities to lighten up the lander's parts. The second approach, the Super Weight Improvement Program, or "SWIP," involved a radical reenvisioning of the lander. A team of engineers second-guessed the entire design, which was 95 percent complete and involved contributions from 20,000 separate contractors. By September 1965, the Scrape project whittled away 45 kilograms from the structure. And by the end of 1965, Scrape and SWIP eliminated 1,100 kilograms of excess weight. A radical change involved using aluminum-mylar heat foil instead of bulky thermal shields. This choice demanded skillful machining and testing, leading to more weight-reducing changes and an oddly angled assembly.

The original Link Trainer was also a radical design from an unlikely source: player pianos. Instead of being born from an existing simulator design, the trainer's organizing force gave rise to new scientific insights into safety and human behavior. Engineering is often deemed derivative and subservient to pure science, even when new sciences are routinely developed from engineering. But the Link Trainer enabled new sciences and organizational ideas for learning, not vice versa. And within a generation, this radical device became an institutional fixture in aviation.

✈

THE APOLLO 11 MISSION—ENABLED partly by the radical design innovations, coordinated efforts across teams, and ongoing training—was a crowning achievement for America during the Cold War era. While the Soviet Union and the United States remained locked in a battle of innovation, not all nations could replicate the engineering feats that launched these nations into scientific history. In 1967, a French politician commented on why Europe lagged in the space race. The deficiency was not so much in brainpower, his reasoning went; it was with "organization, education, and training." In other words, France lacked systems management.

Scholar Stephen Johnson has studied the impacts of systems management on aerospace missions, which invariably involve interlocking practices of planners, engineers, scientists, and bureaucrats. Well-organized processes can markedly improve the performance of launch vehicles, ballistic missiles, and spacecraft. In the decade between the 1950s and 1960s, Johnson points out, the success of such projects following a systems management approach almost doubled, with cost and schedule overruns reduced by 3- to 10-fold.

For adventurers, systems engineering procedures can seem dry and tiresome. But think of it as the tedium of lifesaving flight-safety checks and instructions that precede the thrill of flight. "We can lick gravity," as the sentiment goes, "but sometimes the paperwork is overwhelming." If the concept of flight is about six degrees of freedom, then processes that support it present six degrees of constraint. That's why the aerospace sector lives with a paradox. Innovation and bureaucracy breed each other. The safety-critical products and processes that produce exciting missions must be infallible. "Spacecraft that fail as they approach Mars cannot be repaired," Johnson writes. "Hundreds can lose their lives if an aircraft crashes."

Consider the acronym-heavy, six-step procedure applied to Apollo configuration management, which systems engineers designed to precisely track and control hardware and software modifications. The first step was a Preliminary Design Review (PDR). The second step was the Critical Design Review (CDR). The third step was the Flight Article Configuration Inspection (FACI). The fourth step was the Certification of Flight Worthiness (COFW). The fifth step was Design Certification Review (DCR).

And the sixth was the Flight Readiness Review (FRR). This routine might sound creatively deadening. But these steps were essential to realizing the mission, from parts to the performance and progress reports. People fascinated by the "moonshot" have focused only on the impossible destination and seldom discuss the procedures and discipline that undergirded the mission. No wonder many of today's so-called moonshots often end up as moonshine, only intoxicating advertisements rather than integrated action.

The aerospace sector has relied on systems engineering as "insurance for technical success," Johnson observes. These methods are invaluable when planners have an incomplete picture, with poor objectives, communication, and coordination. Sometimes a mission may combine all these deficits. "Military officers demanded rapid progress. Scientists desired novelty. Engineers wanted a dependable product. Managers sought predictable costs. Only through successful collaboration could these goals be attained," Johnson writes. Success with space rocketry necessitates using systems engineering tools, however mind numbing or time consuming, in the spirit of the Latin expression *per aspera ad astra*—through difficulties to stars.

ONE MAN WHO DEEPLY understood the indispensability of systems engineering—and the tragedy when it is ignored—was Frank Hughes, the former chief of NASA's spaceflight training division. Hughes joined NASA as an instructor in 1966. On the first day of his job at the Kennedy Space Center, the first box of the Link Apollo simulator was delivered. Hughes's career at NASA evolved with the Link simulators through the Mercury, Gemini, Apollo, Apollo-Soyuz, and space shuttle missions leading up to the International Space Station program.

"We were lucky all the time," Hughes told a reporter. "I used to joke that God spent about half of his time just with NASA on those early missions." The early stage of the Apollo program had "no training rhythm to it," Hughes recalled. "The crew walked in the door and set the stage of what they wanted to do. There was no such thing as a lesson plan or objectives or whatever. You trained until you flew, and it just filled up the time. If things delayed, then you trained more, and you just went on like that."

One of the most critical elements of the training was the ability to simulate potential problems that astronauts could encounter. During the Apollo program, the Link engineers updated the simulators with over 1,000 exercises for malfunction scenarios. "The fuel cells could fail, tanks could leak, all that sort of thing," Hughes remembered. Simulations could be sped up and slowed down with advanced visuals and many improved capabilities for every step of the mission rehearsal. The star fields were even more credible. And the simulators now contained the exact measurements of every spacecraft element rather than estimates.

The value of such almost identical features became apparent on Monday, April 13, 1970. Apollo 13, the third lunar landing mission, was 55 hours into its mission and 200,000 miles from Earth. The spacecraft was under the command of Jim Lovell, with John Swigert as the command module pilot and Fred Haise operating the lunar module, all trained as engineers. Meanwhile, after an intense and exciting day, Frank Hughes was driving home from Kennedy Space Center, already thinking about the training regimen for Apollo 14.

Suddenly an oxygen tank exploded in the service module, creating a sequence of breakdowns on Apollo 13. Lovell, Swigert, and Haise couldn't identify the problem. Soon after, the crew heard several more explosions. In an emergency call to mission control, Lovell uttered the most famous, inauspicious words in space history: "Houston, we've had a problem." Hughes raced back to mission control. He and the rest of the team quickly got to work on the simulator, continuing the effort for five restless nights. In the first three hours after the oxygen tank explosion, the crew lost over one-half of the oxygen reserves. Returning the trio safely to Earth was now *the* mission.

With the command and service module *Odyssey* crippled, *Aquarius*, the lunar module, became the "lifeboat" for the astronauts to operate for 90 hours, double the number of hours the module was designed to work in an emergency. As the spacecraft hurtled away from Earth at 2,000 miles per hour, preserving the module's power was paramount. New procedures had to be researched, recorded, and revised. The Link simulator operated nonstop as the ground crew and standby astronauts rapidly reworked their ideas. They tested scenarios based on the remaining onboard resources.

With oxygen reserves depleting, the astronauts' challenge was compounded by the fact that their simulator sessions didn't perfectly reflect their reality. Although high in fidelity, the simulations didn't have the cloud of particles or debris striking the spacecraft that the crew had to contend with onboard. And there were further problems that the simulator hadn't anticipated: maintaining uniform temperatures within the spacecraft. Apollo 13 needed passive thermal control. A tricky maneuver first attempted in the simulator, nicknamed the "barbecue mode," slowly spun the spacecraft lengthwise for even exposure to solar heat. With a disabled command and service-module computer and the lunar module lacking the essential software, the slow-motion roll had to be done manually. Lovell controlled the forward-backward movement, and Haise handled the left-right motion of the spacecraft, rotating the spaceship at three revolutions per hour. This technique saved the spacecraft from being heated on one side and frozen on the other. And finally, thanks to the ground simulator, the crew learned a method to abandon the disabled module, which enabled the astronauts to plummet safely through the Earth's atmosphere. As the astronauts commenced reentry, the procedures were anything but orthodox. With orange-and-white parachutes inflated over the Pacific, they splashed down near Samoa.

For NASA, Apollo 13 was a "successful failure," Hughes noted, and the Link-built simulators out of Binghamton proved that they were the "unsung heroes" of the rescue mission.

✈

"THIS IS THE HALL of Ones and Zeroes," said Susan Sherwood, pointing me to an area in the old ice-cream factory, now the Center for Technology & Innovation, in Binghamton. Sherwood, the center's director, described the workings of vintage IBM computers and printers. "The very creators are now restoring them," she said. Local company uniforms and coffee mugs were also on display. On the other side of the room sat a pair of pianos, one of which was a Link player unit. The coded roll of a "Fats" Waller medley rang out the bubbly "Boogie Woogie" and "Canteen Bounce" followed by the cheery "Fuddy Duddy Watchmaker." The gears turned, the wheels spun, the paper rolled, and the keys mysteriously went up and down. "Makes you feel kind of creepy, don't it," wrote the engineer-turned-writer Kurt Vonnegut in his debut novel, *Player Piano.* "You can almost see a ghost sitting there playing his heart out." The old box had come to life.

The Center for Technology & Innovation is a theme park of antiquated technologies with a vast collection of devices and manuals from IBM, Link, and General Electric, all formerly local businesses in Binghamton. The center's all-volunteer program TechWorks! aims "to showcase regional technology in action—an effort that begins from the inside out." For Sherwood, "what's past is prologue." Along with the volunteers from TechWorks!, Sherwood restores the unsung heroes of the past—machines like the player pianos, IBM computers, and Link Trainers—and reveals the profound lessons these technologies might hold for our futures. "Shall we go see the flying piano now?" asked Sherwood.

Ed Link rigged his first flight trainer from a piano not far from TechWorks! on Water Street. Sherwood took me to the center's working antiquary, a hangar-looking building that doubles as an exhibit hall. The central idea of TechWorks! is operational preservation. Since 2006, Sherwood said, the center had returned three dilapidated Link Trainers to working status. Over two years, volunteers refurbished the 1942 blue-box trainer using new-old stock parts. Repairing the instructor station took another two years. The synthetic trainer returned to an operational state with refreshed servo motors and microcontrollers. Now visitors to TechWorks! can enjoy an "immersive" spin on the device that revolutionized aviation.

Several retired Link engineers have dedicated themselves to preserving these technological treasures in the working antiquary. John Lash recalled working for Link as a career-defining experience. "The company was one big happy family of people," he said, adjusting the belt on his cargo work pants, remembering his daily commute to Binghamton. "From the guy who swept the floors on the second shift up to the president who ran the company, they were all neat people to interact with," Lash reminisced. "Over the years, that changed; it became more businesslike. It all came down to the bottom line."

Lash quickly discussed an electronic component with Carl Mazzini, a fellow Link veteran conducting a circuit test. "Even people living in Binghamton don't fully realize how significant their town is," Lash said. "There are stories behind everything here if you had people to tell them." And Mazzini said, "I saw an old washing machine over there. The company that became Whirlpool started here," referring to the vintage cast-iron Cataract from the early 1900s. "That's something I didn't know."

Lash lamented a long-vanished coupling between a company, its community, and the country. That connection, he argued, inspires meaningful work. "It's one of those things where I think ignorance is bliss. We didn't know we could do it, but we got it done in 10 years," he said, referring to the Apollo program. "Well, yeah, you were lucky," Mazzini said. "Now, you can't build a fighter jet and get an initial operating capability in 15 or even 20 years."

They then discussed the government's influence on defense and aviation. "There's so much revolving door between government and the industry that the defense companies are as inefficient as the government is," Mazzini said. Lash agreed. "I mean, if a defense contractor were to produce a cell phone, it'd be the size of this thing here," Mazzini said, pointing to a big rusted box. "Yeah. We used to have a joke about that," Lash said. "It went like the government wanting a wheelbarrow. By the time the functions were added, and all the organizations got involved, you got this huge device that isn't anything like a wheelbarrow."

Earlin Ward, a peppy engineer, joined the conversation. "We'd get so involved with everything that we'd lose the basic concept and concentration on the problem," Ward said. Even systems engineers sometimes end up sacrificing the big picture for details. A maze of procedures can overshadow the objective. "We got so much into writing the requirements, and adding more and more requirements, that we could never get to a fricking design. We couldn't do it," Mazzini said.

That's why the blue-box trainer was different. "Ed Link didn't need diagrams or systems manuals or anything. He understood his objective, and he understood his resources. And, he understood what capabilities he needed," Ward said. It was the *philosophy* of systems engineering Ed Link fully capitalized on. "This might all sound fairly simple, but go try to build that blue box. Some of that skills and substance are just not available anymore."

At the other end of the antiquary stood Gene Abbey, whose pioneering work with systems integration and testing on the Link simulators contributed to the success of the Apollo mission. With him was Bill Bennett, a retired software-systems engineer at Link, who worked on military flight simulators. They discussed a beam splitter that Abbey was fixing on the Apollo simulator on loan to TechWorks! from the National Air and Space Museum.

Soon after delivering the Link Apollo simulator to NASA in the 1960s, Abbey built simulators for commercial nuclear power plants. Given the risks involved with their work, the nuclear power industry was quick to adopt simulators. After a pause in his discussion with Bennett, Abbey recalled the 1979 disaster in the township of Londonderry, Pennsylvania, at a nuclear plant three miles downriver from Susquehanna. The namesake, the Three Mile Island accident, was one of the worst in history. "The operators didn't recognize the problem till it got out of hand, and then it was too late," Abbey said. "If they had had adequate simulator training, they could have recognized and circumvented that catastrophe."

In Bennett's view, the Three Mile Island case was a scenario with no single output or outcome. "These are not simple calculations to do in your head," he said, remembering his programming for a portion of the Apollo simulator. "The only way you know is the whole thing comes apart." The software had to communicate with 1,500 circuits, each with switching and control functions. "The key was to embrace levels of abstractions and visualize these complex interactions in your mind's eye," Bennett added in his lulling voice. "That's precisely what Ed Link did."

When Bennett started working for Link in 1952, he had heard the company did systems engineering. "But what's that?" Bennett wondered. "It seemed like a word without a definition, you know?" He wasn't the only one who didn't initially understand systems engineering. Jim James, another retired engineer with a long, white biker mustache, echoed his sentiments. "The simulator could be the system with subsystems under that, or the simulator might be a subsystem under a complete training program," James said. "I mean, you know, it's all hierarchical."

In the early 1990s, Bennett wrote a monograph called *Visualizing Software*. He used the example of a simple road map presented in two parts. In the first part, only the dots representing towns are printed with their names, and in the second part, the lines represent the roads that go through the towns. This exercise requires a viewer to simultaneously look at two maps with different sets of information and synthesize them. The key was combining the maps into one mental image—with town names and road names—to see their interactions.

Humans navigate complex sets of information at many levels, prioritizing some details in one instance and another set in another.

Applying this idea to simulation software, Bennett argued for hierarchical design. "Most of the time, it's impractical to show a large piece of software as one single diagram," Bennett wrote. "Some means of breaking up a large software system into many diagrams, like the hierarchical division of the road maps, is essential." This basic multilevel approach underpinned the software development for the Apollo simulator.

MAINTENANCE IS A WAY of life for the seasoned engineers volunteering at TechWorks! This cohort of engineers views retirement as a time for restoration in addition to recreation. It's easy to forget the humdrum yet essential duties of repair, renewal, and reconstruction when we are constantly fed dazzling narratives about tech innovation. "Another flaw in the human character is that everybody wants to build and nobody wants to do maintenance," Vonnegut has written in *Hocus Pocus*. Scholar Debbie Chachra, in a nod to novelist Arthur C. Clarke, has said: "Any sufficiently advanced neglect is indistinguishable from malice." When we valorize moonshots, we might deemphasize maintenance. The former moonshot engineers at TechWorks! are examples of those doing unthanked, unglamorous, unrewarded, and unrecognized care work. Responsible engineers know that maintenance and modernity go hand in hand, not one at the expense of the other. Maintenance is the mother of modernity.

"To say we are underfunded is an understatement," Sherwood said, reflecting a common problem of nonprofits, particularly those committed to stewardship of older technologies. Bob Swarts is another volunteer, who revitalized a general aviation trainer with Carl Mazzini and others. The GAT-1 was Link's fully electronic, fiberglass system from the 1960s—a lightweight, fixed-base, plug-in machine resembling a single-engine aircraft, flying up to 10,000 feet. The restored trainer reproduced the cabin sounds accurately, and Swarts wanted me to take it for a ride.

I climbed, turned, glided, and cut wide arcs through a 360-degree turn as if grooving to Bob Marley reggae. Then, before I could think, I was diving and spinning. The trainer was shrieking. "Watch out. You are stalling," Swarts said. It was clear I had violated every known rule of flying. I stepped off the trainer with adrenaline and sweat as if I had landed on a tilled cornfield.

FIG.1

EDWIN A. LINK
INVENTOR

BY

ATTORNEY

Chapter Five

Wicked Maintenance

I N THE EARLY 1800S, AMERICA WELCOMED STEAM TRANS-
port as technology and public interest galloped beyond horse-
pulled carriages. Since 1820, steam-conveyed cotton, molasses,
and whiskey multiplied cargo 40-fold in 40 years. By the mid-19th
century, over a thousand steam-powered vessels were transporting millions of passengers. While the sizzling steamers "telescoped
a half century's development into a single generation," they also
became death traps, exemplifying why safety isn't possible without
maintenance.

"Speed king" skippers were fast because their mechanics super-
stoked the ships' boilers while shutting off safety valves. This technique ratcheted the steam pressures and effectively turbocharged
the vessels. The treacherous rides thrilled some passengers, even
if it meant that everyone aboard was risking life and limb. One
passenger reminisced about a chancy collision in which two "crack
boat" captains raced recklessly in a narrow channel. As the boats
rushed full steam ahead, their hulls scraped—passengers stood eye
to eye on opposing decks and exchanged handshakes.

In 1833, Andrew Jackson asked Congress in his State of the
Union for preventive and punitive laws against the "evils" of steam
navigation. The president's frequent opponent, Senator Daniel
Webster, proposed a requirement to outlaw racing boats and apply
stringent standards for boilers. Jackson's loyalist, the "Old Bullion"
Thomas Hart Benton, argued that the steamboat owners were

generally "men of the highest integrity." And as a frequent traveler, Benton said, he never encountered any accidents, advising Webster not to meddle in private waters. Benton's attitude enabled him to steamroll Congress and block the bill.

In the years to come, the value of periodic boiler maintenance and the harm that arises from maintenance neglect became apparent. The politics of steamer safety prompted stricter industry practices for navigation, and it all came down to "hard maintenance" of regulated inspections and standards. As with Captain Isaac Perin, a 24-year-old daredevil, the accidents made it clear that maintenance was tied as much to safety as to vulnerability and efficiency.

On March 31, 1838, Perin pulled out a new ship from a boatyard in boomtown Cincinnati named after the engineering prodigy Robert Fulton. The 150-ton *Moselle* was swift and stylish. Three weeks later, on the balmy afternoon of April 25, 1838, Perin took *Moselle* on a trip from Cincinnati to St. Louis via the serpentine Ohio. Its 300-odd passengers experienced a deluxe and delightful voyage as *Moselle* ferried that 750-mile route in a record two days and 16 hours. A mile upriver from the pier, *Moselle* stopped to pick up some passengers with goats and chickens. Perin, eager to flaunt his supership's high speed to a crowd, left the steam levels dangerously high. The onboard engineer panicked and warned Perin. Suddenly, *Moselle*'s four boilers burst with a blaze likened to a "mine of gunpowder." The synchronized blast flung Perin across the river into Kentucky. Over 200 people perished. An eyewitness compared the shattered liner to a tree twisted by a lightning strike. The battered, burned, and bleeding survivors floated on fragments of the ship deck.

An investigative committee blamed Perin's inexperience and ignorance. However, their report argued that greed was the real reason behind the catastrophe. "We are not satisfied with getting rich, but we must get rich in a day. We are not satisfied with traveling at a speed of ten miles an hour, but we must fly. Such is the effect of competition that everything must be done cheap; boiler iron must be cheap, traveling must be done cheap, freight must be cheap, yet everything must be speedy," the report pronounced. The efficiency gains from *Moselle*, both in speed and dollars, was "demanded by the blind tyranny of custom, and the common consent of the community." A local news publisher pleaded guilty for contributing to "inflate the ambition of its captain and owners."

Three months after the disaster, Congress passed legislation for steamboat standards, boiler inspections, and fire safety. Even though the law held ships' owners and operators liable for murder, neglect, and malfeasance, punishable by years of hard labor, the measure was still too weak to prevent accidents. As interstate commerce grew, so did the number of steamboat explosions in the following decades. Between 1807 and 1853, steamer blasts killed over 7,000 people, and defective or ill-serviced boilers were the chief reason.

By 1860, steamers transported over half of all cotton from the South over the Mississippi, perpetuating the thriving slave economy. One member of Congress praised the steamboat as the "triumph of genius!" and affirmed that the "magnificent onward march" of the nation "has received new energy and animation from it." Yet, continued deaths and damages from steamboats compelled the government to confront new notions of liability.

One precedent included an 1845 court decision involving an engine explosion at the defendant's flour mill that killed the plaintiff's horse. The defense claimed the manufacturer was at fault for the boiler's defects. Still, the court ruled otherwise, holding the owner responsible for poor maintenance. This ruling created more questions than answers. How could one establish liability for design and inspection? What's the responsibility of businesses tempted by corner-cutting to skimp on licensed engineers? And how can manufacturers navigate new federal oversight of private industry?

The popular practice of ship operators to knowingly tie down safety valves for surplus speed was designated malpractice. Sometimes, the worst malpractice was when drunk engineers were on duty. Each boiler explosion can be a case study of the dereliction of duty. The government required firms to keep their steam machines well maintained. In 1852, Congress passed a more decisive law establishing boiler operations standards. The law also tightened maintenance standards, thus forming a basis for a separate 1871 authority that issued licenses to engineers, technicians, and inspectors and imposed hefty penalties for owner and operator infractions.

In "hard maintenance" problems such as pressure-vessel explosions, as with the molasses tank failure, straightforward technical fixes and policy statutes solved the problem because the causes and consequences were apparent. Naturally, communications, transportation, food, and biomedical product mandates can have concrete compliance requirements: the problems are obvious, and their effects, preventable.

Now, consider the technological evolution of boilers. Most 18th-century boilers were wood, copper, or cast-iron containers absorbing heat from the furnace. The next generation was riveted wrought iron with valves to ensure safe pressures. But the rolling and pitching motion of the ships created complications for these boilers, including faster corrosion. The increase in steamships created the need for qualified engineers who could do routine hydraulic proof testing to ensure boiler safety. Thomas Jefferson envisioned "a wise and frugal government, which shall restrain men from injuring one another, shall leave them otherwise free to regulate their own pursuits of industry and improvement." But the public soon came around to implementing federal policy to curb steamboat disasters.

Writer Samuel Langhorne Clemens signed up as a steamboat pilot's apprentice in 1857. He was devastated by losing his brother a year later when the steamboat *Pennsylvania* detonated, killing 250. Its engineer was indicted for negligence. A year later, Clemens received his pilot's license. Piloting on the Mississippi led Clemens to his renowned pen name, Mark Twain, a term related to whether a steamboat was sailing at a safe 12-foot depth, or two fathoms. Twain criticized the "lure of technology" to solve complex problems. He felt that as machines increasingly replaced humans, they degraded them. But today, technologies for hard maintenance have extraordinarily high system availability and reliability. In some contexts, power availability and broadband reliability have exceeded 99.999 percent—or "five nines"—equivalent to five minutes of downtime in a year.

The worst disaster in US maritime history, though, happened days after Abraham Lincoln's assassination. The two-year-old Cincinnati steamship *Sultana* was a regular at the port of New Orleans. In the early darkness of April 27, 1865, the four-storied, 260-foot liner journeyed to St. Louis with 300 people and numerous hens, horses, mules, sugar casks, and coal bins on board. It stopped in Vicksburg, Mississippi, a military post that Jefferson Davis called the nailhead holding the South. Nearly 2,000 Union soldiers released from rebel prisons boarded the ship. *Sultana* now carried five times its maximum load. In addition, the boat's chief engineer later recalled poor repair and servicing of the boilers, which led to the explosion. Three boilers blew up a few miles north of Memphis, shredding the sides and blazing the boat. The monstrous mishap claimed nearly 1,800 lives, surpassing the fatalities from the *Titanic* some 47 years later. Others drowned as *Sultana*'s saloons and stacks burned like brushwood.

Calamities like these have become rarer because of technical and regulatory remedies, including, in some contexts, the imposition of penalties on new technologies for breakdowns or shortfalls. Maintenance has evolved beyond risk reduction, which was more relevant to the steam era. Today, maintenance may mean making products long-lasting. Maintenance can also refer to related themes: retain, redeem, refuse, reduce, reuse, repair, replenish, refurbish, reconstruct, remanufacture, repurpose, reform, recycle, and rejuvenate.

While maintenance is at the heart of engineering responsibility, innovation more often gets highlighted. A "sales pitch about a future that doesn't yet exist" is often more attractive than mundane maintenance of existing technologies. However, "the engineer learns most on the scrapheap," as scholar Claude Claremont observed in his 1939 book on bridges, *Spanning Space*. Metaphorical bridges are needed to remind us that most practicing engineers operate and maintain systems rather than pursuing glory-garnering gadgetry. Indeed, most engineers don't chase disruptive innovation; their work *prevents* disruptions. Yet, the skewed focus on "innovation-speak" rather than on care devalues the tenacity of maintainers in public infrastructure. As one scholar noted, "Infrastructure is all about maintenance. Maintenance, maintenance, and more maintenance. It doesn't just get built, like some colossal monument left to stand until natural forces wear it away. It constantly has to be repaired, rebuilt, extended, shrunk, adapted, readapted, continually redefined and reengaged."

The precision of hard maintenance can improve efficiencies and lower environmental impacts with routine upkeep, compliance, and coordination. The many tech companies that brand themselves as innovators spend 60 to 80 percent of their software budget on the maintenance that enables their innovation in the first place. Deferred maintenance has been called a "slow disaster" and a "disaster multiplier." As historian Scott Knowles has chronicled, problems caused by deferred maintenance are "revealed only when catalyzed by enough water and enough wind in too short a time." Chronically delaying maintenance can have unusual complications.

✳

MOST PEOPLE FINISH THEIR business by flushing the toilet, and that's when Pam Elardo's business begins. The former deputy commissioner of New York City's Bureau of Wastewater Treatment, Elardo is a queen of sewer, a "connoissewer." The bureau makes an elixir out of excrement. An average New Yorker uses about 100 gallons of water daily, double the amount used in most of Europe. The 1.4 billion gallons of wastewater precipitated by the nine million residents, workers, and visitors can easily double, or even triple, on a wet day. New York City has 14 wastewater plants, 96 pumping stations, and 7,500 miles of sewer lines, each requiring maintenance.

Wastewater from homes, hotels, schools, and places of work and worship often combines with the runoff from the streets, storms, and snow. Victor Hugo wrote of sewers in *Les Misérables*, "Crime, intelligence, social protest, liberty of conscience, thought, theft, all that human laws persecute or have persecuted, is hidden in that hole." Even though toilets are technically home to only four Ps—poop, pee, puke, and paper—in New York's sewers, one can see wipes, floss, Q-tips, napkins, condoms, and tampons. Nonbiodegradables mix with lumps of fat, oil, and grease ("FOG"), hair, blood, and chemicals from homes, nail salons, factories, and hospitals. The result can be a mammoth, multimillion-dollar problem: "fatbergs." In London, these fatbergs can "grow to the size of a double-decker bus or weigh as much as three elephants," requiring maintainers to dig out the rocky reek and unclog it mechanically.

Flushed materials go through a local collection line down the street. Next, they meet interceptors, the primary pipes that combine flows from several trunk lines and guide the flow to wastewater treatment plants. Gravity, being Mother Nature's fiber, channels a smooth flow. But when the moving gets a bit harder, pumps provide relief. The influent passes through sieves that filter out everything alien to the sewer: cell phones, books, bottles, bicycles, designer watches and sunglasses, engagement rings, wallets, yoga pants, and even guns. That sifted trash is landfilled at a cost that runs into the millions of dollars annually.

Next, the wastewater enters sedimentation tanks. Heavier solids settle in this preliminary treatment, and lighter materials float. The settled sludge and skimmed floats are taken to landfills or processed elsewhere. Then the activated sludge process begins. In large tanks, aeration stimulates microbial reaction, mending the wastewater. Collected sludge used to be dumped in the ocean, until federal law prohibited this. Now dried biosolids are produced for fertilizer pellets and composting. The aerated wastewater is settled and disinfected. This clean effluent is released into the local waters. "We take credit for the whales returning to New York waters because of the work we do 24/7," Elardo said. "Sometimes people propose to get married at our sewage treatment plants. It's a bit geeky."

New York City's sewage-processing systems came to life at the turn of the 20th century with two plants in Brooklyn and one in Queens. Before that, New Yorkers predominantly relied on wells and street suppliers for drinking water. Lack of proper sewage, waste, and trash disposal sparked cholera and yellow-fever outbreaks, and the city constructed new plants in the 1930s. As unsophisticated as the sewer system might appear to the untrained eye, managing the overall enterprise involves scores of "soft maintenance" problems. Hard problems arise in this system, like sewer clogs. When those problems congeal with behavioral predicaments, they create the familiar tangle of trade-offs: how and when to repair and replace assets over their life cycle.

Historically, as with many operating utilities, the New York wastewater system lacked a comprehensive asset-management program. One can delay making essential repairs and save a few dollars soon, but this will lower the lifespan of a part or the system it supports. Or one may prematurely replace an asset without making the most of it. Like soft efficiency planning, asset management is a multicriteria problem; it requires a broader assessment of the past, present, and potential paths ahead. The bureau's wastewater technologies span generations. In some places, charts and clipboards are still the dominant tech, but tablets and touch screens are becoming the norm in others. Historian David Edgerton points out that such a range of technologies suggests a better understanding of society's material constitution and cooperation. In *The Shock of the Old*, Edgerton writes that, in one instance, technology is "the latest technical gizmo, and the next it is the very infrastructure of human existence." So Elardo's job can involve time travel. "Depending on the day, it could be the 1920s, the 1980s, or the 1950s, and each decade posed its particular problems," she said at a Nepali restaurant a few miles from the original Jamaica Water Pollution Control Plant.

Elardo first appreciated wastewater engineering while working for the Peace Corps in Eastern Nepal. "I became an environmentalist in third grade when my family moved next to a wastewater treatment facility," she said. "I wanted to fight pollution. Air, water, whatever." Her high school counselor discouraged her from pursuing engineering, saying it was a "man's job." But Elardo persevered and rose in a male-dominated line of work. "In the 1980s, there were guys with girlie posters in their offices," she said. "While I deeply respect engineering as my profession, I don't necessarily respect some of its attitudes."

Elardo's primary frustration is that policymakers persistently devalue maintenance planning. "Stuff is falling apart," she said, "but we're not going to get enough done because we're so busy responding to emergencies." Before coming to New York, Elardo directed the King County Wastewater Treatment Division in Seattle. There she led reliability-centered maintenance, a stepwise program to proactively identify and prerectify potential failures by learning from the best trade practices of testing. Her results in Seattle showed systemic gains in performance and cost control. But in New York, Elardo initially felt like an anomaly. The bureau spent staff time overwhelmingly on corrective or emergency maintenance over prevention. "We can't see the whole picture," Elardo said. "We don't even know the size of the whole picture."

Even water treatment involves putting out fires. One patchwork narrative emerged in Elardo's general explanation of public infrastructure's frail and failing functions. "Suppose an old pipeline bursts and sprays out raw sewage. It's an emergency. What do we do? We wrap it. The wraps start to leak. Then we wrap it again. Then a new fiberglass wrap comes on, and that can leak. Then, with little choice to keep the plant running, we encase the entire thing in concrete. So, when we think of replacing the pipe, we will spend much time chopping up the concrete," Elardo said. "As a society, we're always doing stuff like this to avoid catastrophes. Oh God, it's hard."

Under Elardo's leadership, the New York bureau started to engineer a new maintenance ethic. For example, they began planning for maintenance on the run time of assets in hours rather than calendar days, which can prove advantageous. They tracked maintenance hours for critical assets more closely, focusing on prevention and cost savings. But having organizations adopt such a strategy and convincing their budget controllers is another challenge. In care for wastewater treatment, as one pun goes, if you aren't part of the solution, you are part of the precipitate.

As writer John Joseph Cosgrove observed in 1909, "A history of sanitation is a story of the world's struggle for an adequate supply of wholesome water, and its efforts to dispose of the resultant sewage without menace to health nor offence to the sense of sight or smell." There's no glamour in dealing with gunk. Yet Elardo's passion for preventive maintenance is manifested in the Hebrew phrase *tikkun olam*, which connotes conscious actions for communal repair and restoration. Maintenance-intensive industries can trail significantly behind new technologies. The Jewish worldly welfare concept could also motivate innovations to improve the maintenance of sanitation. Modern digesters can reduce landfill use by efficiently processing food waste with bacteria. New odor-control systems can relieve communities near landfills. Drones and digital monitoring can forecast and find potential failures. And remote-controlled submersibles and programmable "pipe worms" can help flush out fatbergs.

Maintenance is embedded in the origins of public health, a field founded by engineers rather than clinicians. In the early Bronze Age, Mohenjo-Daro—the "mound of the dead" in the Sindhi language, about 250 miles north of Karachi in present-day Pakistan—was a center of the Harappan culture of northwest India. The ruined remains of the three overlaid Mohenjo-Daro towns, excavated in the early 1920s, indicate that they were developed between

3300 and 2700 BCE. The metal-age metropolis exemplified purposeful engineering: everything seemed preplanned, even competently managed. The streets ran north and south, east and west, supplied by 700 wells; the houses were built with well-burned, standard-sized bricks, and the system of weights and measures was consistent. While little is known about these people, we can learn a lot from their sweeping social vision for sanitation, which many parts of "modern" India lack millennia later.

Every house in Mohenjo-Daro had a washroom and water-based latrine; waste was conveyed to an extensive sewer system by sloping and tapered terra-cotta tubes to intermediate settling sumps that reduced clogs. Harappan engineers included cunettes, or smaller conduits for outflows of routine discharge, saving the entire channel capacity for wet weather loads. Similarly, the city of Dholavira, located in the Indian state of Gujarat, was known to have created water reservoirs within the brick-and-stone city walls, an act of heroic hydraulic engineering in a desert. Dholavira's rectangular stepwell was three times larger than the Great Bath of Mohenjo-Daro.

Around 2600 BCE, the Egyptians knew where to place channels and spouts, often crafted as animal-headed gargoyles, to divert the rare rainwater into the trenches so it wouldn't flow into the temples and damage the painted walls. In the Minoan civilization that prospered on the island of Crete between 2800 BCE and 1100 BCE, engineers laid dual ducts, one for wastewater and another for rainwater. This informed future designs for divided drainage. The latter-day engineers of the Mesopotamian Empire designed vaulted gutters. They ensured that the rainwater efficiently flowed into baked-brick drains and cisterns sealed with asphalt. The Achaemenid Persian Empire renovated the wastewater system after seizing control of the region from the Babylonians in 539 BCE. They introduced strict laws to protect the rainwater and runoff from contamination. In the Pompeii of 500 BCE, streets had small elevations to guide rainwater.

In the late fourth century BCE, after the people of Etruria were absorbed by the Roman Republic, engineers rebuilt the Etruscan sewers. They pioneered "crowned roads," higher in the middle and slightly sloping toward the shoulders for water runoff. Then, the Romans developed underground sewers to stop sanitary waste from going into surface-water drainage. The "Greatest Sewer," Cloāca Maxima, "large enough to allow the passage of a wagon loaded with hay," may have been initiated two centuries earlier. It helped drain a local swamp that became the Forum, the nucleus of Roman life.

Almost a millennium later, the Edo-period engineers of Japan conveyed water from a local spring in exposed aqueducts, then distributed it across town in wooden pipes that metal ones replaced at the end of the 19th century. They regulated water use, privileging the samurai of high status to draw water from the central system and allowing others to use water from the wells. A beer company in the early Meiji era directly tapped into the main water supply, and water surplus from excess rain one season led to the creation of the Shinjuku Gyoen, the national imperial garden. The rivers were regularly flushed out to prevent pollution, and in the mid-1600s, toilets feeding into the rivers were demolished. The city dwellers were forbidden from dumping waste into the rivers, and the local farmers "recycled" sewer waste into fertilizer.

In the mid-1700s, the maladies of mechanization began to loom large in Europe, launching a "new" version of public health engineering parallel to the Industrial Revolution. The 1831 cholera outbreak debilitated Britain during a time when people believed "foul air" caused the calamity. But it took the Victorian civil servant Sir Edwin Chadwick to chronicle links between the unsanitary living conditions and health. In 1842, he published a report on the "diseases of the civilization" triggered by urban development. He instigated parliamentary inquiries that eventually led to the 1848 Public Health Act. Later, Chadwick worked through the British bureaucracy advocating for long-term drainage inspection and improvements, mainly a comprehensive water and wastewater circulation system. Parliament was skeptical of this "systems stuff." The government instead appointed his rival, Joseph Bazalgette, an engineer, who in the 1860s employed an intercepting sewer network with pumping stations to move the outfalls downriver.

Bazalgette's ideas proved influential even across the Atlantic. The audacious Ellis Chesbrough adapted them to engineer the first comprehensive American sewage system in Chicago, lifting the city out of its low-lying grounds with jackscrews and relieving the metropolis of cholera and dysentery. Chadwick's ideas may not have succeeded against Bazalgette's during the Great Stink from the Thames in England. Still, his global vision inspired essential engineering works that the military often calls SWEAT, an acronym for sewer, water, electricity, and telecommunication. Shortly before he died in 1890, Chadwick acknowledged that deaths from poor sanitation were reduced by one-half because of engineers' hidden sweat and toil.

The *British Medical Journal* reported that public protection enabled by sanitation engineering had been the most significant medical advance since 1840, followed by achievements in antibiotics, anesthesia, vaccines, and genetics. When put into scale—1.4 billion gallons of wastewater along with 16 million pounds of trash and nearly 10 million pounds of recyclables every day just in New York City—one can appreciate the work of maintainers, whose routine and repetitive labor blesses us with passive civic immunity against illness. The sewer is the "rendezvous of all exhaustions," Victor Hugo eloquently quipped, calling it a city's conscience. Everything in it converges and confronts everything else. A sewer tells the truth; there are no lies or secrets, only confessions. And the graceful grunt work of cleansing deserves our gratitude.

In his *Essay Concerning Human Understanding*, British physician and philosopher John Locke wrote, 200 years before Chadwick, "It is ambition enough to be employed as an under-labourer in clearing the ground a little and removing some of the rubbish that lies in the way to knowledge." From the practical wisdom of Locke, Chadwick, Elardo, and the workers at New York's treatment plants, it's clear that maintenance is an essential enabler of health and knowledge. But it can also be pricey. Costs are incurred at every layer of bureaucracy, and distractions and delays quagmire our systems, especially regarding large-scale climate complications. "Everyone's got their views and their little boxes, and those boxes don't intersect," Elardo lamented. "We need to be breaking down those boxes if we are serious about the harsh realities of climate change." Cognitive boxes still constrain civic systems engineering.

Sanitation should remind us that maintenance is the most underrated of our civic duties and that caring for our constructions is as sacred as the act of construction itself. Engineers, much like nurses, are professional care providers. Nursing care is often particular and focused on individuals, whereas engineering responsibility is general, often infrastructural. However, being a good engineer in a technical sense doesn't necessarily mean being a good engineer in a moral sense. Scholar Sabine Roeser is among those calling on engineers to develop a better moral and emotional sensitivity toward the technologies and systems they design that directly affect the seen and unseen, now and in the future. Technical development is outsourceable; moral development isn't. Moral and emotional caregiving is too often invisible; it doesn't appear in financial statements, nor are there market incentives. But the future has a way of issuing warnings to the present. We just need to care about them.

✳

ONE UNDERAPPRECIATED WAY TO understand the world is to see how it falls apart. It's in its nature to break, scholar Steven Jackson has written. In a "broken world thinking," repair and renewal begin from fragilities. But the feats of fixing and forestalling those faults recede from our view. Maintenance is taken for granted, and it isn't easy to conceive our experiences without it, much like electricity and eyeglasses. The idea that care can be invisible yet integral to life animated Anne O'Neil's work for the New York City Transit.

O'Neil grew up in rural Connecticut. Her parents deftly maintained their 200-year-old house, passed down through generations. The upkeep didn't require special powers, just ordinary grit, an attribute that launched O'Neil's engineering career. In the 1990s, she was in the technical teams that laid high-voltage cables to Nantucket Island off Cape Cod, upgraded the electrical control systems for the Boston Sumner/Callahan Tunnels, and designed a four-lane underpass system for the Toronto airport expansion.

With experience in intelligent transportation systems, applying sensors, analytics, and connectivity to improve mobility, O'Neil joined New York City Transit in 2002. One of O'Neil's initial projects within their subway-expansion program was to make new software and sensors communicate with legacy hardware. The communication systems of cables, controls, fire-and-life safety, and customer information had to operate continuously, posing different maintenance needs. O'Neil's team had to establish what repair and replacement regimens were appropriate.

The 7 train to Flushing was a 1.5-mile, two-station extension of the existing line, and the Second Avenue subway was a new 8.5-mile, 16-station line. The Fulton Street Transit Center in lower Manhattan required major reconstruction to connect five subway stations serving nine lines, and the South Ferry station needed a new terminal with a fresh interior design. Delivering each subway project, old and new, required a concept of operations to inform procurement, cost, and schedule. Systems integration proved challenging for O'Neil's team; the stations' tech came in many makes and models, and critical decisions required stakeholder engagement. As with wastewater management, teams understandably conducted quick and dirty fixes on the assets in the past, but those tactical tasks became strategic setbacks.

Systems engineers term this *technical debt*: the difficulty arising when one satisfies immediate requirements at the expense of longer-term solutions. Whether this occurs unthinkingly or under duress, passing the burden on rather than promptly dealing with it can worsen the ultimate responsibility. As in financial debt, compounding interests make repayment arduous. "Maintenance became monstrously significant," O'Neil said. "Typically, teams put off longer-term analyses because of immediate project pressures."

O'Neil gained a different perspective on technical debt when appointed the first chief systems engineer for New York City Transit. In this role, she became an advocate for improving transit operations and customer experience using systems engineering. But first, she had to create awareness across the agency about why these methods were essential. O'Neil recalled her experience working in a fragmented culture where the different divisions didn't necessarily collaborate or even communicate. "Even the managers were very skeptical of the 'systems stuff,'" she said, "but they gave it a chance and ultimately benefited."

The Metropolitan Transportation Authority manages over a trillion dollars in assets, from tracks to tunnels, buildings, and buses. It's the most extensive American public transit system, 10 times bigger than the second largest, Chicago's. Its services continue 24/7, regardless of New Yorkers' mood or Mother Nature's. In software design, a saying goes that for every 10 percent increase in problem complexity, there's a 10-fold increase in the solution complexity. And typically, maintenance consumes up to 80 percent of software costs. Transportation problems are messier than software alone: each year, the transit authority needs to deliver on its multibillion-dollar capital programs on time, and as is usual with infrastructure projects, it incurs significant technical *and* financial debts. O'Neil recalled that the projects missed capturing critical operational requirements or were too late to recognize them. The teams didn't collaborate effectively in the early project shaping, and the executives couldn't decipher the root reasons for the delays. "These were all fundamentally *systems* issues," O'Neil said.

Keeping New York running while new large projects are introduced is a messy maintenance problem. While it's composed of solvable hard problems and resolvable soft problems, the *"systems issues"* that chiefly create delays are shared across organizations. Building large projects can be technically and politically captivating and have significant rewards and risks. Constructing ports, airports, hospitals, highways, stadiums, cargo logistics, and even tax and census infrastructure often consumes far more time and resources than anticipated. These "megaprojects," typically a billion dollars or more, are getting bigger and longer, with "no end in sight."

Scholar Bent Flyvbjerg has analyzed projects where the complexity of the scope has multiplied. Over eight decades, one can see a 160 percent increase in building height in the skyscraper business. In 1930, New York's Chrysler Building stood tall at 1,046 feet. By the early 2000s, skyscrapers had more than doubled in elevation, with Dubai's Burj Khalifa skyrocketing 2,722 feet. In the same period, bridge spans have extended by 260 percent. The lines of code in Microsoft Windows increased 16-fold from 5 million lines to 80 million lines between 1993 and 2009. In Flyvbjerg's estimate, project sizes estimated by dollar value have doubled two or three times every century. One example from the UK epitomizes the problems caused by increased scope.

In 1998, two firms, Railtrack and Virgin Rail Group, embarked on a project to renew the aging West Coast Main Line, first created in the 1830s, a 400-mile corridor between London and Glasgow that links to Birmingham, Manchester, Liverpool, and Edinburgh. This essential upgrade was to introduce new high-speed Pendolino tilting trains. The improvements focused on the tracks, the stations, 60 tunnels, 2,800 signals, and 10,000 bridge spans serving a medley of commuter and freight rails. However, the program quickly went off the rails. Costs exploded. The schedule was off-kilter, and teamwork was absent. While one specialist group was tasked to modernize new rolling stock, the other zoned in on the signaling system, and so on. Each group focused on its own deliverables, but no overarching program could ensure a unified outcome. Soon, the entire project was derailed.

In 2002, the project's expense soared from £2.5 billion to an absurd £14.5 billion. By then, Railtrack had delivered only one-sixth of the project scope. Facing bankruptcy, Railtrack pulled out of the project, and the public sector intervened. The nonprofit Network Rail inherited Railtrack's assets and took over the project management. Network Rail created requirements with a renewed focus on outcomes, monitoring, and accountability—every team had a common, connected set of goals. Further delays and cost overruns were unacceptable. In 2007, nearly a decade after the project began, clearly redefining the requirements turned the project around, and teams successfully delivered the first two phases. Results showed increased train speed and frequency, better punctuality, and decreased journey times. Nevertheless, the billions of pounds lost from previous, disorganized efforts were avoidable.

Similar cost catastrophes are evident in the construction of the Suez Canal at 1,900 percent over budget, the Sydney Opera House at 1,400 percent, and the Denver International Airport, Boston Big Dig, and Panama Canal at over 200 percent. If Pharaoh Khufu had gotten a cost estimate before starting work on the Great Pyramid in 2250 BCE, that figure would have been unreasonably optimistic. In ongoing, large-scale projects like the New York City Transit, designers and installers don't necessarily talk to the maintainers at the outset. Plus, people who worked on the current systems may not be in the system anymore or may have held temporary contracts. These problems aren't simply features of today; they were evident when the subway system was first conceived.

In 1800, when London and Tokyo's populations exceeded a million, New York's population was only about 60,000. The city's population diligently doubled every two decades, reaching 3.5 million a century later—and by 1900, Manhattan's streetcars and omnibuses, or the "horse railways," crowded and congested the city. Traveling in New York was "modern martyrdom," with "dis-

comforts, inconveniences, and annoyances . . . almost intolerable."
When London's Underground began running in 1863, New York
couldn't ignore it. But New York's underground transportation
faced a hard problem: a forbidding geology.

Courtesy of the Precambrian continental shift, Manhattan sits
atop bedrock of marble, gneiss, and schist, with the last variety
proving to be particularly difficult, susceptible to decay, and dangerous for an underground railroad. But 35-year-old William Barclay Parsons, named the first chief engineer of the Rapid Transit
Commission in 1894, was ready for this rocky affair. To people who
doubted his experience, the precise and practical Parsons coldly
responded: "Success doesn't depend on age nor does it depend on
will or enthusiasm. It depends on the rigorous analytical methods
of a trained and educated mind." Fond of Archimedes and his lever
to move the world, Parsons was "city-born, city-bred, and city-
minded." He came from a politically influential New York family,
lived in a brownstone with a ballroom, studied in elite European
private schools, graduated from Columbia University, and chaired
its board of trustees in later years. He was a bit serious and humor-
less; his friends called him "Reverend Parsons." At 27, a year after
he founded his engineering firm in 1885, he wrote *Track: A Complete Manual of Maintenance of Way*, addressing railroad problems. His memoir, *An American Engineer in China*, described his
1898 journey, traveling as "the first foreigner ever seen" to chart the
1,000-mile railway line through the Hunan province.

Meanwhile, New York's bureaucracy stalled the subway project. "We have the worst transit problem in the world," Parsons
said. "Are we to be the last to act?" When the project finally kick-
started in 1898, Parsons found a solution for the hard problem of
Manhattan's schist: a shallow subway. Unlike much of London's
Underground, which bored deep through soft clay, Parsons chose
the "cut-and-cover" method instead of conventional tunneling
through the bedrock. Workers dug shallow trenches through existing streets, removed the utilities, laid tracks, and covered them back

up. Beyond the engineering challenges, Parsons had to combat the political and psychological pushback. Public officials and private financiers argued about the ideal passenger and what people should pay for their rides. One politician said, "New Yorkers will never go into a hole in the ground to ride." Another contractor argued that the subway "will give them more time, more ease" and "change their very lives." Over the next six years, people derided the project as "Parsons's Ditch." Still, his ingenious design enabled passengers to access most city transit by stairs rather than by the deep elevators and escalators that are more common across the London Tube.

On October 27, 1904, the subway opened for service with a celebration. A bishop said prayers and consecrated the new machinery that would define New York for generations. The "mayor-motorman," George B. McClellan, received a silver ignition key from Tiffany's, made exclusively for the festivity. The train jerked out of City Hall, thundering on a single nine-mile track with 28 stops, zipping via Grand Central and Times Square up to 145th and Broadway in the Bronx. Much as in Karl Marx's view on industrial society, the subway became an object for the subject, the modern urban passenger, and a subject for the object. The subway was a form of "social media," mixing people from different classes, races, ages, and identities. As scholar Stefan Höhne points out, with its own governance, the subway network converted unruly commuters into more obedient ones, increasing their discipline and dependency on the system. The subway turned into a "mega-machine."

But Parsons's mega-machine requires mega-maintenance. Years later, he told the graduating students of Columbia University to seek the "spiritual side" of engineering and "to understand its ideals." Yet, modern infrastructure projects are hardly like the ones Parsons imagined after researching Renaissance renegades like Bartolomeo Ammannati, Filippo Brunelleschi, Domenico Fontana, and Leonardo da Vinci. The reality is messier; the paperwork, onerous. Try completing a comprehensive environmental review today to launch a new project or upgrade an existing asset. It may take years and millions of dollars to complete, requiring many stakeholders to consider the proposal. In some cases, there are several hundred stakeholder groups to engage.

"NIMBY"—or Not In My Back Yard—opposition tries to block any project people don't want for their community. Who wants a new wastewater treatment plant on their beachfront, an entertainment arena, a sports bar, or a homeless shelter in their neighborhood, each with its problems? Then there are "BANANA" citizens, who instinctively oppose any private development that may harm local businesses or damage the environment: Build Absolutely Nothing Anywhere Near Anyone. Considering diverse stakeholder views is imperative to engineering design but can also aggravate it. How would one build a perfect fighter jet? As one colonel said, "The perfect design would have contractors in each state and a part made in each congressional district." Engineers and developers grumble about large infrastructure projects as "death by a thousand duck bites."

Whether it's Parsons's original groundbreaking work or New York City Transit's ongoing maintenance work, one cannot ignore *future neglect*. We can build social systems that outlive us only if we invest care and consideration. "Infrastructure problems confound us because too many creep up in ways we don't or can't properly recognize," O'Neil said. When one completes concept development and enters preliminary design in a new subway line, much of the life-cycle cost is already committed to that work. Capital costs are now foremost in delivering on the system's performance requirements. For example, all subways in New York are four tracks, two for express and two for local services, except for the 7 train, which has a third "swing" track switchable for express service. The Second Avenue line has only two tracks because of the original cost constraints. "That's the investment decision we're going to be living with for the life of the subway," O'Neil said. "There are maintenance implications at many different levels beyond just physical equipment." That's why developing a strong concept of operations is crucial for planning large projects. "When you're thinking about introducing systems engineering to an organization, remember, it's a journey," O'Neil said. "If the best place to start is not at the beginning, that's okay."

✳

THE WORD "MAINTENANCE" COMES from the Latin expression for "hold in hand." Mierle Laderman Ukeles ascribed value to this meaning in her radical 1979 artwork *Touch Sanitation Performance*. In it, Ukeles spent a year of long shifts, or "sweeps," tracking down New York City's 8,500 sanitation workers. She shook their hands, saying, "Thank you for keeping New York City alive!" The photographs in *Handshake Ritual* conveyed how vilified the "sanmen" felt: "I'm invisible, I don't count, I'm part of the garbage." Ukeles refuted the "invisible hand" of economics and replaced it with the human touch. "The sourball of every revolution," Ukeles wrote, is this: "after the revolution, who's going to pick up the garbage on Monday morning?"

Maintenance in wicked systems should begin with understanding an absence: What would the world be like without maintenance? We are familiar with many forms of innovation that turn valuable resources into waste. Give that waste to maintainers, and they will industriously refashion the rottenness into a life-giving resource. Maintenance achievements are hardly celebrated in society, even though our highest ideals—life, liberty, and pursuit of happiness—are about creations of care and conservation. As the novelist David Foster Wallace observed, "Actual heroism receives no ovation, entertains no one. No one queues up to see it. No one is interested."

In engineering, heroism is often depicted as design and innovation. Engineering is "perhaps the quintessential revolutionary activity," offers historian Ken Alder, describing "a simple, but radical assumption: that the present is nothing more than the raw material from which to construct a better future." Engineers typically collapse "*is* into *ought* to build (or promise to build) a world more consonant with human desires," as they understand it. With this responsibility, it's tempting to hype engineering design and innovation and obscure the slog of engineering maintenance the world truly relies on. Writer Iain Sinclair pinpointed the "dangerously volatile relationship between actual human shit and the flights of promotional bullshit plastered around major holes in the ground."

In wicked systems, the awareness that maintenance can create new knowledge, synthesis, and values can be profoundly helpful. Indeed, no innovation will pay off without due regard to the maintenance of the underlying platforms. Ordinary miracles can stave off extraordinary menaces. A tight synergy between innovation and maintenance enables system longevity. It allows us to introduce new measures unavailable when the system was developed. The codependence also makes it clear that innovations that ignore foreseeable future maintenance can and do hamper systems operations. Ultimately, the economics of innovation should hew to the ethics of care. Undervaluing maintenance, because it's less exciting, is like dismissing our need to sleep because we would instead do something else. The world of sleep, with its phases of repairs and nourishment, prepares us for the onslaughts of the waking world. This self-renewing surprise provides daily rebirth.

Many ancient traditions have incorporated restorative practices for generational benefits. Medieval Sanskrit scriptures point to the importance of rejuvenation over replacement with the life-instilling ritual of *jīrṇoddhāra*, which means digesting and progressing. Consecration ceremonies of South Indian temples occur over several days as an invocation of the holy elements. They involve bathing the crown of the temple with waters brought in a sacred vessel, praying to the earth, and reciting hymns around the fire altars offering grains, honey, and spices. Such practices revivify life; what's now present as a loan from the past is digested and forwarded to a usable future. "Conservation, repair and maintenance have been key to Indian thought," one scholar has written, allowing for a different imagination of "adopting and transforming the old, rather than through slash-and-burn methods that try to wipe out the past in order to announce one's arrival." This centuries-old practice of ritual renovation envisions maintenance as a worshipful work.

As we have seen in this chapter, caring for existing systems is just as important as creating new capacities, and one shouldn't emphasize the latter at the expense of the former. The aging of systems is shaped by our environments and many known and unknown forces we encounter. Retirability is a crucial attribute of systems maintenance, quite different from market-driven obsolescence that prematurely sends otherwise good products to the graveyard of gadgets. Any underlying logic for retiring a system relates to why it must be discontinued and how one values degradation and depreciation. But too often, decision-makers are tempted by the notion that the systems may hold out just long enough to become someone else's problem.

We have two silently conflicting personalities that sway our consumption: a "civil self" and a "consuming self." The civil version is other-oriented: we help neighbors, we care about the environment, we take our bags to the grocery store, and we consciously separate the recyclables. The consuming version is self-oriented: we dream of new products, preorder them, play with them, and once the product is even slightly out of date or breaks, we don't dwell on it but start looking for a new one.

The consuming self is at play when a manufacturer retires a dishwasher model; there aren't spare parts in the market, or the available service fee exceeds the price of new equipment, and one decides to try something new. Corporations have increasingly tuned in to these aspects of human consumption. An elevator manufacturer is now both a producer and a long-term service provider. The incentive for a jet-engine manufacturer is to keep maintenance costs low for the airliner and keep it profitable for the seller. So, what's then left of the civil self? To go beyond the thrill of buying and using products and explore how they end.

Designer Joe Macleod advises firms on "gaming out their end games." Macleod uses the shorthand "endineering" to refer to the practice and philosophy of creating customer off-boarding experiences of a product—the opposite of onboarding, the process of getting customers hooked, which companies invest heavily in. Endineering is about responsible consumerism. "Although it is not anti-business," Macleod notes, endineering "recognizes clearly that our current pace, approach and philosophy around consumerism cannot continue."

Taking maintenance seriously also provides an opportunity to validate and recognize the historical contributions of individuals and organizations that are largely invisible, even ignored. As an example of maintenance giving rise to innovative design, let's look at the evolution of long-range jetliners. Four engines used to be the default in these aircraft, but now the newer varieties come with two engines. Although four engines are safer than two, upkeep and upgrades with the twin engines are more manageable in an economic sense. A collective maintenance experience informed the innovation, but that fact is seldom highlighted.

In aviation, as it was for steamboat safety, laws require proactive maintenance. So much so that the US Air Force recently decided to phase out some attack aircraft, surveillance units, and tankers and cut thousands of flying hours to fund modernization. With new engines, fuel efficiency, and USB ports, the service life of the upgraded Cold War–era Stratofortress bomber B-52—or the Big Ugly Flyin' Fella—would be 100 years in 2050. Centenary airframes aside, in recent years, the Pentagon has committed about 40 percent of its budget to systems maintenance and operations, which surpasses its investments in weapons, bases, and the people who support them. In contrast, a new development with uncrewed aircraft is low-cost vehicles with limited life that are highly reusable but ultimately expendable. These "attritable" designs, as in outfoxing the enemies in a war of attrition, above all seek minimal maintenance, analogous to the expendability of use-and-throw plastic cups on a sliding scale between the disposable single-use paper cups and reusable crystal glasses.

Historian Steve Shapin points out that innovators can understand technology elements better than their operators. Still, operators can provide insights that may have never occurred to the innovators. Beethoven may have understood his piano better than its manufacturer, perhaps with more insights into its practical maintenance needs. "The piano is one thing to a pianist, another to a piano tuner, another to an interior designer with no interest in music, and yet another to a child who wants to avoid practicing," writes Shapin. "Ultimately, the narrative of what kind of thing a piano is must be a story of all these users." Inaudible maintenance narratives are everywhere, from the story of a deaf composer to a world tone-deaf to its civic priorities.

The word "innovation" in Latin is rooted in renewal and restoration. Yet, innovation has come to mean newness by default. One literature scholar reminds us that novelty "tends to evaporate almost at the very instant it is recognized." Hence, a more grounded vision for innovation—and the governing ethos around it—can help us deliver more effective maintenance regimes for wicked systems. Disruptive products can be less reliable than the quieter quests for the care and conservation of vital systems. And indeed, the durable creations of engineering are often based on the repeatable, reliable, and routine. The results may be grand, but all that is genuinely grand—even the usefully ungrand—emerges from the diligence of the daily grind, as the old Zen axiom goes: "Before enlightenment, chop wood and carry water. After enlightenment, chop wood and carry water."

Engineers like Elardo and O'Neil are like building cranes and scaffolds. They never get much credit, but they are essential and stand taller than the shiny buildings they enable before gradually disappearing, securing us from disruption.

Refrain

Sway

A 1957 PROFILE OF ED LINK IN *SPORTS ILLUSTRATED* described him as "part fish and part fowl . . . a restless sort of aquatic bird with a habit of disappearing frequently into the sky or sea." Ed Link had spent a career with his gaze fixed on the heavens, but his interest in the sea was relatively new.

To offset the stress of his wartime duties with the blue-box trainer, Ed Link took up fishing in the remote wilderness during the 1940s. He flew his Grumman Widgeon floatplane to the Great Lakes and the Canadian bush country during his off time. To solve the problem of transporting an unwieldy fishing boat on a small floatplane, he rigged up a collapsible canoe that could be disassembled and fit into two large suitcases. The sections could be reassembled into a watertight, functional canoe. Ed Link called the commercial version the Linkanoe, and it sold in the thousands. Evidently, he was ready for new engineering challenges.

Over the years, Ed Link's hobby evolved from weekend fishing in lakes and rivers to weeks-long excursions in the sapphire waters of Florida, the Bahamas, and Cuba. Shortly after World War II, he purchased the *Blue Heron*, a 43-foot yacht, to hone his sailing skills. Ed Link was able to transfer his navigational expertise from flying to sailing, melding information from navigation charts, radio guidance, and local weather forecasts. But as a newcomer, he was less bound to seafaring traditions. Ed Link shocked the judges in a St.

Petersburg–Havana race with an unconventional technique. When all the other competitors went south, he sailed west—and won.

Once Ed Link sold his aviation company, he had more time to focus on sailing. With his wife, Marion, and their sons, William and Clayton, as exploration partners, sailing became a beloved pursuit for the whole Link family. Although some could have seen this as a drastic change of focus, applying his expertise to a different context was a welcome challenge for Ed Link. He once said about his switch from the skies to the seas, "water and air are both fluids. One is just thicker than the other." He even designed a submarine to look like a helicopter. While rockets fired off from Cape Canaveral, just to the south near the Florida Straits, Ed Link mused about the underwater frontiers.

IN JUNE 1951, Ed and Marion Link accepted a friend's invitation to explore shipwrecks in the Looe Key of Florida, named after a small 18th-century British warship grounded in the War of Jenkins' Ear. Accompanied by professional divers and marine archaeologists, Ed Link plunged into the museum of the ocean floor to recover part of the wreckage, made of ivory. Hours later, he helped raise a heavy cannon from 1617. The experience solidified Ed's conviction: he would be a treasure hunter. "I took up golf once and put it back down very gently," Ed Link recalled later. "I don't like to get my hands dirty, but I like to get them greasy. Wreck hunting seems to have the right challenge."

Soon after his first shipwreck expedition, Ed Link convinced Marion to learn to dive. He wore an aqualung and fastened the mask over Marion. Then they slowly dipped into the crystalline world beneath the waves. "A beautiful sea garden stretched out on every side, a world of waving sea grasses, fantastic coral formations and lacy traceries of sea fans," Marion wrote. "The spiny creatures frightened me, and I backed away, lost my precarious balance and wavered helplessly toward another pit, which was liberally dotted with more sea urchins." Ed and Marion Link were instantly addicted to the world of sunken relics. "When we made those first exploratory dives, we were as green as the moray eels which curled malevolently within the jagged coral reefs," Marion wrote. She had imagined the wrecks as "hoary hulls still intact upon the bottom, and swarming with fish—dangerous and predatory fish, writhing octopi and huge rays with flapping wings."

For Ed Link, the engineer, the dives were a technical challenge. He needed to upgrade the *Blue Heron*. He purchased a 65-foot shrimp yawl, which he modified and christened as *Sea Diver*. The capacious workboat became their second home. Under the tepid blue waters of Italian islands, Ed and Marion Link found ancient Greek pottery, collected jewelry and coins, and retrieved antique kitchen utensils and cannonballs. "I was able to swim about this strange undersea ravine, floating up the steep sides of the coral rocks, peeping into eerie chasms and then gliding down to examine a nearly buried cannon, scarcely distinguishable under its disguise of coral," Marion wrote. She spotted the anchor of a sunken warship, coral-covered and almost hidden under a jutting rock shelf, and swam through its colossal ring. The shank, half buried in the coral, weighed over a ton. "I hovered over it enthralled," she wrote. "I was no longer fearful of this strange environment. I was at last home on the bottom."

In 1953, Ed and Marion Link found a smoothbore cannon likely used by Christopher Columbus. Enthralled by this find, they searched for the wreck of the *Santa Maria*, which led them to question where exactly Columbus had landed when he "discovered" the New World in 1492. The Links partnered with Captain Philip Weems, Ed's previous collaborator on the Celestial Navigation Trainer, to probe the dominant narrative surrounding Columbus's voyage.

Ed Link studied Columbus's journals, explored the north coast of Haiti, and aerially retraced Columbus's route to America. "I found out that no one even knew what a league was in Columbus's day—it could have been one mile or three. It seemed to us the only way to prove anything was to sail the islands at the same speed we calculated that Columbus did, checking the time elapsed, as Columbus recorded it, against our own, and against the landmarks Columbus saw." Ed Link challenged the accepted theory of Columbus's first landfall. Writing with Marion, he offered that the "Great Navigator" did not land on the swampy San Salvador but on the Caicos Islands, 200 miles to the southeast. When the Links detailed their findings in *A New Theory on Columbus's Voyage through the Bahamas*, published by the Smithsonian Institution, they instantly sparked controversy among historians. Some dismissed the Links as amateurs or told them to stick with collecting ivory tusks and picking over long-sunk pirate ships. In the *New York Times*, one Pulitzer Prize–winning Columbus expert at Harvard scoffed at the Links' findings, ascribing them to a "sailor's imagination." Nevertheless, Ed Link stuck with his conclusion.

In 1956, the Links also found their way to Port Royal, a Jamaican port city, home to the indigenous Taino population and once notorious for the sins brought by its settlers: piracy, prostitution, and the plantation economy. In what some saw as a divine decree, in 1692, a tsunami struck this "wickedest" land and killed thousands. The Links were astonished that despite the repeated disasters that the city endured—fires, earthquakes, and hurricanes—Port Royal was described as "a place of spectacular beauty rising in majesty from the sandy sea bed." They had to see it for themselves. Five years after Hurricane Charlie, the Links' crew anchored close to Port Royal's reefs. The water was murky, with barely any visibility for their dives. Silt from streams and muck from latrines lining the shore flowed into waters infested by barracudas and sharks. "It took guts and persistence to descend day after day into these unpleasant waters," Marion Link wrote. They recovered scores of rum and drug bottles; many were preserved flawlessly with coral coating.

Ed Link's crew airlifted portions of the brick walls presumed to be the ruins of Fort James, a fortress built by the British during the

18th century to guard the city against French invasion. The divers expanded their search, lured by the prospect of finding chests of gold cakes and coins. After a day's search, they returned with only a copper medal, clay pipe, broken porcelain, and beer bottles. When a storm suspended their search, the divers rigged up a wire cage suspended by an airplane to catch and sift their recovered debris. A small sparkling object from the airlift prompted Ed Link to dive deeper, but he was disappointed to find that it was only recently minted Jamaican pennies. After a week of fruitless exploration, Ed Link's crew was disheartened and ready to give up. Then, on the final dive on the final day, they uncovered an artifact long hidden from view: a British cannon weighing over 5,200 pounds. Ed Link's extraordinary finds off the coast of Port Royal were momentous for underwater archaeology. Over the following decades, extensive missions led by others with insights from Ed Link's early excavations helped make the "catastrophic site" into an incomparable UNESCO World Heritage Site with universal cultural value. The Port Royal experience only heightened Ed and Marion Link's appetite for further expeditions.

In early 1957, Ed Link retired *Sea Diver* after a stretch of wild weather and waves in the Florida Keys wrecked its platform and broke its anchor chains. His 1959 upgrade was a custom-built hundred-foot *Sea Diver II* with big booms and winches. The 168-ton, double-hull, and double-diesel steel cruiser was capable of 10 knots. It had two electric generators, two high-pressured compressors, two radars, six cabins, two large freezers, a fireplace, a four-burner stove, a dishwasher, a washer-dryer, a machine shop, and the most advanced technologies found in ocean liners: radios, radar, sonar, loran, echo sounders, air hoses, and airlifts. The space between the hull walls became tanks for oil and water. The vessel's underwater viewing chamber was a portal for the camera-ready, neon-colored vistas. It had an on-deck decompression chamber and two cruising boats—the 19-foot *Reef Diver* and the 15-foot *Wee Diver*. A smaller, battery-operated submarine, *Power Diver*, also supported more elaborate expeditions.

In the summer of 1960, Ed Link steered *Sea Diver II* into waters off the Mediterranean coast near Tel Aviv to explore Caesarea, an ancient seaport sunk by an earthquake in 130 CE. Herod the Great constructed Caesarea in Roman-ruled Judea—later Syria Palaestina, then Palaestina Prima, and now Israel—over 12 years in honor of "Octavian" Caesar Augustus. When the Greek philosopher Apollonius of Tyana visited the port city, he described Caesarea as "excelling all others there in size and in laws, and in institutions and in the warlike virtues of ancestors, and still more in the arts and manners of peace." The boat's underwater searching devices got to work: a magnetometer to detect metals, a fathometer to locate debris, a forceful jet hose to vacuum away the detritus, and an underwater television camera to record the dives. Caesarea was an engineering spectacle with colossal shrines, citadels, hippodromes, watercourses, and a columned promenade. The city was also an important port for ships ferrying between the trade centers of Egypt and Lebanon. As the Book of Jonah describes, when storms struck, "the mariners were afraid; and every man cried out to his god, and threw the cargo that was in the ship into the sea, to lighten the load." The twisty tempests, too often, took down the ships themselves.

Relying on the writings of the first-century historian Flavius Josephus, Ed Link's archaeological expedition charted the size and scope of the ancient port. Using a combination of aerial maps, water levels, and underwater measurements, Ed Link recalculated the size of specific structures in Caesarea. He wondered if Josephus exaggerated the size of the stones or if the book translation was incorrect. Aided by his calculations, the probing dives from the *Reef Diver* recovered structural elements, including fallen pillars, large wooden beams, and marble columns, as well as long-hidden artifacts, including amphoras, commemorative coins, pieces of sculpture, and Roman and Byzantine lamps brought on trade vessels over the Sea of Galilee, the crossway of civilizations. The crew of 11 also retrieved two exquisite Roman cooking pots from the first century CE and primitive stone anchors from even earlier. Ed Link entrusted these relics to the Israeli government, and they were widely exhibited at museums.

These diving expeditions gave Ed Link a fresh perspective. There were worlds underwater that were yet untapped by human civilization. Treasures like the Port Royal cannon and the Caesarea ruins were found over months of laborious back-and-forth between the land and sea. But Ed Link wondered, what other artifacts would have emerged if teams could stay underwater for the entire expedition? In his mind, living at the great depths of oceans for extended periods was a similar challenge to human flight in the air. As with the early aviators, Ed Link's ambition presented many unknowns, but these were known unknowns.

THE NOTION OF UNDERWATER living has appeared in many different forms throughout history: from ancient Indian scriptures, in which the Hindu god Varuna resides in the seas and ascends to land on a crocodile, to myths about Alexander the Great in which he spent time underwater in a glass chamber, looking at a colossal sea monster that took days to pass by. Engineers have long sought to create submergence vessels inspired by the possibilities and mysteries of underwater living. In 1620, Dutch engineer Cornelis Drebbel demonstrated a leather-wrapped wooden submarine in the Thames for the British Navy. It was never deployed for combat. Then, in 1776, David and Ezra Bushnell, brothers from Connecticut, worked with clockmakers to develop a wooden submarine called the *Turtle*. Though the vessel sank in the American Revolution against British warships, future president Thomas Jefferson brought publicity to the effort in 1798. Two years later, Robert Fulton prototyped the practical submarine *Nautilus* for the French Admiralty. Fast-forward to 1934, when the first deep submergence was demonstrated in a bathysphere, a spherical diving chamber lowered to 3,000 feet in the Caribbean. Twenty years later, inventor Auguste Antoine Piccard designed the first bathyscaphe—a self-propelled deep diver—that reached 13,300 feet in the waters off Dakar, Senegal. That record was broken in 1960 at the Mariana Trench by Piccard's last bathyscaphe, *Trieste*, which reached almost 36,000 feet, a depth exceeding the peak of Mount Everest by over a mile.

As Ed Link's interests evolved beyond underwater archaeology, he believed he could engineer the means for humans to stay on the bottom of the ocean for extended periods. He imagined that underwater living would unleash new opportunities from agriculture to national security. But the technical challenges that water posed were daunting. Water weighs much more than air and exerts greater pressure. Hence, staying on the ocean floor for extended periods means battling harmful ambient pressures. One atmosphere of pressure (ATM) equals 14.7 pounds per square inch. At 33 feet underwater, one experiences 29.4 pounds of pressure, or 2 ATMs. Every additional 33 feet underwater adds an ATM, or 14.7 pounds of air pressure. So, at 66 feet, 2 ATMs of water pressure and 1 ATM of air pressure combine to produce 3 ATMs of ambient pressure. Once humans descend deeper underwater, the blood circulation resembles a highly pressurized soda can. Crack the can open too quickly, and the contents burst out. For this reason, decompression, the careful process that returns humans to surface atmospheric pressure, can take several days.

Ed Link realized that deep submergence was better achieved with a submersible decompression chamber. The decompression chambers aimed to provide divers a safe place to return to ambient pressure while deep underwater. The chambers were to prevent two deadly effects of diving: "getting bent" and "getting narc'd." The first complication, the bends, causes sharp pains and vertigo because of bubbles in the bloodstream, which, without decompression, can lead to paralysis and even death. The second complication, nitrogen narcosis, is a form of intoxication from absorbing excess inert gases while deep underwater. Ed Link designed a one-person aluminum cylinder that could be lowered from the ship and hauled up like an elevator to afford more safety and comfort during the lengthy decompression ordeal. During his Port Royal expedition, he first tested the chamber—11 feet long and about 40 inches in diameter. In 1961, Ed Link met the famed Jacques Cousteau in Monaco. They agreed on a joint venture to use the decompression chambers to transport people to Cousteau's proposed seafloor houses.

In August 1962, the 58-year-old Ed Link, saturating himself with a helium and oxygen gas mixture (heliox), entered the diving cylinder. He descended 60 feet below his ship, anchored near the French Riviera, and remained there for eight hours. His 21-year-old son, Clayton, delivered macaroni and cheese to a relaxing Ed Link inside the cylinder underwater. A month later, under Ed Link's supervision, the Belgian underwater archaeologist Robert Sténuit became the world's first aquanaut. Sténuit spent 26 hours in the submersible decompression chamber 200 feet below the *Sea Diver* off Villefranche. These preliminary demonstrations led to successful tests in batches of mice and a seven-month-old goat named Caroline, who was treated to a "big meal of cabbage" after spending 16 hours at 400 feet deep. The team was awarded a grant from the National Geographic Society, officially kick-starting Ed Link's "Man-in-Sea" program.

To bolster support for his project, Ed Link wrote a prospectus to the US Navy describing how his program could simultaneously advance physiological studies, instrumentation development, and operational testing in deep-sea trials. The initial phase of the Man-in-Sea program focused on tests at 400 feet, a depth beyond the capabilities of standard scuba gear. The emphasis was on engineering inflatable underwater suites called Submersible Portable Inflatable Dwellings. The SPID was a bottom-moored setup with breathing-support systems. Another design was an igloo-shaped enclosure for workers to perform undersea pipeline-repair work. While testing these suites, the decompression chamber served two functions: a conduit to the SPID and the IGLOO and a conveyance back to the *Sea Diver* for deck decompression.

During this time, Ed Link also created Ocean Systems, which quickly attracted the US Navy's interest. The company's vision was to enable humans to live and work at depths of 1,500 feet, opening up an underwater region the size of Africa for research and exploration. "We cannot afford to neglect any longer the great resources of the oceans with man living and working directly in its environments," Ed Link wrote. "Wake up America, we are on the verge of a new era." He had a reason for optimism, given his tremendous success with his underwater developments up to this point. But Ed Link's dreams of deep submergence were disrupted by a dark disaster in 1963.

ON THE MORNING OF April 10, after a flurry of commands and checks, the USS *Thresher* began descent for a safety-check dive 220 miles east of Cape Cod. With 3,500 tons and 3,000 silver-brazed piping joints, the nuclear-powered hunter-killer was acclaimed for its stealth and silence in the great depths. The submarine and its crew of 129 people reached its assigned depth. Fifteen minutes later, the troubles began, but the sub's emergency call to the rescue ship *Skylark* was garbled. *Thresher* had attained its crush depth, at which a watercraft collapses under pressure. Death and destruction were instantaneous. The navy's closest rescue equipment worked only to 800-odd feet and couldn't go deeper. They brought in the bathyscaphe *Trieste* for the deep search.

Finding the wreckage in the saline shadows of the North Atlantic was agonizing. The failure was mysterious and the following congressional inquiries contentious. What about the ship's 20-ton nuclear reactor and its long-term impacts on the oceans? And what about the safety of other active submarines? Admiral Hyman Rickover testified that the *Thresher* accident "should not be viewed solely as the result of failure of a specific braze, weld, system or component, but rather should be considered a consequence of the philosophy of design, construction and inspection, that has been permitted in our naval shipbuilding programs." He argued that recent advancements "may have forsaken the fundamentals of good engineering." Eventually, investigations revealed that metal failures from corroded pipes contributed to the submarine's destruction.

The US Navy created the Deep Submergence Systems Review Group, tasked with advising the development of rescue vehicles for submarine crews and planning for future missions. Ed Link was perhaps the most practically accomplished engineer on this panel, and he proposed ideas for lighter-weight submarines. With the weight savings, he said, "it could be possible to double the striking force as well as even increasing the safety of the crew and possibilities of going to greater depths than our Polaris submarines can go today."

Ed and Marion Link docked their *Sea Diver* at the Washington Navy Yard and began their yearlong work on the design. But after President Kennedy's assassination rocked the nation, the Deep Submergence Systems Review Group received little publicity. Their final report sank without a ripple. This policy experience was a departure point for Ed Link. He had a fuller view of the problem than others, not that different from the flight-training mindset only two decades before. His mission went deeper: to strengthen the foundations of ocean engineering.

JOE MACINNIS, a 26-year-old Canadian doctor, read Ed Link's Man-in-Sea articles in *National Geographic* and was determined to work for him. "I went to his office and knocked him over with my enthusiasm," MacInnis said. Ed Link hired him that day and committed to a fellowship to support MacInnis's training in diving medicine from the Link Foundation, established a decade earlier to support work in aviation and ocean engineering.

MacInnis asked Ed Link about the *Thresher* tragedy. "We should not have lost all those men," Ed Link told him. "It confirms how little we know about the depths of the ocean." Then, Ed Link took

MacInnis around the sun-warmed *Sea Diver*. "For our next dive, we're going to place a small station on the seafloor at 400 feet," Ed Link told MacInnis. "This chamber will carry two divers down to the station and return them to the surface. It's going to take a lot of hard work to get it done." Ed Link appointed MacInnis as the medical director for that project.

On an ultrablue morning in June 1964, Ed Link and Joe Mac-Innis stood on the *Sea Diver*, ready for a test near the Berry Islands of the Bahamas. They watched Robert Sténuit and Jon Lindbergh, the cave diver and son of the legendary aviator, suit up. Sténuit and Lindbergh entered the submersible decompression chamber. They were winched 430 feet underwater, where they intended to spend a record-setting 49 hours in the SPID. A long hose fed heliox gas to the eight-feet-by-four-feet rubber station. This blend of breathable gases rendered the divers' voices indecipherable on the intercom, akin to the high-pitched squeaks of Donald Duck. As a solution, the divers used Morse code to communicate from the SPID to the *Sea Diver*—similar to how instructors once spoke with pilots training inside the blue box. Sitting just inches apart, the divers wrote messages to each other, which Ed Link could read as he supervised them on the coarse closed-circuit television.

Sténuit and Lindbergh went on outings, making friends with an extroverted 200-pound grouper that hovered near the exit hatch. The divers enjoyed good meals: canned hash, lunch meats, bread, fruits, and a rich lamb stew for dinner. While the divers could withstand the extreme pressures, their cans of food looked like a truck had run over them. "Living in the depths, I have become a creature of the depths, adapted to their pressures," Sténuit wrote about this experience in *The Deepest Days*. "I have always found joy in dangers lucidly accepted and prudently overcome." A triumphant Ed Link wrote in *National Geographic*: "Our mission was accomplished. It proved that humans can adapt to the inhuman conditions of the deep sea."

By 1965, Ed Link further improved the design of his submersible decompression chamber. He contacted John Perry, a high-energy Palm Beach entrepreneur, former army aviator, and submarine inventor. As Perry's business partner, Ed Link sketched the world's first lockout submersible. The *Deep Diver* had 21 acrylic viewports and could cruise at three knots on battery power. It carried a crew of four in two separate chambers, the forward and lockout chambers. The forward chamber remained at atmospheric pressure, providing the pilot and an observer with a "shirt sleeve" environment, requiring no special clothing or equipment. The lockout chamber could be pressurized to match the depth outside. A floor hatch allowed the two divers to enter and exit the sub once on the seafloor, tethered to the sub by long umbilical-like cords that provided breathing and communication support. The dimly lit eight-and-a-half-ton submersible could stay underwater for 24 hours. Ed Link designed the *Deep Diver* for maximum maneuverability, adapting the central feature of his flight trainer. Designed to operate at around 1,300 feet, it could maneuver up, down, and sideward and rotate on its axis with six degrees of freedom. "Living in the sea is good movie talk but a hard way to live and it isn't necessary," Ed Link said. But with *Deep Diver*, crews could "do their work and get back to their sub for dry clothes, a charcoal steak and some rest."

To Ed Link, the *Deep Diver* was just the beginning for submersibles. By the mid-1980s, he wrote, when "it will be commonplace for man to enter and work freely in the seas, we can foresee elaborate engineering developments scattered upon the ocean floor— underwater tunnels such as that contemplated beneath the English Channel, networks of telephone cables and forests of oil derricks with interconnecting pipelines all available for ready servicing." Humans could "live in the sea for a month, even up to six months if necessary," he added. He imagined underwater hostels for enthusiastic skin divers to vacation in. He dubbed the aquatic rides "the Volkswagens of the deep," noting the operators of such vehicles would require the equivalent of flight training.

Months before Apollo 11, while Neil Armstrong, Buzz Aldrin,

and Michael Collins were training in the Link simulator for their unprecedented journey, Ed Link was setting a record of his own. He reached down 700 feet in the Tongue of the Ocean, a region of incredible depth in the Bahamas. In an intense, six-week expedition, Ed Link's crew completed over 40 dives and spent hours on the ocean floor near the Great Stirrup Cay studying exotic marine life yet to be named. In the twilight glow of the ocean floor, Ed Link's small steps, in their way, were making a giant step for humankind. But despite Link's contributions to deep-sea engineering, the world of submarines remained extremely treacherous.

IN 1968, FOUR NUCLEAR submarines disappeared. The *Skipjack*-class USS *Scorpion*, Israel's *Dakar*, France's *Minerve*, and the Soviet Union's *K-129* lost 315 crew members in total. The circumstances of each disappearance were different, but the exact causes of each disaster remain mysterious. At the time, very little was known about the properties of materials used to construct these submarines and how these materials behaved at extreme depths, temperatures, and in contact with one another. Following the four submarine disasters over a year, new research characterized submarine alloys' durability and susceptibility to failure at great depths. Ed Link built the *Deep Diver* from "plow steel," or carbon steel, a special heat-treated alloy that he knew to be highly durable. But the new research led Ed Link to the costly conclusion that he must decommission his vessel. These disasters also bolstered his recommendation that the navy panel designate ocean engineering as a formal field of study.

Despite the costly loss of the *Deep Diver*, Ed Link forged ahead and fabricated a lighter and highly maneuverable submersible. The 23-foot-long, 10-ton submersible featured panoramic views and two hulls: the forward hull, with a four-inch-thick, transparent, acrylic sphere like a helicopter, and the rear hull, which was a cylindrical pressure vessel made of marine-grade aluminum alloy. Each hull could fit two people. The sturdy submersible contained sonar, telephone, intercom, echo sounders, magnetometers, and Doppler navigation. It also featured wide-angled, low-light cameras and other critical life-support systems. Ed Link credited the project's financial backer, Seward Johnson Sr.—son of Robert Wood Johnson, the cofounder of Johnson & Johnson—in the submarine's name: the *Johnson-Sea-Link*, many of whose mechanical and electrical parts were made of scavenged airplane material. The first model was commissioned in 1971, operating at a depth of 1,000 feet and later upgraded to triple that depth. The sub was primarily suited for search and recovery, archaeological applications, and delicate sampling of tiny creatures. Fifteen years after its debut, Disney World's *The Living Seas* exhibit deemed this design "futuristic."

But soon, Ed Link endured a calamity very close to home.

JUNE 18, 1973, WAS unremarkable at first. Four men were ready to make a routine dive off the Florida Keys to collect research specimens on *Johnson-Sea-Link I* under Ed Link and Joe MacInnis's supervision. But when they reached the seafloor, the submersible's cables became fatally entangled with a scuttled World War II destroyer. Because the dive was intended to be brief, the crew was ill-equipped for long-term survival. The navy dive teams were summoned but arrived late. They never reached the sub due to the drag forces from strong surface currents acting on their air-supply hoses. The outwardly calm Ed Link on the *Sea Diver* blinked tears, keeping the walkie-talkie pressed against his ear. Finally, a torpedo-recovery vessel managed to snag the sub and winch it to the surface. The lives of the pilot and observer had been spared due to the thermal insulation provided by their acrylic hull. But the two men in the rear chamber weren't as fortunate. They were in their swim trunks, expecting only a brief dive, and the aluminum hull afforded them no protection from the cold. Due to hypothermia and the carbon dioxide buildup in the chamber, they slowly lost consciousness and died. One of them was Clayton Link, Ed Link's 31-year-old son.

MacInnis couldn't sleep that night. He went to a pub, double rum straight up. "At this moment, I hate the goddamed ocean," he wrote. "I hate its size and its force. It is a malevolent son of a bitch whose currents and cold have just taken a good man's lifetime of effort and compressed it into tragedy." Even in grief, Ed Link concluded that the navy's rescue measures were inadequate. "This loss need never have occurred," Ed Link wrote to the Coast Guard. "The sole remedy—rapid, effective submarine rescue systems— must be developed and kept in strategic locations."

Ed Link's response was to launch a new breed of remotely oper- ated vehicles called CORD, the Cabled Observation and Rescue Device. The CORD overcame the problems of previous rescue systems. Unlike the unwieldy hoses and cables that prevented the navy rescue team from reaching the sub in distress, CORD's slen- der umbilical cord cut through the fiercest currents. It also had an airplane-type joystick that offered more precise control and advanced instruments that helped rescuers "work blind" in the ocean depths. Once again, Ed Link's familiar techniques from the early days of blind flight could be applied to negotiate a foreign envi- ronment. "It made us realize how helpless we really were," Ed Link said, reflecting on that dreadful day. "But now we have developed five different types of rescue systems, and it couldn't happen again."

ON A CITRIC-WARM FEBRUARY morning in Florida, Marilyn Link asked me to follow her car. The 93-year-old half sister of Ed Link steered with the alertness of a race driver. "I'm demanding; a stickler for details," she once told a writer. Marilyn Link received her pilot's license in 1946 and earned proficiency entirely on the Link Trainer. She couldn't become a full-time pilot, with flying jobs rarely available for women. She had a graduate degree and held some teaching roles, an administrative position at the Smithsonian, and an executive job at the Hughes Aircraft Company. Ed and Marion Link were her constant mentors. When Ed Link started a research institute with Seward Johnson on an abandoned sand mine on Florida's east coast, Marilyn Link became the organization's first managing director.

Marilyn Link and I pulled into the parking lot of the Link Port, a former swamp, now home to the Harbor Branch Oceanographic Institute, part of Florida Atlantic University. Gabby Barbarite, a marine microbiologist and science communicator at Harbor Branch, received us at the entrance. Harbor Branch has about 200 scholars and students focused on technical topics such as finfish aquaculture, cancer-cell biology, physical-biogeochemical ocean observation, and modeling. Among them are scholars in ichthyology, benthic ecology, phytoplanktonology, and zooplanktonology, who have devoted their careers to improving the health and preservation of aquatic life. "Our seas are doomed unless we act quickly," Ed Link said when serving as Harbor Branch's director of engineering in the 1970s. "Pollution and depletion of the rich resources of the ocean bottom are taking their toll. Even with a complete turnabout in man's attitude . . . it will take many years to reestablish the essential underwater environment that will preserve the oceans for use of future generations."

Ed Link condemned the oil industry for its egregious pollution risks. The companies "haven't taken severe enough steps to provide backup systems" for oil spills and other ecological disasters. Ed Link leveled some of his strongest criticism against Operation CHASE, a controversial toxic waste disposal program administered by the US government until early 1970. Operation CHASE, an acronym for "Cut Holes and Sink 'Em," intentionally sank ships in different locations to dispose of unwanted munitions. Some of these ships carried deadly chemical weapons or nerve gas. Ed Link thundered that the government was taking a "devil of a chance" with this program. "There is ignorance at all levels of government of the problems of pollution."

At Harbor Branch, Barbarite explained the workings of aquaculture and highlighted the latest research on harmful algal blooms.

Then she opened a hangar-like room that held the *Johnson-Sea-Links I* and *II*. In their 40-year underwater service, the two submarines completed nearly 9,000 dives retrieving vital artifacts from engineering history. In 1977, they helped recover the wreck of the Civil War ironclad USS *Monitor*, the long-lost armored ship. In 1986, the *Johnson-Sea-Links* helped locate scattered parts of the space shuttle *Challenger* after it exploded in flight and took seven lives. For the *Challenger* alone, they completed a total of 109 dives. "Were it not for CORD, the solid rocket booster pieces couldn't have been found quickly, the *Johnson-Sea-Links* wouldn't have recovered the failed O-ring seal, and the space shuttle program would most likely have been discontinued," ocean engineer Andrew Clark said. "Ed Link not only had the foresight to recognize the need for a new field of education but the fortitude to actually launch it himself." Even before he knew anything about Ed Link, Clark wanted to be underwater. As a teenager, he read issues of *National Geographic* wherever he spotted them to monitor developments in undersea technology.

Clark, now in his 60s, calls himself the black sheep of his family. His mother was a nuclear radiologist, his father a nuclear physicist, and his three older brothers were National Merit Scholars, but Clark dropped out of high school. At 16, he left home and hitchhiked to Louisiana, seeking a job as a diver in offshore oilfields. In the early 1970s, he worked for Sun Oil Company on the platform supporting the first crewed oil-well completion on the seafloor. "In the offshore business, we were beginning to apply in practice all those methods envisioned and proposed by Ed Link's Man-in-Sea program," Clark said.

But the school of hard knocks could only take Clark so far. "Everything I saw out there I wanted to do, the older fellas would tell me, 'Son, you need college for that.'" After being rebuffed several times, Clark relented and returned home. His father patiently taught him calculus and physics evenings and weekends. Clark went to the Florida Atlantic University for his bachelor's and master's in ocean engineering, a program Ed Link helped create, before completing a doctorate at the University of Hawaii. He joined Harbor Branch in 1979 for a Link Foundation internship. A decade later, he followed Ed Link's footsteps by serving as the institute's engineering director.

Clark's work has ranged from developing large crewed and uncrewed vehicles to testing small instruments for exploring the oceans in three dimensions. He has designed and deployed seafloor sensor networks for monitoring tsunamis and seismic activity, and broadband satellite communications to transmit data and images in real time from maritime vessels, buoys, and platforms back to shore. "Few major fields of study can be so clearly traced back to one defining event. Even fewer can be traced back to the imagination and initiative of one remarkable individual," Clark wrote. "It wasn't Ed's nature to simply make a recommendation and then leave it to others to implement."

Since working as the medical director on the SPID missions, MacInnis has studied leadership in extreme environments, with Ed Link among his prominent case studies. "Ed Link was the essence of an alpha curiosity, an action-driven curiosity, a solution-seeking curiosity," he said. In the 1980s, MacInnis was part of a French-led expedition to the *Titanic*, whose wreckage lay 12,000 feet deep in the waters south of Newfoundland. His subsequent expedition in the 1990s filmed the search and inspired James Cameron's 1997 epic. And in 2012, MacInnis advised the Hollywood director on his seven-mile-deep solo mission into the Mariana Trench in the *Deepsea Challenger*, a "vertical torpedo" sub. Ed Link's template of courage and commitment, MacInnis said, "has guided me and kept me safe."

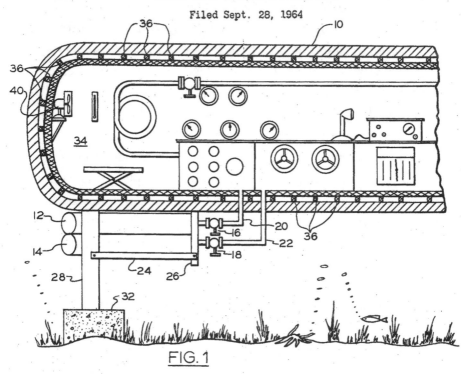

FIG. 1

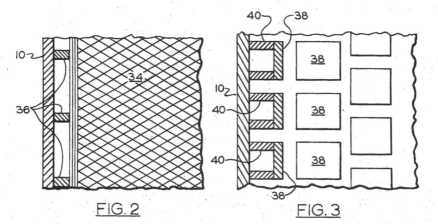

FIG. 2

FIG. 3

INVENTOR.

EDWIN A. LINK

BY

ATTORNEY

Chapter Six

Wicked
Resilience

F OR THE ROYAL NAVY, THE FORMULA WAS SIMPLE. GOOD
timber led to sea power, and sea power led to good timber.
Before oil, timber shaped geopolitical fates and futures. Feeble
timbers were a more ferocious threat than the empire's enemies.
Shipworms damaged the oak keels of King Henry VI's discovery
vessels. Even Columbus's dread of ruin from "fouling organisms"
surpassed his worry about impassable waters. Dry rot and marine
borers sank ships; fist-sized fleshy mushrooms feasted on wooden
panels. Sailors protected the hulls with pine pitch and tallow, an
upgrade from arsenic and tar sealants.

By the early 17th century, Britain was desperate for wood. Seven-
eighths of the nation's virgin forests were depleted. Construction-
quality oaks took 80 to 100 years to mature, but a ship constructed
out of them typically lasted 25 years. As British ambitions soared,
their navy built "seventy-fours," double-decker majesties named
for the number of guns they carried. The Royal Navy had over 100
seventy-fours, each engineered with stern elegance and staffed by
hundreds of crew members. With its dwindling timber supply, the
British Empire needed other wood sources. As did France, Spain,
and Holland.

These maritime rivals considered North America an "endless
frontier" of new resources and logging routes. Boston, New York,
and Philadelphia quickly became shipping centers for spruce, cedar,
maple, and white pine from New England and oak, mulberry,

cedar, and laurel from the mid-Atlantic. Each seventy-four was a floating forest furnished by far-flung supply networks. Mainmasts were from the mountains of Maine, topmasts from Ukraine, and hulls from Sussex. Small spars and planks were made from snaky stock that floated down Polish and Norwegian rivers.

As one marine historian pointed out, the credit for the success of these ships belonged to engineers as much as admirals. These engineers had to generate solutions to a perplexing question: How can one make timbers waterproof, rotproof, and stressproof? The notion of resilience provided essential guidance, particularly the observations of pointy-nosed English engineer Thomas Tredgold.

In 1802, 14-year-old Tredgold began honing his woodworking skills as an apprentice to a cabinet maker. Then he worked as a carpenter for an architect who renovated royal residences. In 1813 Tredgold moved to London, hoping to make his mark as a technical author. As new bridges spanned the Thames, cast iron became a popular construction material. But Tredgold's faith in the wood didn't wane. He proposed that understanding timber's resilience would enable its better use, especially for building ships that had to wait long for lumber.

In 1818, Tredgold wrote a short paper on the comparative strength of beams for the *Philosophical Magazine*. In it, he suspended different weights from six varieties of wood—two species of red larches, two kinds of English oaks, and Memel and Riga timbers. He compared their stiffness, stretch, and strength to determine how the wood bent or broke. He dropped weights from different heights on inch-thick wooden pieces in another experiment. Tredgold found that while larch wood was inferior to Memel or Riga timber in stiffness, it surpassed oak in strength and "in the power of resisting a body in motion (called resilience)."

A Scottish nobleman who had introduced fast-growing European larches in his estate invited Tredgold to study them further. He analyzed the "work" these deciduous conifers performed, recommending them as sturdier, stress-resistant substitutes for tall oaks. Tredgold whittled these insights to decrease the drying times of timber, a hard efficiency tactic for cost savings with ship frames and masts. Then he arrived at a question engineers often face: How can I make this success sound serious in theory? In 1820, Tredgold published *The Elementary Principles of Carpentry*. This dense, dry treatise proved durable; new editions were published until 1946, 117 years after he died in poverty.

Tredgold didn't invent the concept of resilience. In the Bhagavad Gītā, for example, Krishna advises a vulnerable Arjuna during the *Mahābhārata* war that the essence of self-resilience—and therefore, self-realization—is control of the senses. Equipoise stems from *sthita-prajña*, a state of steady wisdom unaffected by emotions, undisturbed by desires, and unshaken by adversity. A similar Zen insight contains a visual: "In a storm, the grass bends with the wind while a mighty tree is blown away." One must be supple, like grass, through the winds of delight and disappointment, delusion and dread, to retain balance. Seneca the Elder used the word "resilience" to mean "to leap." For Ovid, resilience meant "to shrink or contract." Cicero defined resilience as the capacity to "rebound," and Quintilian, to "avoid." Saint John Chrysostom, Archbishop of Constantinople, wrote: "An arrow never lodges in a stone: often it recoils upon its sender."

This concept has a long philosophical and religious tradition across cultures. Tredgold also didn't invent the word "resilience" as we have come to use it colloquially. Contract law in Middle French denoted the term "to cancel." The word migrated to 1520s England when King Henry VIII annulled—or "retracted"—his marriage with Queen Catherine of Aragon. In his 1625 book *Sylva sylvarum*, Francis Bacon used "resilience" to describe the intensity of echoes. In the early 1800s, polymath Thomas Young, known as "the last man who knew everything," considered resilience a performance measure for bulky beams. Later, Samuel Taylor Coleridge applied it to express emotions in his 1834 "Hymn to Earth in Friendship's Offering."

However, what Tredgold did was to make resilience operational as an attribute to engineer "a species of strength." His ideas about resilience contained both physics and philosophy. They entailed a "just economy of materials," achievable from "judicious combinations." But they were also about the value of elasticity: "Trees planted in the worst soils, and most exposed situations, have thriven the best." And indeed, the engineers of civilizations far before Tredgold must have taken resilience seriously. In 30 BCE, for example, the Roman engineer Marcus Vitruvius promoted volcanic ash–lime mortar for crack-resistant construction. This self-reinforcing "concrete" is an early case of a material being engineered to become stronger under stress, as with the structural integrity of the Colosseum and the Pantheon, which have spanned millennia.

Tredgold's conception of "hard resilience," with its solid bounds, referred to a material's elastic memory to handle swift and severe loads without failure. With hard problems, resilience can be quantified as the opposite of brittleness. Engineers can simulate various stressful scenarios (called the performance envelope) to see how their products and services rebound and adapt from hardships when possible. For example, during the unexpectedly stressful state of the 9/11 attacks, AT&T effectively deployed mobile backup systems using semitrailers. Today, AT&T's response system includes equipment for rapidly deploying "mobile cell sites and mobile command centers, such as Cell on Wheels (COWs), Cell on Light Trucks (COLTs), and Flying Cell on Wings (known, of course, as Flying COWs)."

Tredgold's insights for hard resilience problems have motivated solutions for better designs that make products less vulnerable. But the insights may have also fueled the efficiency motive of environmental exploitation. Engineering has logged vigorous profits from virgin forests to build "seventy-fours" and a surplus of flimsy furniture. The results of the rampant deforestation included disaster and famine. In the early 20th century, forests were "a lumberman's carnival" and "a great reservoir" to be "drawn upon almost at will and in any quantity to meet expanding requirements." As with hard problems and their solvability, the timber problem had a solution: ironsides.

In the 20th century, resilience gained greater cultural prominence, and the basic idea was, after all, appealing. Why can't society be as resilient as a steel beam is resilient to pressure? In 1973, scholar Crawford "Buzz" Holling presented crucial insights from conservation ecology that magnified and modernized the idea of resilience. Engineering, he argued, gravitated toward "the quantitative rather than the qualitative."

Holling's work distinguished between engineering resilience and ecological resilience. Engineering resilience relied on a system robustly restoring its equilibrium after a perturbation, as with Tredgold's timbers. Ecological resilience, however, required something more from a system: persistence, cohesion, and adaptation in the face of extreme disturbance. The former typically relied on balancing steady states; the latter was about regime shifts, as in a clear lake now conquered by algae. Holling arrived at this distinction while studying commercial harvesting in forests and fisheries. The hard efficiency motives of "maximum sustainable yield" created messy vulnerabilities in certain ecosystems, including the chance of species collapse and extinction. For Holling, the concept of robustness, as Tredgold practiced, was too limited in its scope. An equilibrium-centered view, he wrote, "is essentially static and provides little insight into the transient behavior of systems." Concerns about pollution and population weren't adequately described "by concentrating on equilibria and conditions near them."

Holling's scientific argument was significant. It boosted resilience as an integrating force to understand how human societies can coevolve with natural systems as an organic whole. But such a totality required pursuing a vital and vague concept: socioecological resilience at multiple levels, to encompass the entire biosphere. And how was one even to visualize this resilience? It took an infinity graph inspired by a goat-footed god.

Pan was a wild trickster in Greek mythology, the patron god of meadows and music. He had goat horns, legs, ears, and a tail. If disturbed during his naps, he aroused "panic" in others. But Pan also meant "all," as in all-encompassing, which inspired Holling and his colleague Lance Gunderson to create a resilience framework called "panarchy." Social-ecological systems, they argued, are made of self-organizing cycles connected at multiple levels across space and time. Each cycle has four interlinked phases to convey how a system adaptively grows, stabilizes, collapses, and regenerates.

If you link these phases on a panarchy graph, it looks like the infinity symbol. One shape-shifting infinity is nested within another infinity at a different level. Each infinity loop affects another, and a lack of resilience at one level can weaken another. We saw an engineering example of interactive multisystems, multilevel abstractions in soft vagueness. Next, we'll see another version for "soft resilience." Panarchy as a concept helped apply resilience from engineering to the context of ecologies. Total resilience requires constant engagement with our world. One scholar observed: "Just as 'the price of liberty is eternal vigilance,' the price of resilience is a restless mind."

In the prologue for his carpentry book, Tredgold wrote that a builder should understand geometry and mechanics but cautioned against hyperspecializing to avoid a final result with "neither solidity nor beauty." Tredgold alluded to a systems view of the world even if "this kind of excellence seldom secures praise." While serving as the secretary of the Institution of Civil Engineers in London, he wrote his famous words about engineering: "Art of directing the great sources of power in Nature for the use and convenience of man." Tredgold knew that engineering was always politics by other means. But this fact is seldom advertised or acknowledged in engineering practice today. Anyone who refuses the political nature of engineering negates its roots. Another relevant perspective from art that may be true for engineering comes from novelist Toni Morrison. *"All good art is political! There is none that isn't."*

In one of his final letters, Tredgold wrote, "Each self-important creature thinks for himself alone, till at length all ties will be broken and all will fall in one common ruin." This is true in the case of Tuscan wood's transverse stress and in the cases of the toxic stress of modern work, which can rot personal and group harmony. Unfortunately, Tredgold's vital work in engineering didn't provide him with dependable income. His only wealth was his book collection. Near destitute, Tredgold died in January 1829. Left to fend for their children, Tredgold's wife sold one of his encyclopedias to the Institution of Civil Engineers. They gave her £40.

✳

CHRIS GLAZNER CALLS HIMSELF a nerd on retainer for the government. He grew up in Channelview, Texas, a refinery town, and the subject of the 1993 film *The Positively True Adventures of the Alleged Texas Cheerleader-Murdering Mom*. In his youth, Glazner was drawn toward Native American traditions. An elder in the Ponca tribe of Oklahoma told Glazner that there are many paths, with different perspectives, to arrive at a common destination. He chopped wood, learned their songs and dances, listened to folktales, and partook in powwows. "I was highly atypical," Glazner said with a shy smile while wearing beaded moccasins.

While studying engineering in his 20s, Glazner felt that something seemed off-balance. "A part of me was missing," he recalled. "I was solving problems devoid of context." He kept asking: How does this derivation change the world? And an answer flashed. He signed up for a second major in philosophy.

Glazner went to the Massachusetts Institute of Technology for graduate school to study technology and policy. In his first problem set, a professor asked Glazner's class to create a fair system to allocate kidneys for people on a waiting list. Each team tried to impress the professor with their algorithms during the presentations. "You all think you are so smart," the professor said. "You are so confident in your program. You followed the same approach but couldn't agree on what fair meant." With results differing widely, the consequences of their code became clear. "Your system decides who gets to live and who gets to die," the professor said. Fascinated by these early insights into the social impacts of technology, Glazner enrolled as his PhD student, focusing on the simulation of complex systems.

Now at MITRE Corporation, Glazner directs systems engineers developing complex models and simulations for the government. "Twenty years ago, if you asked me what systems thinking was, I'd probably have said something pejorative," Glazner observed. In the eyes of a stone-cold specialist, some elements of systems thinking can sound a bit woo-woo and bereft of numerical delight. These comments can stem from the typical insecurity about "real" engineering and its ostensible rigor. But just because a model doesn't use formulas or generate fixed-form solutions doesn't mean it's a lesser form of engineering. "The key is to make sure people feel part of the solution, and their mental models are brought into what's developed," Glazner said.

Soft problems are like forced-perspective art, which sways the viewer's perception of space and time. The comparative issues of soft problems must be pinched and zoomed in and out to generate different systemic views. Recall that soft models don't predict the future states of complex systems but can produce evolving insights. With both soft and messy problems, a challenge is to agree on the scope and keep getting agreement. Most of the intractability, as Horst Rittel and Melvin Webber wrote in their 1973 paper, "is that of defining problems (of knowing what distinguishes an observed condition from a desired condition) and of locating problems (finding where in the complex causal networks the trouble really lies)."

Some years ago, Glazner got a call that would test everything he knew. It was a federal government assignment, a top priority for the White House. The goal was to eliminate homelessness among veterans. The Department of Veterans Affairs asked MITRE to develop a model that reviewed the progress over the past 25 years to determine what didn't work and could be done differently.

Glazner's team started with the historical data and found that government money spent didn't correspond to less social displacement. He then pulled together a workshop on systems analysis for government officials, policy analysts, scholars, and social workers. In a large conference room surrounded by flip charts and sticky notes, Glazner visually represented the different perspectives of the people in the room as they interpreted the problem. He displayed those views side by side, so the group could simultaneously get a visual grasp of the complexity.

Glazner didn't want to complicate his draft model too hastily. He probed: Why do people become homeless? How do they overcome this? What are the critical points when people go into or out of homelessness? What is the influence of criminal justice or VA policy on social displacement? The workshop participants were divided into subgroups to focus on different subsections of Glazner's model. The group reconvened and drew a much-expanded model from Glazner's whiteboard diagrams. The participants could now advance a simulation prototype from Glazner's draft that distilled their shared understanding of the system of homelessness.

Over the next several months, Glazner worked closely with scholars and social workers. Their model became more nuanced as each group provided a different outlook and action plan. The model became a coordination device to express everyone's logic coherently. Glazner's model didn't necessarily solve or control something in a traditional engineering sense. Instead, it was a collaborative tool for the participants to engage with the system, model it, explain what works and doesn't, and intervene.

Soft resilience problems like homelessness aren't amenable to hard solutions, which in this example might look like a quick financial fix for the vulnerabilities that fuel social displacement. Indeed, it's unrealistic to define what robustness means for social displacement and for whom and who is willing to pay for it, a concept we encountered back in soft efficiency. Robustness makes sense only in an ongoing analysis of trade-offs against the overall performance and what costs and consequences one is willing to accept. Further, one may rashly assume that problems of homelessness are about situations in an individual's control and not institutions. While some of those situations may be temporary, their effects are long-lasting. Breaking the cycle of homelessness requires a combination of public policy, public money, and public buy-in.

A youth experiencing homelessness can face feelings of despair, doubt, and disconnection. An adult living in poverty may maintain an optimistic hope for the future while adapting to their circumstance. Each situation requires enormous psychological resilience that differs from a timber's physical resilience to overcome adversity. As a participant in a study of individuals experiencing homelessness said: "So, you have to go through those struggles every day, and when you're homeless, you don't have a lot of really good days. You have days where you're . . . you get by, compared to what used to be, and you have to change your . . . your barriers and your parameters as to what success is. And those are adjustments that again, some people do well and some people give up."

Some individuals remain positive. "And you have to be able to . . . to forgive yourself for the mistakes you've made. You make 20 mistakes a day every day, every one of us, and we don't even know we're doing it. And then it's, you know . . . you wake up the next day and try to be positive and work at it again. And it's hard when you suffer from depression to do that. But you have to forgive yourself, and you have to realize that there are better days and you can get there. And the only way you're gonna get there is trying to have a positive attitude." Some are tenacious: "I just can't give up . . . I haven't given up and I'm not gonna." Some, worried: "Well you feel yourself swirling down the toilet." Some hopeless: "I am still stuck in here. When is this hell going to end?" Others, appreciative: "You don't forget, you know, what you have. I appreciate every day now." And some are afraid: "My fear is being found on the street, but no one knowing how to help me or who I am." One can relate resilience from these sentiments to the essence of survival, a life altered by adversity and a potential hope for a better tomorrow.

Glazner knew the problem was at once personal and political, much like the case studies his PhD adviser taught him. When Glazner's team delivered the results, the agency leaders were surprised. They were expecting the usual text-heavy PowerPoint presentation. Instead, his team presented an interactive systems-dynamics model of causal thinking. It was a "flight simulator" that one could get into to test different policy scenarios without crashing public investments. The model's many perspectives made it a valuable training tool, not just a data presentation.

The model's chief insight to policymakers was simple: move away from temporary shelters and provide permanent supportive housing while focusing on preventing the flow into homelessness. The recommendation wasn't particularly novel, but there was little real-world evidence for its efficacy at the time, and no one wanted to take a risk on something new. The assumptions were typically kept opaque, even for those familiar with this approach. As more decision-makers engaged with his model, the more flexible it became, which made the problem more approachable and less unmanageable.

Glazner explained the benefits of thinking about the system as a bathtub. How are the flows into and out of the tub (and the recirculation, as in people moving in and out of homelessness) influenced by policy and other system elements? What information was available to support different hypotheses? What could explain a system's behavior under certain conditions? Glazner's team coded these questions into the program and tested different assumptions. The critical engineering consideration was that the system couldn't and *shouldn't* be made resilient only from one stakeholder's perspective or optimized on a single attribute, such as efficiency.

As in the case of chronic homelessness, policy interventions intended to ease can instead exacerbate soft problems of resilience by temporarily making things better without tackling the underlying causes. Without a fix that addresses the causal underpinnings of homelessness, any improvements are fragile. These issues do not have solutions to create magical transformation because it's impossible to know what factors make people resilient. A community's resilience is not a total of personal resiliencies. Even if that were the case, the total would expose the problem of what constitutes social good and how it is valued, which is at the crux of many soft problems. At a minimum, engineering for soft resilience, as with soft vagueness and soft vulnerability, requires different views and abstractions of the system. This is a principal reason why systems engineers develop multilevel models. Multilevel models focus on blind spots and perceptual factors to generate gains that would be impossible through specialized metrics.

Resolutions for soft problems, and dissolutions for messy problems, shouldn't be considered in isolation; their effects can reinforce and lead to increased resilience. In the case of homelessness, focusing on long-term health, such as support for mental health, substance abuse, and even dental care, can reduce recidivism. Such an approach becomes even more effective when permanent supportive housing is in place. According to the VA, implementing permanent supportive housing has helped lower veteran homelessness by nearly one-half since 2010. Still, permanent supportive housing contradicted some people's notion of fairness. As decision-makers became mired in discussions about entitlement and untested efficacy, Glazner observed how each participant steadily advanced their local and personal interests. Each perspective became adversarial; a puzzle where the pieces were competing.

The VA took the working model to the US Congress and the Office of Management and Budget and had *them* assess it. These interactive framing sessions were revealing. The decision-makers felt that the model's results went against their initial feelings on fairness. Glazner realized that his model had become a conversation facilitator. He gave them a visual vocabulary to share perspectives on what they see or choose to see about the system. By visually explaining the dynamics backed with different data, different perspectives could be supported or refuted. "I can't say solving differential equations is what led to us reducing veteran homelessness," Glazner said. "We brought stakeholders together in a way that had not been done before and gave them tools to share and test their mental models."

We know resilience can help a system work against a negative influence. This, too, is a function of how one perceives resilience. What's supportive for one constituency can be detrimental to another. In the case of reducing homelessness, this can be seen with the traditional shelters across the United States. These provide a temporary helping hand. They are easy to donate to but don't address the underlying causes of homelessness. Problems of soft resilience emerge from this complex grouping of struggles.

Scholar Donald Ludwig used a boating metaphor for resilience. When floating on water, the boat will oscillate if you abruptly throw some weight on it. Then the weighted boat will eventually settle in a new state that may feel the same as the previous unloaded state. The boat is now stable. However, suppose you gradually load the boat by hanging weights below. In that case, the boat will sink deeper, failing to balance the gravitational force. If you load the boat from the top, it may suddenly flip over and sink. Whether the boat sinks slowly or suddenly depends on how the system responds to the external weights that disturb it in this resilience test. Ludwig used the idea of the "domain of stability," a set of actions that prevent the boat from tipping. Of course, the boat may also tip if the occupants move around, which is a source of internal disturbance.

There's a soft resilience problem even in this boating metaphor. The boat's resilience "cannot be determined outside of its social and institutional context," Ludwig notes. The boat's occupants have different goals and rights and will weigh their risks differently, just as its owner would. Then more considerations enter: Who decides on the boat's loading and occupancy? Which government agency regulates boats and their waterways? Are there interest groups that would rather not see the boat on the river? As Ludwig wrote, the final fate of the boat depends at least on its physical characteristics, the environment in which it operates, and the surrounding social and political structure. One has to resolve the boat's resilience rather than solve it.

Glazner knows a thing or two about stabilizing boats. He volunteers as a paddle-sports instructor, teaching white-water kayaking to wounded veterans at Great Falls Park. The swift waters of the Potomac may look chaotic for the fresh initiates, and there's a thrill in the turbulence, jagged rocks, and crosscurrents. But those that observe long enough will see patterns in the chaos. Sharp maneuvering is shaped by training, whether our reflexes or resilience.

Glazner's model didn't eradicate veteran homelessness; it's not a solvable hard problem. But because the model helped reformulate and resolve certain critical decisions to a timely policy requiring funding, he shared his accomplishment and gratitude with a Facebook post of a news article about the resulting changes at the VA. One of Glazner's kayaking students, in her late 20s, saw it and sent him a message. She was a veteran recovering from injuries and post-traumatic stress. She shared the secret that she had been homeless for a year. She lived in her car and showered in public facilities. A day earlier, she had received a voucher for a place to live, a voucher provided by a program shown to be effective in Glazner's model. It was her own apartment for the first time in her life. She thanked him. Glazner cried.

"Never in my career have I had something like that happen to me."

✳

AFTER SCALING MONT BLANC, the frostbitten romantic George Mallory marveled, "Have we vanquished an enemy? None but ourselves." It's unlikely that the alpinist was alluding to the garbage dumps of the world but instead aspiring for the eventually tragic attempt to conquer Everest. However, the dizzying trash peaks of humanity's refuse do rise in many places. Historian James Carse declared waste an "antiproperty." No one wants to see it or own it. The junk mountains of the world don't get named after explorers. They are, in Carse's words, "the emblem of the untitled."

In December 2015, a landslide of construction waste in Shenzhen, China, swamped 33 buildings over an area the size of 60 football fields. The rescue operation required 4,000 people and dozens of excavators; over 70 died. The regulator in charge of the construction plant jumped off a building in suicide. In March 2017, a 50-year-old mound of discards gave way in Addis Ababa, Ethiopia. The avalanche killed over 115 people in the nearby shanty settlements. A month later, an unstable heap of demolition debris in Colombo, Sri Lanka, loosened and rumbled like thunder. The landslide killed 26 people, destroyed hundreds of homes, and released befouled liquid.

Cities in India are home to giant waste hills, the badlands of decay, dead and deathless. Once lauded as the Garden City, Bengaluru became a Garbage City recently. The *Times of India* reported that the 200-odd tons of daily detritus in the year 2000 swelled to over 3,700 tons by 2015. Bengaluru's trashy triumph mushroomed around the time its information technology business surpassed all the megacities of India except New Delhi. With this waste explosion came an unmanageable workload for the city's sanitation workers.

Bengaluru's decades-old centralized system coordinated waste collection and disposal from various localities. The waste was dumped on the city's outskirts. Adjacent villages soon began to feel the effects. Thick leachate seeped into groundwater, wafting a choking stink. People couldn't eat, and they threw up when fresh hospital waste arrived. Activist petitions led the Indian high court to order a new, decentralized system. Residents were required to segregate their waste. Wet waste was composted. Dry waste was collected in each city ward, and scrap merchants recycled a fraction. The local government paid the expenses. Still, the rest of the responsibility was left to citizens and the informal sector of street sweepers, ragpickers, and waste dealers. Since door-to-door waste collection is inadequate in India, collecting and segregating trash translates to higher costs. And people unwilling to pay would dump their garbage elsewhere, including in the streets. Therefore, much depends on the informal sector and solid-waste gathering through dry-waste collection centers.

One analysis reported that in six major Indian cities, an unofficial industry of 80,000 people handled nearly a fifth of the total waste processing. These workers lacked visibility because of the entrenched Indian caste system. Trash picking, street cleaning, dealing with acrid odors, and the most menacing work, manual scavenging, have been assigned to the "lower caste." A clean community depended on the success of these collection centers and, in turn, on the revenues they reaped. The more trash the residents segregated, the more functional and profitable the dry-waste collection centers were. This decentralized approach meant less dumping in landfills and fewer consequences of leachate and pollution.

A nonprofit in Bengaluru consisting of engineers, social scientists, and journalists collaborated with activists, waste cooperatives, and the municipality to confront this literally messy problem. The organization wanted to "create more informed citizens, a small step toward creating more conscious citizens." The group eschewed the idea of a simple technological fix and the usual view of seeing people as "users." They built a trilingual board game called Rubbish! The prime actors in the game Rubbish! were the waste producers themselves. The game was pliable and "in a state of constant becoming," the principals write in *We Are Not Users: Dialogues, Diversity, and Design*. Participants can engage in multilevel abstractions during gameplay by zooming in and out, entering or eliminating information. The developers compare this process to choosing meal ingredients; different combinations can yield different recipes.

The game's central feature is to model the waste collection and segregation supply network. Its goals are simple: make people aware of the importance of waste segregation and the complexity of creating a new system of trash management where no such system existed before. The game lasts about two hours and can accommodate up to 12 players. Each player can impose their view on the system and insert subtleties and conflicts that affect other people's perceptions of the problems. These features enable a networked view of the problem and each decision's inherent costs and trade-offs.

One game scenario allowed the dry-waste collection centers to profit, and another focused on the citizen's role in improving waste-management systems. The latter required more citizens to segregate their waste at the source and use dry-waste collection centers. Yet another game scenario was to change citizens' apathy, worsening the hygiene problem. The game's goal, after all, was to illuminate to the participants the vast implications of their actions and inactions. Each neighborhood in Bengaluru is a different constellation of consequences. More affluent localities produce more diverse waste blends than the low-income areas and vegetable markets, separate from the shopping malls and wedding halls. If one group complained about dry-leaves collection, others complained about the broken coconut shells. And often, people from the informal sectors were more mindful of the environment than educated professionals. But everyone leaves the game agreeing that segregating waste is vital for the system to perform effectively.

Unlike a typical game, Rubbish! has no winners or losers. The game is an artificial situation for gaining insights. Players collaborate or compete and follow the rules. A simulation, in contrast, is typically judged by its fidelity to realism. A game need not be as realistic, high-tech, or immersive like a modern flight simulator to be effective. One community-engagement game developed to advise on improving the Dutch railway system's performance and quality used "low-tech" alternatives: pen, paper, sticks, and sponge to depict the trains, passengers, timetables, and infrastructure. The organizers of the railway game reported that the low-tech sessions gave them conclusions similar to those of the high-tech sessions.

The high-tech tools may have given them more data, but "little difference was found in the actual use of these games." The US Marine Corps routinely employs low-tech tools to teach tactical decision-making. Even with a multibillion-dollar annual training budget, the marines reliably use low-tech terrains of dirt, sticks, strings, chalk, cotton, paper, paint, and a few props when appropriate. One observer noted that more expensive training doesn't necessarily mean it's more effective. "Technological advances notwithstanding, there will always be a nearly limitless supply of sand."

Rubbish! uses a paper map. Like a good systems model, it serves as a conversation facilitator on a topic most prefer to disregard. But there's no point in cities installing "smart" receptacles if the bulk-waste producers are not committed to trash segregation, let alone cleanliness. This scuzzy problem may seem unique to India, but global waste production is dramatic—approaching 30 billion tons per year by 2050, nearly a third of it coming from India and China.

Scholar Noah Sachs notes that the trash on India's streets reveals something disturbing about America's trash problem. The Indian trash is explicit, but American garbage hides. The average American generates up to seven times the amount of trash that a resident of Bengaluru might produce and two to three times more municipal solid waste than a Japanese resident. The lives of US sanitation workers are complicated and dangerous. For example, in New York City, their fatality rates are twice that of police officers and about seven times that of firefighters. The American garbage footprint goes beyond landfills. Sachs also points to disposable plastics, fossil fuels, and carbon emissions of US citizens. These hundreds of millions of tons of waste each year have far more grievous social costs. Then there's waste from the wealthy nations shipped to the poor. These developments show our actions aren't isolated; when we toss our trash, we spark a vast supply network of activities.

While the mass of humans is only one-hundredth of the total biomass, the human output of objects has doubled every 20 years over the past century. Engineered products such as bricks, concrete, metals, glass, plastic, coffee cups, sneakers, laptops, and mobile phones will soon exceed the wet biomass. The resulting waste surpassed the planet's total dry biomass in 2013. Another analysis estimated that chemically produced new substances, newer forms of existing substances, and modified life forms had multiplied 50-fold since 1950. Their volume may triple again by 2050 compared to the 2010 levels. The chemicals released from such novel entities and the related plastic pollution have transgressed the "safe operating space" of the planet. The UK Royal Academy of Engineering highlighted that almost one-quarter of the world's municipal solid waste goes uncollected. Worse, over a quarter of the collected waste is routinely mishandled; a considerable fraction is disposed of by open burning. In sum, about a billion tons of garbage yearly degrade our environment and health, with the toxins from trash killing one person every 30 seconds in developing countries.

Waste production is now a central phenomenon of life, stemming primarily from our pursuit of hard industrial efficiency. However, messy resilience can also stem from problems of hard resilience. Composite materials engineered for greater robustness and performance may be harder to process. For example, plastic with metal coating can't be processed the same way as plastic or metal. It may inevitably end up in a landfill or an incinerator. And the technology to properly process it may be available only in developed countries that won't provide access to it without intellectual property licensing.

"Stare into dumpsters long enough and they stare into YOU," wrote the dumpster diver John Hoffman. And waste and disposability have grown from an "unpleasant but marginal problem" to being "deliberately generated as the very metabolism behind economic growth," as scholar Justin McGuirk observes. "Messy resilience" problems of hygiene and health aren't straightforward; as with local traditions, they influence messy efficiencies. A bird flock is simple to explain but complex to describe. But waste management, as with homelessness, is simple to describe and complex to explain. It must address resilience by preventing excess waste and its spread, recovering from the waste generated, and overcoming people's persistent habits, with the change ultimately being social. Resilient practices require addressing all of these problems, making it difficult to create public understanding and buy-in. Games unravel the layers of a problem by embedding the participants in the middle of it. They can educate and deepen debates. The "trash talk" of Bengaluru helps reformulate how different stakeholders view their everyday problems.

As we saw earlier, one function of resilience is to resist forces and another is to recover from them. Yet another form of resilience, more relevant to the trash problem, is to adapt to accommodate those forces, like Donald Ludwig's boat trying for stability. Yet, from a traditional engineering sense, the last variety is challenging to design for because such transformations need to occur perceptually. Low-tech games such as Rubbish! provide examples to guide the participants to a viable future through changed views and habits. Everyone thinks about the consequences of the waste one unthinkingly leaves behind. The futures discussed in the game aren't commanded and controlled; instead, they are meaningfully codesigned by those who are often left out of the conversations.

One game like Rubbish! or a systems-dynamics model for homelessness may not necessarily tame a messy problem. However, each makes apparent the factors that aren't considered, the first step to reframing a problem. Rubbish!, then, is a "language game," to borrow from Ludwig Wittgenstein, "a whole cloud of philosophy condensed into a drop of grammar." The participants who played the game defined the vocabulary with which they engaged with one another and framed the problem. And language, Wittgenstein wrote, is a "labyrinth of paths. You approach from *one* side and know your way about; you approach the same place from another side and no longer know your way about." Sometimes jumping into the language of fiction can help us dissolve a messy nonfictional reality.

Much like George Mallory, another wanderlust naturalist went whenever the mountains called. After a hike on the Wrangell range in Alaska, the one-time engineer John Muir wrote that the world was continually making itself. Each morning was an act of creation, and the "mountains long conceived are now being born." The same is true for our trash.

*

WHEN LANCE GUNDERSON AND "Buzz" Holling created their panarchy framework, they were tapping into the ancient folklore of tricksters. Pan was a flute-playing god, and so was Krishna. Both were boundary crossers, and their paths were both a spirit road and a real road. They were adept at moving between heaven and earth, lifting good things from heaven and delivering them to humanity, as Prometheus the trickster did by stealing fire. As a child, Krishna liked to steal butter. His foster mother, Yashoda, would instruct him to stay away from their butter. But Krishna broke the clay pots, coating the fluffy cream over his dark face. Krishna cleverly told Yashoda that he was saving the butter from the ants or claimed that he wasn't really stealing since the butter was homemade. The young butter thief was also a thief of hearts, charming everyone. Scholar Lewis Hyde has written about tricksters like Pan, Prometheus, and Krishna, who weren't *im*moral but *a*moral. They were a "mythic embodiment of ambiguity and ambivalence, doubleness and duplicity, contradiction and paradox." A trickster is an agent of change, disrupting the bounds of good and evil.

Varied as they are, wicked problems are like the tricksters of folklore. They surprise us and profoundly affect our culture. Rittel and Webber must have grasped the inherent trickiness with such problems to name them "wicked." Resilience for wicked systems is a case of the collective more than the particulars. However, the need for collective resilience doesn't often appear in a company's action items or, for that matter, a country's. Stating resilience as an explicit organizational goal can create unwanted complications.

One physician observed that the idea of absorbing and recovering from harmful conditions that were avoidable in the first place "makes resilience a dirty word." For example, in the burgeoning crisis of clinician burnout, the blame is often placed on the individuals. Few openly acknowledge that medical care is over-politicized, ill-allocated, understaffed, and disorganized. Add to it the consequences of overwork, compassion fatigue, and deprived work-life balance contributing to this crisis. Resilience becomes a function of people voicing their input to make prudent choices. In a rain-short region of Zimbabwe, a community might prioritize the growth of drought-resistant cereals, like millets, engineered to thrive with scarce water. Their survival strategy is tied to the community's ability to trade off nonvital assets to cultivate the pearl grains better.

An equivalent practice of trade-off exists in the realm of systems engineering. Resilient design is influenced by agility, adaptability, extensibility, quality, reliability, repairability, flexibility, changeability, supportability, and versatility. Engineers call these "ilities" the nonfunctional features beyond the attributes such as safety, maintenance, and efficiency required for mission-critical performance. One of these nonfunctional features, reliability, duly satisfies its requirements over varying conditions. If an internal agent changes the system, it's adaptability; if it's by an external agent, it's flexibility. At the same time, extensibility is about accommodating new features after design, and versatility involves achieving new features without sacrificing form. Each trait is part of resilience. Engineers often use "ilities" to quench society's thirst for further features and scale. From those new features come the maintenance challenges, and from that scale, arise waste. And, of course, achieving resilient design may become impractical when perfect quality or supportability becomes expensive.

There are many sides to resilience. Recovering from a romantic heartbreak and recuperating from the tragedies of rogue microbes, terrorist attacks, or hurricanes are different starting points. Ultimately, our relationship with resilience is much like what James

Baldwin wrote: "People are trapped in history, and history is trapped in them." If being resilient was easy, it wouldn't be genuine. But "resilience," as we now know, can be many things in one, an aerosol word easily sprayed. Scholar Andrew Park captured it as "paradigm creep—the use of buzzwords far beyond their original sphere of application." The word's meanings mutate into uselessness. Their excessive use "short-circuits critical thinking about the actual challenges for which they were supposed to be convenient metaphors." This means we must consider resilience as a mindset rather than a metric. As we've seen throughout this book, the same can be said for efficiency, vagueness, vulnerability, safety, and maintenance. The collective effect of the six concepts can make our approaches to wicked problems far more resilient.

The vaccine pioneer Jonas Salk understood this well. In his 1975 article, "The Survival of the Wisest," he suggested a visual metaphor for societal transitions. Salk presented an S-shaped curve to identify two phases, much like the beginning of an alternating-current sine wave. He called the rising part of the curve Epoch A, where humans are more individualistic and independent, with opportunistic and short-term rewards. In Epoch B, the flattening upper part of the curve, humans are more collaborative and engaged. They recognize the ever more interdependent need for prosocial design and long-range systems approaches. Epoch A is a period with a zeal for efficiency, while Epoch B favors resilience. If Epoch A is "just-in-time," then Epoch B is "just-in-case." If governments and businesses in Epoch A dodged disasters, then in Epoch B, they will need to circumvent catastrophes. From the case studies in this book, we have seen how parochial and patchwork approaches can create a confluence of health, economic, natural, and political calamities. Yet, resilience is the multivalent vaccine against the strains of inevitable surprises.

Refrain

Heave

S IMULATED LIVING IS EVERYWHERE. WE "PROBABLY received simulators during the first days if not hours of our lives." Scholars William Moroney and Michael Lilienthal refer to pacifiers that serve as a *simulator* and a *stimulator*. "It simulates the nipple of a mother's breast or a feeding bottle and stimulates sucking and rooting reflexes, thus improving infant muscle tone." From pacifiers and model toys to imitation butter, simulated objects often script specific behaviors—nursing, running a toy train on tracks, or spreading butter on bread. Some early simulators were games that bonded entertainment with strategy. One example, *petteia*, a board game from fifth-century Greece, involved a strategic competition played with different-colored stones. In the sixth century in central India, the game *chaturanga* was developed. The name refers to the "four parts" of the Gupta dynasty—infantry, cavalry, elephantry, and chariotry. Future versions of this game, notably the *shatranj* of Iran, laid the foundations for modern chess, from which players developed tactical role-playing games that simulated potential conflicts. These *kriegsspiele*, or "wargames," would later ensure crucial victories on actual battlefields in the Franco-Prussian War. Wargames also had a long tradition in the United States, beginning in Naval War College under Theodore Roosevelt, then the assistant secretary of the navy.

In the early 20th century, wargames became routine, used to strategize the 1941 Pearl Harbor attack, the Nazi invasion of Belgium, and many campaigns during the Cold War decades later. Tabletop exercises called BOGSAT—an acronym for a "bunch of guys sat around the table"—have literally shaped human history. Wargames shrink an "approximation of war itself onto the surface of a table," notes scholar Jon Peterson. Because a wargame "was always just an approximation, it would never tame war, never control it—but it became an ideal vehicle for commanding, in the words of H.G. Wells, 'a game out of all proportion.'"

In 1974, Dungeons & Dragons was born, germinating a game genre. The description was simple: "Rules for Fantastic Medieval Wargames Campaigns, Playable with Paper and Pencil and Miniature Figures." The players could create their characters and act as their surrogates. The D&D quests entailed problem-solving, partnership-forming, treasure-seeking, and knowledge-creating. This form of role-playing "no longer simulated the experience of command—it simulated the experience of being a person who did many things other than commanding." The realistic but imaginative techniques of wargaming and role-playing that proved valuable in gaming the unthinkable would soon be applied to simulators.

ALTHOUGH QUITE UNLIKE LEWIS CARROLL, Jack Thorpe also believed in six impossible things before breakfast. Thorpe was a program manager for the Defense Advanced Research Projects Agency, or DARPA, whose singular mission is to imagine and invest in seemingly impossible technologies.

In the 1970s, Thorpe had his sights set on an area of national security needing transformation: military simulators. While simulators provided realistic practice scenarios, Thorpe contended that they did not recreate the intensity of close combat. As such, they were at risk of becoming rote. In 1978, Thorpe wrote a paper called "Future Views," presenting a 20-year vision for superior aircrew training. He proposed synching individual simulators that would plunge pilots into unscripted scenarios to rehearse for collaboration during combat. This approach was revolutionary: in Thorpe's vision, hundreds, perhaps thousands, of individual simulators would be "networked" to create a virtual world that participants entered and navigated together.

Thorpe was eager to transition from simulation devices to simulator networking. In SIMNET, trainees would no longer fight computers. Instead, teams of soldiers fought other teams of soldiers on the network. Each simulator within the network was stand-alone, with distributed controls and capabilities for real-time role-playing. For example, if one simulator was a tank, another served as a bomber, and a third was the command post. Although performing different functions within the simulation, each stand-alone simulator provided soldiers with the same highly realistic battlespaces to navigate. Many of the terrains for SIMNET were created from footage of real places—picture the opening combat on the beaches of Normandy in *Saving Private Ryan*. Each simulator came with its image generator and terrain data. They provided the operator with a unique line of sight, providing high contextual complexity and less display complexity.

Thorpe compared SIMNET to "Alice's looking glass" because it created a portal to the network that brought pilots, gunners, and drivers together as a team. Through this medium, individuals came together to hone their collective abilities and integrate their skills. "Instead of communicating with thick proposals or lengthy briefings," Thorpe argued, "government officials and legislators can live the weapon system in combat conditions." SIMNET's designers opted for selective fidelity rather than complete realism; the essential functions were included and the trimmings left out. The resulting prototypes of the simulated battlefields resembled early arcade games. Developers placed greater emphasis on the network's ability to train a team. Thorpe defined this focus on the network and deemphasis on realism as the "sixty-percent solution." It extracted maximum value from a 60 percent finished design, allowing for midcourse corrections rather than a completed design. Thorpe said, "Fast, approximate, and cheap was better than slow, deliberate, and expensive."

Eventually, SIMNET incorporated the kinds of panoramic graphics that we have come to expect from simulated environments. The system's graphics now recalled the visuals of action-adventure films, video games, and war itself. In a way, SIMNET's networked warfighting resembled a tryst between the Pentagon and Hollywood. Consider "flying carpet," a radical stealth visualization system on SIMNET. The technology swept its participants into the heat of combat. They could fly, zoom in and out on areas and persons of interest, and try god's view with the latest intelligence. The commander could customize and control the combat zone at will.

Modern warfare has integrated many aspects of simulation and become much more automatic. Advanced surveillance technologies the size of a housefly can provide an exacting eye-in-the-sky situational awareness to outwit adversaries. The Pentagon increasingly relies on autonomous weapons systems—devices like automated drones, robot sentries, and even "self-driving" submarines—to carry out hazardous missions. Soon, automation might infiltrate all combat elements, not just weaponry but decision-making, intelligence gathering, and commanding. But is this all "just war"? The feelings of moral abstraction or dissociation that these simulated guiltless environments might provoke in their "users" leave serious underlying questions unresolved. The ethics of remotely piloted drones and uncrewed aerial vehicles are not merely new concerns. These modes of war elicit unique feelings of inculpability, captured in the concept of "distant intimacy"—a hostility radically distanced yet close.

One SIMNET exercise involved reverse simulation—graphics reconstructing an actual event from accounts compiled after the fact. In early 1991, Thorpe's team recreated a panoramic battlefield of the Gulf War as though fought entirely in simulators. The "Battle of 73 Easting," named after the grid coordinates on the map, simulated combat in a blinding sandstorm. The programming synthesized information from actual battlefield tours, accounts of survivors, missile trails, satellite images, and army database reports about the explosions and casualties. This computer-generated exercise based on concrete reality trained soldiers' actions and emotions. Simulation now properly meant anticipation and, from it, an essential coherence. Scholar Patrick Crogan calls this "gameplay mode," in which the "reorientation is paired with a disorientation, just as reanimation is with a deanimation of what had come to life previously." The resulting rehearsal archetypes, such as common synthetic experiences or the joint simulation environment, enabled increasingly lifelike impressions for crew training and weapons testing.

"War is the province of uncertainty," General Carl von Clausewitz wrote. "Three-fourths of those things upon which action in war must be calculated are hidden more or less in the clouds of great uncertainty." The military philosopher understood that war was rife with psychological uncertainty and that no amount of technological exactitude—maps, graphs, or geometry—would simulate the true experience of war in a game. SIMNET vastly improved upon the early wargames and the stand-alone simulators of the 1970s. Yet, SIMNET posed a host of conundrums. In the words of one Hollywood producer, when simulators offer an environment that blends entertainment and education, "What's the difference between fighting Saddam Hussein or fighting Klingons?" Sometimes, even technologies with clocklike precision can come with cloudy moral challenges.

WHEN A 1973 OIL embargo increased fuel prices, the US military substantially reduced its open-air flight hours and substituted them with simulator hours. Flight simulators' sophistication and use grew throughout the decade, competing with the likes of aircraft, "light as a feather, though weighing billions of tons," in the words of Walt Whitman. Training on a simulator was not just safer, it was compellingly cost-effective. When the air force switched to a simulator, hourly training costs for one earlier bomber fell from $600 to $1. Consider the simulator for the B-52 Stratofortress bomber. It weighed 43 tons, with over 5,000 circuit boards and 13 computers. The 220 instruction manuals for the bomber alone weighed over 1,500 pounds. This simulator did everything a B-52 could: precision bomb sighting and following over 530,000 instructions. The only thing it didn't do was fly.

For the 50th anniversary of the invention of the Link Trainer, in 1979, Ed Link attended a commemorative conference at Piccadilly in London's West End. He and Marion Link visited all the major railway stations in the UK. They even ran a scaled-down steam-powered train through Harbor Branch Oceanographic Institute to show the potential of steam energy. Ed Link had been considering growing food for laboratory shrimps and oysters from sewage waste. He sketched a paddlewheel that used water to produce electricity.

One could imagine Ed Link retiring to his upriver Spanish-style stucco house with the operettas of Victor Herbert, regaling visitors with tales of barnstorming and underwater treasure hunting. But instead, after a beloved theater in Binghamton closed, Link purchased from the theater the pipe organ he had handmade and installed there in the 1920s. From then on, Link used his retirement to travel back to a largely forgotten period, filling his garage and small garden shack with a small organ factory. "I've had a fun career, I guess. First pipe organs, then aviation, then the bottom of the sea, and now I'm back to organs," he told a reporter.

Ed Link reignited his dormant love and resurrected the art of organ building that had given rise to the first Link Trainers. Professional organ builders were largely extinct, but Ed Link spent two years gathering the required parts. He sourced over 2,000 metal pipes from Germany, the Netherlands, and the US Northeast. Ed Link's financial adviser, Doug Johnson, remembered when the inventor turned up at the bank in his denim overalls with a suspicious request. Ed Link urgently wanted $10,000 in 20-dollar bills in a paper bag. He had located an original blue-box trainer in some barn and intended to repurchase it for parts to restore the organ. Ed Link assiduously revoiced the dusty, damaged pipes by adding new tubes for flexible tonal groupings from the Baroque and Romantic eras. The ensemble of precision controls, combination pistons, and wind-pressure adjustments produced a miniature symphony orchestra. The "string" instruments from the left side blended with the "wind" instruments of the right and unified with the center chimes and percussion.

True to his lifelong ethos of maintenance, when Ed Link complained that his wheelchair was poorly designed, he found the tools to rebuild it himself. His other passion, for his hometown of Binghamton, stayed with him to the very end. "Don't you ever believe any rumors that I'm moving anywhere. I love it here," he told a reporter. "I wouldn't live any other place."

Ed Link died in 1981, leaving behind an incomparable legacy of engineering.

THE SCOTTISH ENGINEER THOMAS TELFORD, nicknamed the Colossus of Roads, argued that engineering projects could have benefits beyond infrastructure, namely providing access to better education, jobs, agriculture, and health. He termed that process of building social capital through infrastructure a "working academy," through which "the moral habits of the great masses of working classes are changed."

In the 20th century, Binghamton had its own versions of the working academy. One such civic leader was George F. Johnson, a co-owner of Endicott-Johnson Shoes—or EJ. At its peak in the 1920s, the company employed over 20,000 workers and sold over 50 million pairs of shoes yearly. One 1934 advertisement bragged that these were "nifty shoes for the chap who is aiming to be somebody." The employees at EJ were predominantly European migrants who had landed on Ellis Island and gone straight to Binghamton. In return for their labor, EJ offered employees practically interest-free home loans, subsidized their food, and paid for their medical, dental, and childcare. Local philanthropy by the leaders at EJ soared, as did community pride. Wielding his progressive aims for the community, Johnson assailed the local chapter of the Ku Klux Klan active in 1920s Binghamton—a city once labeled the "Klan Capital of the North"—rooting the group out in a few years. The impetus was to improve the lives of his immigrant employees targeted by bigots and xenophobes.

Johnson liked to repeat that there were no rules in the company "except ordinary decency in human conduct." This ideal was part of EJ's Square Deal, which contended that company-wide profit sharing is as simple as "cutting melons or dividing plums." But this Square Deal was not purely altruistic. It was rooted in a "negotiated loyalty," whereby employers offered benefits to placate workers' demands. As historian Gerald Zahavi points out, EJ's welfare expenses were dubbed "efficiency expenses." In line with other welfare capitalists of his era, like George M. Pullman and Milton Hershey, Johnson cast himself as a "friend" of his workers. He sought to create orderly, wholesome communities to prevent any worker unrest. EJ Shoes' proactive paternalism fostered worker allegiance and stability, and profits. Indeed, this progressive industrialism made EJ and Binghamton a "veritable beehive of industry."

After learning about Ford Motor Company's successes and potential weaknesses, George Johnson organized EJ around his distinct community instinct. Ford's labor practices were "at a tremendous disadvantage," Johnson and his managers believed. "Their people do not live around the works. Mr. Ford does not live with the people—he goes into the works but seldom—they do not know him personally—it is all handed down to them through the medium of a lot of hired people—devoted people, good people, hard working people, but, still, hired people."

Between 1915 and 1950, EJ built thousands of houses for its workers. The "EJ Homes" designs followed a template, priced under $3,000. Johnson discouraged "EJ Families" from owning cars, persuading them to live within walking distance of the factories. He built parks, churches, and playgrounds in the triple cities of Binghamton, Johnson City, and Endicott. "There can be no security, there can be no guarantee of prosperity and industrial peace except through homes owned by the plain citizens," Johnson wrote in a letter urging a local banker to finance the homes. "I believe myself that the home is the answer to Bolshevism, Radicalism, Socialism, and all the other Isms," he added. "You will find that the home is the basis of all security."

In 1916, the company celebrated creating the 8-hour workday, the first major employer to reduce it from the standard 10 hours. During the Great Depression, EJ decreased his staff's working hours instead of furloughing them. Dubbed the "godfather of sports and recreation," Johnson made amusement and athletic facilities in the 1920s within walking distance of EJ homes. He gave pep talks to boost the energy and morale of his employees. And he shaped his identity as a "workingman's advocate" for relaxation and pleasure, encouraging weekend movies and horse racing. "The working people are no different than you and I and others," Johnson wrote to an associate. "We must be occupied, and we will be occupied." All this, of course, led to good business. In the summer of 1934, as a show of gratitude, over a thousand children, dressed in their best clothes and making festive music out of kitchen utensils, swarmed over Johnson's lawn to present him with a bouquet. Johnson had installed the sixth carousel for their enjoyment. The price of admission? One piece of litter.

Through the 1960s, EJ remained a dependable, respectable feature of the Binghamton community. "My father's family was an EJ family, and many of my aunts and uncles worked for the company," one resident recently said. "The really lucky ones got to work for IBM. That's how it was, you wanted to work for EJ's, but you wanted your kids to work for IBM." As this perspective suggests, Johnson's Square Deal would eventually lead to the Watson Way, and the focus of Binghamton's labor market would shift from the feet to the head.

$$\hspace{2em}\textbf{\large +}$$

THOMAS J. WATSON SR. didn't pay much heed to statistics. He preferred pithy, persuasive slogans. "Don't sell machines, sell results," he often reminded. "Emphasize applications, not hardware, the *why*, not the *how*."

Watson was born in 1874 in a small lumber-and-flour-mill town called Painted Post, an hour west of Binghamton. Keen in business, Watson sold sewing machines and pianos for a local family at country fairs. Then he joined National Cash Register, where, in 1911, he bellowed at a humdrum sales meeting: "The trouble with everyone of us is that we don't think enough." From that outburst came his soon-to-be-famous, one-word maxim: THINK. It defined not only Watson's identity but also the persona of his future megacorporation. This particular brand of simplicity and persuasiveness would one day make him America's Number One Salesman.

After he was fired from National Cash Register, Watson joined a holding company called the Computing-Tabulating-Recording Company in Endicott. A decade later, Watson renamed the company International Business Machines. IBM was first formed as an amalgamation of four firms, including the Bundy Manufacturing Company, renowned for its timekeeping devices. In 1892, one publication commended the Bundy time recorders: "No errors can be made in booking time, no disputes can occur as to accuracy, no jealousy is possible between time keepers and employees, neither can there be collusion between them." By the 1920s, time recorders became integral to the widespread efficiency movement of the era. Their underlying principles later informed IBM's "largest bookkeeping job" for the Social Security Administration. IBM's tabulating machines mutated into accounting machines.

IBM and its model of precision intelligence became a mindset. With the motto THINK, "I mean take everything into consideration. I refuse to make the sign more specific," Watson once explained. "If a man just sees THINK, he'll find out what I mean." The word was printed in giant block letters on the factory walls of IBM, emblazoned on coffee mugs, calendars, and concrete, and at one point, spelled out with red tulips on the company lawn. The steps leading to IBM Endicott welcomed visitors with the words "Think. Observe. Discuss. Listen. Read." IBM had its corporate songs and symphony orchestra. In one verse, choruses joined:

> *Our products now are known in every zone,*
> *Our reputation sparkles like a gem!*
> *We've fought our way through—and new*
> *Fields we're sure to conquer too*
> *For the EVER ONWARD IBM.*

Watson's approach had undeniably religious tones. Company meetings in later years opened with a prayer delivered by an IBM pastor. People who didn't buy into Watson's spiritual enthusiasm for work were let go. For Watson, working at IBM meant a commitment to intense loyalty, as his biographers put it in *The Lengthening Shadow*. "Loyalty is the great lubricant of life. It saves the wear and tear of making daily decisions as to what is best to do," the IBM newspaper mused. "The man who is loyal to his work is not wrung nor perplexed by doubts, he sticks to the ship, and if the ship founders he goes down like a hero with colors flying at the masthead and the band playing." IBM emulated EJ's family spirit to boost employee loyalty and built a country club. In 1946, IBM helped found a college that became part of Binghamton University.

Like EJ Shoes, IBM believed itself to be the employee's friend, even if this friendship was one of unusual control and formality. Watson was always dressed in dark suits, starched white shirts with stiff collars, and striped ties, no matter the temperature. "He always shaved twice a day and bathed and changed his shirt at least as often," his biographers wrote. And like George Johnson, Watson loved the pomp of a good ceremony, chorus, and celebration. His staff, in return, displayed excessive esteem, once arranging an impromptu dinner for 2,000 employees. This "spontaneous tribute" included a colossal cake with 50 pounds of frosting. In the early 1950s, for Watson's 40th anniversary with IBM, the corporation spent over a half-million dollars in pageantry, "giving three hundred and fifty tribute dinners in fifty-seven countries, with an attendance of well over fifty thousand." Such extravagances ended when IBM entered a new era under the leadership of Watson's son. With a more relaxed demeanor, Tom Watson Jr. was more of "a conservative rebel" who said IBM "needed wild ducks, not tame ones." And unlike his father, the younger Watson believed that loyalty entailed "people caring about the company because it was their company, and the company caring about them."

BECAUSE OF HIS FAMILIARITY with EJ and IBM as a Binghamton resident, Ed Link also believed that loyalty to his company was "one of the finest experiences" of his life. "We don't have to simulate cooperation," a 1960s Link company newspaper ad said. "In Broome County we have the real thing." In the late 1940s, a Link employee visiting England for business recounted his experience. "Many English workers have never seen their managing directors," he wrote. But this lineage of Binghamton leaders, including Johnson, Watson, and Ed Link, made it a personal priority to be attentive to their employees and communities.

"Many employers like EJ, IBM, and Link enriched the community with their cradle-to-grave benevolence," said Tom Kelly, the retired business school dean and vice president of Binghamton University. Kelly suggested that before corporate social responsibility became trendy in business schools, Johnson and Watson pioneered its earliest forms in Binghamton.

"When the nation got pneumonia in the 1930s depression, Binghamton got only a slight cold," Kelly said. By the end of the

1960s, the Binghamton region was home to some 200 firms, from the world's largest furniture producer to the oldest photographic supplier, as well as suppliers of clothing and cosmetics. Subsequent decades saw defense contractors moving into the area, and these strong trends continued into the 1980s.

On a crisp September afternoon in 1984, President Ronald Reagan visited Binghamton. He celebrated the "Valley of Opportunity," which represented, in many ways, America's story. "The computer revolution that so many of you helped to start promises to change life on Earth more profoundly than the Industrial Revolution of a century ago," Reagan said to a cheering crowd in an outdoor school stadium. His dream for America was "to see the kind of success stories in this valley multiply a million times over," he told the thousands gathered. "When those immigrants came to our shores and said, 'Which way EJ?' they were asking which way opportunity, which way peace, which way freedom."

In the decades that followed, Binghamton's experience belied Reagan's optimism. The region saw a tremendous population outflow and some of New York State's lowest economic growth rates. Jobs moved offshore, layoffs followed, factory buildings lay vacant, and property values plummeted. Once an industry capital and a role model for community building, Binghamton has faced stiff economic headwinds. In a 2012 well-being survey, it ranked second among the most overweight cities nationwide.

Baby boomers imagined working for one company their entire life with the expectation of a comfortable retirement. "They were confident that their children, my generation, could go even further," writes Kristina Wilcox in her study of the factory town. "Many saw these dreams, expectations—even 'givens' crumble before their very eyes."

IBM had 1,300 workers in 1914, then grew almost 10-fold by 1960. By 2008, the number of IBM employees grew globally by 30-fold. But all the while, the Endicott plant was being downsized. By 2014, the once-prominent buildings of the community were abandoned. The IBM Country Club was deserted; it became a victim of vandalism and was described as one of the area's "biggest eyesores." Universities and hospitals have replaced IBM as anchoring institutions.

Kelly suggests a cycle or pattern that has existed since Binghamton's river hub made colonial trading possible. "There was a kind of inspiration and inventiveness that enabled Binghamton's progress," Kelly said. "That led to new types of industries, new types of innovations, and new prosperities." As pressures of globalization adversely affect many American industry towns, Binghamton's opportunity, Kelly said, is to do what Ed Link did industriously well—routinely inventing and reinventing himself. One lesson from Binghamton's illustrious and ill-fated history is clear: prosperity is not permanent. "Remember: the past won't fit into memory without something left over," the poet Joseph Brodsky wrote; "it must have a future."

Not long ago, about 90 former employees of EJ got together. Some traveled long distances to Endicott and brought memorabilia from the company. "We kind of grew up together," one attendee, who worked for the company for 26 years, told the *Binghamton Press & Sun-Bulletin*. "It sounds trite, but it really was like family."

TOM KELLY AND I met during a busy lunch hour at the Lost Dog Café, a vibrant eatery in downtown Binghamton housed in a former cigar factory. Our view included the historic bluestone Victorian Gothic church, which rose into the bright June sky. Kelly recalled his own life experiences with simulation growing up in northeastern Pennsylvania. During his school years, students performed air-raid drills and practiced duck-and-cover in case of a foreign attack. The Binghamton region had been one of the prime targets in the United States during World War II and the Cold War, with all the espionage work underway, especially IBM's work for NASA.

In his recollections of Binghamton, Kelly lamented that companies don't make the kinds of long-term investments into the community that they once did. Even latter-day GE would be affectionately called "Generous Electric" for its investment in employee development in Binghamton. "Corporations are renting the talents of young people and putting them to work for short periods," he said. "I think it's a problem. We'll have to develop new kinds of lifetime learning and corporations enabling that." The industry loyalty culture that profited Binghamton now seems a vintage idea. Job-hopping and the gig economy have given rise to quicksilver career development, and community development has become less of a priority for employers and, perhaps, employees. Some may wish to invest themselves in the community but can't because of the existing job market. And some of these trends may point to a generational issue, or maybe not, but fickleness has displaced fidelity.

A block away from the Lost Dog Café is a one-time movie theater promoted as "absolutely fireproof," now converted to a brewery. Across from it is an old department store. Kelly and I walked into the adjacent dreary parking garage—concrete-gray and dusty, as it might have been at the Link Piano and Organ Company, which once stood on this location. Dulled wall art, a neon green background with fluorescent, *Matrix*-type font, read: "Welcome to the Birthplace of Virtual Reality." We walked down the ramp to the lower level and saw a fading mural on a corner wall behind oil-stained parking spots. This was where clocks became business machines, and the pianos reached for the clouds. Big block letters read: "On this site, Ed Link invented the flight simulator which transformed how pilots learn to fly 1929–1934."

We stood in silence. Between the piano factory and the parking lot, it felt like *The Twilight Zone*. Or, more precisely, "a dimension of sound, a dimension of sight, a dimension of mind," as the Binghamton boy Rod Serling usually opened his blockbuster series with the iconic four-note guitar riff. "You're moving into a land of both shadow and substance, of things and ideas." The garage ceiling was fractured, and the ribbed concrete bars were exposed and rusted. The floor was grimy.

I DROVE OFF THE parking ramp onto the asphalt quilt of Water Street. An Art Deco bridge relayed me across the Chenango River. "$Cash Paid$," a billboard read, soliciting Victorian furniture, old signs and toys, soda makers, pedal cars, pottery, and glassware. Then I drove down Clinton Street, the Antique Row. In the 1940s, this street buzzed with stores and bars, as writer Ronald Capalaces has recounted. Conversations in Slovak, Polish, Russian, Lithuanian, and Yiddish blending together with Italian filled one's ears. Restaurants enticed customers with hot pies, spiedies, and banana splits and shakes. While waiting for their prescriptions, ladies in the drugstore sprayed Evening in Paris perfume samples. The adjacent EJ shoe showroom had a peep-down fluoroscope to look at the foot—the sales clerk threw out the kids who were naughty with it. A nearby theater showed silent matinees accompanied by the music of a Link pipe organ. And during the city-ordered blackouts during World War II, people "walked up and down sidewalks in the pitch dark with cigarettes hanging from their lips," Capalaces has written. "It looked like fireflies glowing in the night and the cold of winter with snow covering the ground, the dancing dots of burning cigarettes made it seem like summer in winter. It was magical."

Today, the Antique Row has become antique. As if on an immersive simulator ride into the past, I took a deep breath of the time that no longer existed. The street is punctuated with onion-domed churches, funeral homes, coin-operated laundries, and a hip vegan joint. The old encountered the new, or as photographer Berenice Abbott said, "the past jostling the present." One can still spot the four square EJ company-built homes across the west side of Binghamton. Making my way through tree-lined streets, I reached Binghamton's Recreation Park, or simply the Rec Park, a few blocks from where Ed Link lived most of his years. A couple were walking their dog on the soft green grass with shade from the oaks and maples. A father and son played baseball in the adjacent field as a student idly strummed his guitar on the park bench. A teenager skateboarded in the parking lot.

The park's centerpiece is a 1925 Allan Herschell carousel in a large cupola. George F. Johnson donated the machine as free entertainment. The carousels were built by immigrants—ornate, naturalistic scenes and a "playful storybook" quality, as Herschell preferred. During its prime, the carousel operated 2,000 rides on any summer day. Although now painted with scenes from *The Twilight Zone*, this merry-go-round still maintains its 60 original hand-carved horses galloping to the original Wurlitzer military band organ.

I sat on a gallant horse with its head up and flowing mane. The operator rang the bell. And slowly, the carousel spun to a tune from 1926, "Take in the Sun, Hang out the Moon." This carousel was a system of gears, motors, and hand-carved dreams and delight. While technological revolutions of many kinds were happening in Binghamton, the carousel was also a revolution, both lilting and literal.

The operator rang the bell. One journey finished, and the next began.

The carousel whirled again for the Beatles' "Here Comes the Sun."

I was set free.

ABOUT FOUR YEARS LATER. Interstate 81 North, crossing into New York's southern tier from Pennsylvania, combines Endless Mountains with endless maintenance. The industrial park outside Binghamton once housed the wartime Link Aviation Devices and its future owners. The Court Street corridor to downtown is lined with architectural works of the master builder Isaac Gale Perry from the mid-to-late 19th century: the New York Inebriate Asylum or "Castle on the Hill," the Phelps Mansion, the Broome

County Court House, and the Perry Block, with a distinctive cast-iron structure.

Near the aerodynamic Art Deco bus station rises an imposing carved marble building. It's named after Sylvester Andral Kilmer, a bald, flamboyant, gunslinger-mustached physician who developed the famous Swamp Root formula, an alcohol-rich herbal concoction of the late 1800s. The brew claimed remarkable cures for human ailments until the 1906 Pure Food and Drug Act curtailed its sales. A comb manufacturer in the 1890s excelled with efficient automation a few blocks from the Kilmer building. In 1904, they sold a record 17 million combs made from elk horns and cow bones at 10 cents apiece before Bakelite and celluloid materials won the day. Now the former factory is an electrical substation bordered by empty buildings. The 1900-built Lackawanna Train Station, almost lost to neglect, has been revived and is home to small businesses and the last remaining of four radio towers. The Nobel-winning engineer Guglielmo Marconi erected this 97-foot steel structure for telegraphy in 1913. He successfully tested wireless communication to trains moving at 60 miles per hour between Binghamton and Scranton.

Tom Kelly and I met for lunch after the lean years of the Covid pandemic. It was a cold-swept January day with a wintry mix ready to rush in. "How quickly things change!" Kelly said at the Little Venice Italian restaurant, with Armando Dellasanta's impressionism and "Funiculì, Funiculà" in the background. "The Binghamton area has once again reinvented itself." Kelly was referring to the improving city economy and the rising cultural diversity of students and young professionals in the area. Binghamton University's downtown incubator has helped spur new technology businesses. The state and federal governments have invested significantly in a new, sizable lithium-ion battery manufacturing plant, where some of the original IBM buildings once stood, to create thousands of jobs. A new three-mile walking and biking trail stretched between downtown Binghamton and the university, and the local mall was turned into a "commons" with a sporting megastore and multi-purpose venue. And that dreary garage near the department store, once home to the Link Piano Company? It was razed, paving the way for a commercial complex.

I went to the Gorgeous Washington Street, where I lived as a graduate student in an apartment above a tattoo and piercing studio. The building next to it, once the historic headquarters of the Votes for Women Club, which led the local suffrage movement, is now a Chinese take-out joint. Past more restaurants, student housing, and the city's entertainment arena is the university's downtown center. The building's archaeological exhibit, *Our Invisible Past*, takes one to a different time. "Daily life is not usually the subject of written history, which tends to focus on significant events and prominent people," the description read. "However, everyone contributes in some way to a community's history." Binghamton is no single thing; it's simultaneously the bedrock, the people and the buildings that have come and gone, what it is now, and what it might become. The historic lenticular truss bridge a couple of blocks south, spanning the Susquehanna, was slick with frost. The South Mountain and Ingraham Hills appeared monochrome, starkly serene. I stood at Confluence Park, imagining the first communities that settled here. The Chenango splashed with the Susquehanna, quivering in the embrace.

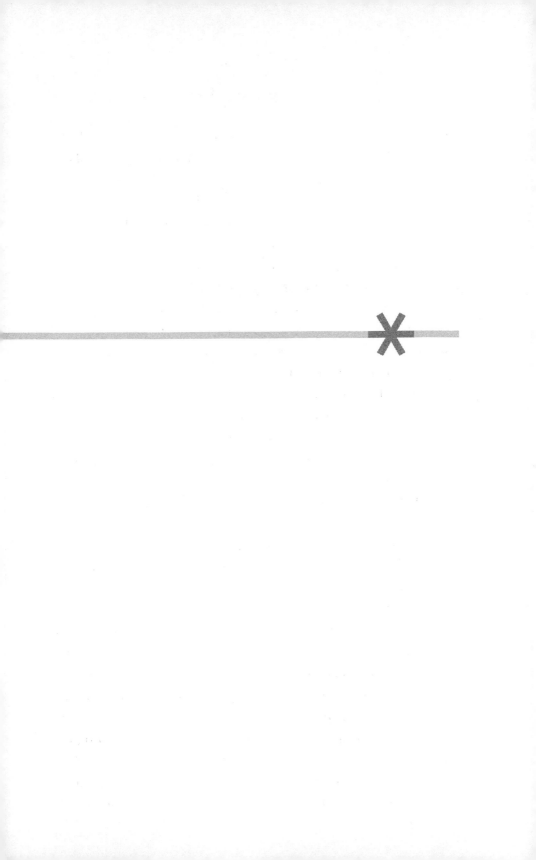

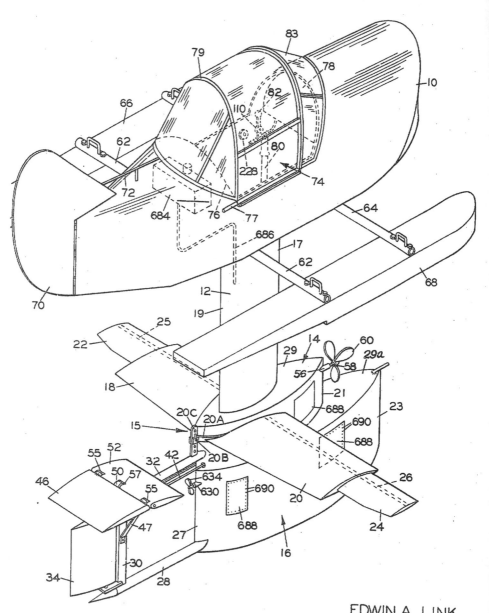

FIG. I

EDWIN A. LINK
INVENTOR.

BY *Donald F. Heller*
Philip L. Hopkins
ATTORNEYS

Civicware

The descent beckons
as the ascent beckoned

—*William Carlos Williams*

The many fictional inner journeys available to us—those that
unfold in imaginary places—also come equipped with maps.

—*Margaret Atwood*

Let's start from the beginning again, Jeff. Tell me everything
you saw—and what you think it means.

—*from Alfred Hitchcock's* Rear Window

T HE EDWIN A. LINK FIELD, A FORMER COW PASTURE ON
a Scottish settlement, is home to the Greater Binghamton Airport. With the heyday hurly-burly long gone, the airport is a small
operation. The parking lot was vacant and grim in late autumn as
distant lightning zigzagged. Muffled thunderclaps brought a brief
rain shower. Airline counters were empty, with no humans in sight.
The flight schedule on the electronic display showed two round-
trip services a day. Automated security announcements overhead
blended with an FM pop mix.

An easy-to-miss corner houses a display of Ed Link's Pilot
Maker. After a lifetime of hectic hissing and huffing, it now stood
with an air of reassurance, like a commanding bumblebee. The

wingless wonder didn't describe knowledge; it deftly transferred it, committed to the idea that no single interpretation of the problem will suffice. And under the hood was an intimate grandeur of darkness that unified the real and the plausibly real.

If Karl Popper formalized a distinction between clocks and clouds, Ed Link nimbly coupled them without leaving the ground, one forming and informing the other. The Link Trainer work was far ahead of today's deep-learning devices. Yet, as a "depth technology," Ed Link's blue box recognized multiple dimensions related to one another. Perceptions become renditions only through relationships. As scholar Judith Roof has noted, "It always takes at least two—two eyes, two objects, two fields, two images."

Modern headsets and goggles now connect us with immersive worlds that span everything from concerts to competitive gaming. Even our politics operates in virtual worlds, concerned more with perceptions than facts. "It exists for only the fleeting historical moment, in a magical movie of sorts, a never-ending and infinitely revisable docudrama," one observer put it. "Strangely, the faithful understand that the movie is not true—yet also maintain that it is the only truth that really matters." In Ed Link's engineering, though, virtual reality conveyed an operational mentality, a code of conduct. Nothing looked more commonplace than this minimalist masterwork. It's a cage to liberate us from our cognitive cages.

✳

ENCOUNTERING UTAH'S DELICATE ARCH, Edward Abbey wrote what could very well apply to the nature of wickedness: "Suit yourself. You may see a symbol, a sign, a fact, a thing without meaning or a meaning which includes all things." Our modern lives are such that driving on a freeway without GPS navigation can seem a wicked problem for some, as can spending a few hours without cell service or Wi-Fi. Many of us can't explain how electric kettles operate, let alone electric cars or electric grids. Nor can most describe the workings of dishwashers and the delete function on the keyboard, or how the neuromuscular system throws a ball. We seem comfortable with certain forms of ignorance. Yet the wicked complexities of climate change, construction of better cities, population-health policies, natural resource management, and work cultures demand more from us. As the world becomes more segmented and specialized, expecting informed citizens to keep up with details seems unreasonable. Even engineers looking at the same data reach different conclusions. New designs will supplant the old, always leaving us out of date. And no one can ever be an expert on all aspects of complex systems. Such colossal capabilities emerge only in the commons.

Our competencies are often limited by what we can conceive. This condition is called hypocognition, and it's pervasive. It differs from the tendency to unthinkingly apply familiar frameworks to issues even if those ideas are inept and irrelevant to the challenge. When Americans introduced ice blocks to Martinique in 1806, the locals couldn't comprehend the meaning of ice and let the unsold shipment melt. Rather like the vegetable merchant in the Indian parable who used eggplants to value the diamond, unfamiliar conditions test our cognition. Our bounded worldviews became apparent when clocks were first introduced. And again, when zero was invented, music was notated, clouds were classified, compound interest was calculated, dictionaries were developed, and portable music players were marketed. Making decisions based solely on what we know and expect to happen is a form of ignorance.

As presented in this book, flexible concepts of operations can allow us to conceive different futures and test different cognitive and cultural flight paths. With ConOps, community organizers with a single cause may consider issues beyond social justice—say, the efficiency and safety of their actions. Corporate managers could begin to explore the potential impact of short-term profits on the broader vulnerabilities and potential resilience within their business. And city and county planners can test their assumptions with inherent vagueness and indispensable maintenance in their budgets. Combining efficiency, vagueness, vulnerability, safety, maintenance, and resilience may appear forbidding. But we also saw how maintenance is intertwined with safety and vulnerability, ultimately affecting efficiency and resilience. It's inevitable that when these six concepts are put into practice and refined, additional considerations will present themselves. As with Ed Link's original imagination, it's prudent to start with a few comprehensive concepts rather than a confusing vastness.

Imagine using systems engineering concepts to engage with elected representatives in town halls. Or to guide federal budget discussions or national climate targets and business strategies. And how about approval processes for medical and consumer products or funding taxpayer-supported research programs? Any policy decision is inadequate without a healthy concept of operations. We risk disasters when a messy resilience problem, or even a messy efficiency problem, is treated as a hard efficiency problem. Wickedness, as discussed, can emerge not only at the interface of hard vagueness, soft vagueness, and messy vagueness but also from the interaction of hard efficiency, soft safety, and messy maintenance. There's an efficiency argument for maintenance, a resilience consideration in safety, and vulnerabilities lurking within vagueness. And similarly, maintenance and safety contribute to efficiency, just as their deficiency can affect vagueness, vulnerability, and resilience. Yet again, the tendency to excessively emphasize hardness—and related problem-solving—can fester into softness, messiness, and, ultimately, wickedness. We'd be better off confronting such complexity up front, as with the calculated torture from the Link Trainer, than dealing with the costs and complications later.

Many constraints and trade-offs complicate each civic scenario, as they do in engineering. Interacting with wicked problems without the essential elements of cooperation and competition, and constraints and trade-offs, is like flying an aircraft without instruments. Above all, civics is a multiplayer game. Just as engineering transcends the thrills of technology, civics is more than civility. We can cultivate a broad—and broad-minded—civic understanding of common conditions by approaching a problem through the six general concepts. Their grouping can enable multiple views, viewings, and viewers, allowing fixed and free-form approaches to guide policy requirements. Building such a "civicware" will take time—as it should. Asking for patience might sound preposterous in a time-starved, screen-obsessed, distraction-dosed, thumbs-up-thumbs-down culture. But consider the evolution of our earliest tools from the Lower Paleolithic. To advance from sharp stone tools shaped from river cobbles to the teardrop hand axes of the Late Acheulean took two million years. Prehistoric stonework required skills and standards while slowly evolving our motor skills for perception.

The immersive process of reworking, reinterpreting, and reflecting on wicked problems in the modes of a Link Trainer can be dizzying, triggering "simulator sickness." Still, helpful disorientation is part of this training. Such disturbances emanate from the problem conditions rather than the systems engineering concepts themselves. Pilots cannot overcome the complexity of weather, but their training allows them to approach it more carefully and cogently. The key is to balance the relationships among the six concepts, like the forces operating on an aircraft. Merging a flight-training mentality with a concept of operations for civic engagement may not be quantifiable. Indeed, many of our quantification activities have no connection to reality. Even the notions of time and money critical to social transactions are meaningful only to humans. Such metrics mean nothing to viruses and volcanoes.

Despite our faith in numbers, we can't simply substitute an aircraft's altitude for speed, direction, and fuel readings. Worse, crushing and compacting those dimensions into a single metric would be suicidal in flight. Yet, we fly our economies and enterprises as if they are single-point measures: GDP, SAT scores, cost-benefit ratios, scholarly citations, unit sales, and so forth. Public policy and business planning contain separate and simultaneous goals. Our survival depends on collectively making sense of these indicators like pilots interpreting a flight dashboard. The world's problems can and do appear hopelessly wicked. Still, we all share the burden of responsible engineering. Interesting territories can come into view once we adopt a systems engineering approach: to consider multiple complex systems at work and how they overlap and intersect, to make familiar problems fresh and alive to new assumptions, and to explore and discover civic vigor and variety.

✳

EVERYTHING IN OUR LIVES has engineering, but engineering isn't everything. In this book, we saw how excess rationality can misdirect engineering to mindless, mechanistic goals. Consider the anxiety the words "social engineering" ignites. The idea implies sinister acts and setbacks to progress. Comparatively, look at other activities with the prefix "social." "Social media" connect people, "social innovation" empowers strong civil society, "social capital" fosters healthy communities, "social entrepreneurship" brings positive change, "social businesses" win Nobel Prizes, "social physics" shows how ideas spread, "social justice" strives for equal rights and opportunities, "social equity" promotes justice and fairness, "social costs" propose the sacrifices a society should be willing to make, and "social sciences" should be prioritized. If the term "social engineering" generates angst about potential harm, then engineering that's asocial—and not attuned to its consequences—should be recognized as far more pernicious.

Two decades before presenting on clocks and clouds, Karl Popper wrote about "piecemeal" social engineering. He argued for open-ended reforms over utopian blueprints. A piecemeal approach is evolutionary and begins by realizing that facts are fallible and contexts change. Yet, such increments require caution. Piecemeal responses can cancel one another out when not coordinated by an overarching principle or guided by a standard set of concepts. And obviously, you cannot optimize a system by optimizing its parts separately. Because wicked systems cannot be planned from the top down, they require an evolutionary approach to selecting and replicating improvements to civic welfare. The concept set of efficiency, vagueness, vulnerability, safety, maintenance, and resilience can facilitate such conscious cultural evolution.

A metaphor from Greek mythology reminds engineers of their close contact with reality. Libyan giant Antaeus enjoyed strength so long as his feet touched the earth; it vanished when he was up in the air. Engineering, much like civic engagement, is a mash-up of metrics and markets, the municipal and the moral. Yet, consider how young people are usually introduced to kinds of engineering promoted as "changing the world" through seductive "tech" and reductive "innovations." Equating technology to engineering is like claiming that a spreadsheet is an economy. And again, emphasizing problem-solving as the supreme professional identity evades the role engineers play in producing such problems in the first place and not acting to correct the long shadows of these failures.

Even as our environments become increasingly human-made, the natural parts are no less critical. Writing about this in 1974, Russell Ackoff was concerned about environmental degradation reaching a point beyond which it is irreversible. But "there's little agreement, however, on what should be done to reverse deterioration of environment, who should do it, and how it should be financed," a problem that persists today. Ackoff observed that failing to account for complex interactions isn't a sin exclusively committed by politicians; even scientists and engineers who should know better do so embarrassingly and recurrently. "They too are inclined to look at immediate effects and not consider later consequences," he lamented. As a cultural practice, engineering has long been a source of market power. Now is an opportunity for engineers to grow and guide more responsible forms of pronatural and prosocial uses.

But the deep structures of engineering education and the motivations of engineering enterprises inhibit such contemplation. Social issues are still set aside as inaccessible topics in many engineering firms. At the core of the engineering enterprise lies a contradiction, notes scholar Darshan Karwat. "Engineers in Texas figure out how to extract ever-harder-to-reach fossil fuels, while engineers in Florida develop urban adaptations to rising sea levels caused by global warming. Engineers for gun manufacturers refine assault weapons, while other engineers devise medical instruments to treat gunshot wounds. Engineers design addictive social media apps, while others design apps to manage that addiction," Karwat notes. Engineers are hardly conversant on such issues because market motives often control them. Worse, many engineers reflexively shirk moral complexities, often in training or work environments. It's a state of affairs reminiscent of revenge Westerns with stereotypical set pieces: Hollywood heroes in wide-brimmed Stetsons lassoing and lashing villains.

Frequently, engineers feel that they are servants of capital and hired guns for industrial interests. When clients demand swift solutions and a corporation is held hostage to market forces, engineers tend not to prioritize the input of the communities that their projects will impact. In this working reality, engineering teams have fostered a template of so-called projectification. Whether it was Noah's Ark or the Newark International Airport, the term "project" can be all-encompassing from an engineering sense. At least one-fifth of global economic activity now occurs through projects. In China, projects approach one-half of their gross domestic product. In India and some advanced economies, one-third of their economic activities are project work. The concept of a "professional service company" is project based, as is much of the freelancing in the gig economy. Engineering firms outsource billable work and compete on contracts and subcontracts. The bulk of engineering now relates to cost centers, daily huddles, and weekly reports to manage workflows, milestones, and action items. As one project ends and another begins, broader reflections on the human conditions these projects aim to address can fizzle in the face of fixed targets.

Now add the processes engineers must comply with: one can lose sight of the overall system in a maze of color-coded charts, road-maps, interacting loops, intersecting circles, influence diagrams, solid lines, dotted lines, and acronyms. One current standard that guides architectural approaches requires engineers to address 6 processes, 45 "required activities," and 416 "recommended tasks." Systems engineers also employ design heuristics, 600 of them by recent counts. And such mental shorthands can lead to a growing list of cognitive violations and biases. Consider "intertemporal choice," through which failure is feared more in the short term than over time, or the "narrative fallacy," the tendency to fit data to a marketable story. Because formal processes might bake in such biases, designers need to consider how to avoid them explicitly.

Engineering education has long been trapped between culture change and upholding the innate "rigor" that is the currency of professional recognition. All these matters have further separated engineering from public concerns. Scholar Carl Mitcham has argued that engineering "does not provide its own justification for transforming the world, except at the unthinking bottom-line level, or much guidance for what kind of world we should design and construct. We wouldn't think of allowing our legislators to make laws without our involvement and consent; why are we so complacent about the arguably much more powerful process of technical legislation?" The idea of bracketing out cultural, ethical, social, and environmental issues from engineering has become a creed that can—and does—silence conversations about equity and inclusion in professional settings.

Researcher Erin Cech has argued that engineering training often underemphasizes civics in professional consciousness, including community engagement, racial understanding, and peacebuilding. Cech has found that such public welfare considerations among engineering students in some leading American universities significantly declined throughout their education, extending into their early employment years. A common regret about such civic disengagement is that developers with technical tunnel vision may frame "any 'non-technical' concerns such as public welfare as irrelevant to 'real' engineering work." Indeed, as a student, when Cech asked her engineering professors questions about access and inequality, she notes, "they not only didn't know the answers, but didn't always think the questions were relevant and pushed me to pursue other fields, particularly sociology."

Broadly, though, Cech's work positions civic disengagement in engineering education as a depoliticization that frames science and technology as "pure" spaces. Hence the still prevalent idea that political and social concerns in engineering are not merely optional but omittable. Indeed, the decisions about what's to be studied, designed, scaled, delivered, and financed are inherently political choices. "The electrons don't whisper to you what about them you should study," Cech says. "These are decisions we make in part through the social power hierarchies we are embedded in." As more supposedly nontechnical concerns are expunged, existing power structures in engineering are silently strengthened. As Cech puts it: "If depoliticization means that it is illegitimate to talk about inequality issues, except within explicitly defined diversity and inclusion seminars, how are we to make any headway?"

A chief symptom of this problem is the deep belief in a "technical optimum" that's detached from public debates—as we saw with the idea of hard efficiency. Indeed, Big Tech firms have been criticized for their pride in precision and profits, seemingly rational but seriously risky. The Nobel-winning Friedrich von Hayek described this impulse as "the religion of engineers." A drive to produce quick fixes can promote a social order that engineers cannot "detect by the means with which they are familiar." Social media are the latest case. Technologies that once promised to unite the world have instead deeply divided it with algorithmically amplified distrust and disgust. The result is "demosclerosis," in which chronic greed and gridlock choke our political arteries and prevent their ability to adapt. Our click patterns on tech platforms have become a proxy for civic practices. And profit maximizing has fueled fact-free factions of secrecy, shitposting, and slurs as daily forms of online "civic engagement."

What does this all have to do with wickedness? As we have seen throughout this book, wicked systems don't present with clean-cut issues and expect answers under a deadline with a trophy to follow. Prosocial forms of engineering involve solutions, resolutions, and dissolutions of often unoptimizable, as-yet-undefined wicked problems. Combining the concepts of efficiency, vagueness, vulnerability, safety, maintenance, and resilience can give us a start. Engineering focused on rigor and technical purity but shorn of reflective practice and disengaged from civic welfare cannot be resilient and ultimately effective against wickedness. After all, standards must be higher for a profession whose results are perhaps far more influential than democracy in everyday life. To engineer civicware is to first be civic aware.

✳

"GREAT EVENTS ARE PERHAPS so only for small minds," the French poet and critic Paul Valéry wrote. "For more attentive minds, it is the unnoticed, continual events that count." Special events like the Wright brothers' first flight or the Apollo 11 landing are celebrated in the history of flying machines. They are headline episodes in the annals of aviation. However, the engineering revolution that Ed Link fomented in relative obscurity deserves greater appreciation. The Link Trainer democratized an activity previously available only to an adventurous few. It enabled broader participation for women in an arena defined by men's interests, giving those women vital roles of instruction, obscured yet integral.

After Kitty Hawk, the Wright brothers rose to an exalted status. Some characterized them as "an instrument in the holy hands of God." They transformed the world by wresting monopoly over the skies from birds. But without leaving the ground, Ed Link was equally transformative in imitating that reality. His work remains a quiet triumph in a world of thunderous takeoffs and show-offs. The point is not to create arbitrary fame contests between the first practical flight and the first practical flight trainer. There's hardly any controversy over who engineered what, when, where, and how. But a more profound issue lurks: how society assigns fame, manufactures eminence, propagates popularity, and ultimately judges one individual over another. "Due recognition is not just a courtesy we owe people. It is a vital human need," writes scholar Charles Taylor, describing how our identity is partly shaped or misshaped by recognition, with its presence and absence. In a relentless ratings-and-rankings culture, someone will always get more publicity than someone else, whether worthy or not. Economists often invoke moral hazard, a product of individuals behaving differently if their risks are protected. For example, insurance coverage might encourage reckless driving. Likewise, in civics, far more consequential is *laurel* hazard. In competitive fame-and-fortune seeking, we discount the dull, diligent, dutiful, and deserving.

The synthetic capabilities enabled by the Link Trainer now underpin the entire aviation enterprise. Today, they have democratized access to education, combining psychology, sociology, and technology to produce knowledgeable and skillful crews in many areas. Just consider the collective cost of World War II without flight trainers, the sheer feasibility of the original moonshot, the viability of the modern airline business, and even the global commercial and cultural connectivity. If the Wright Flyer was an extraordinary first, the Link Trainer was an even better first. Flying a prototype a few times to succeed in a 12-second demo project and making it last 59 seconds doesn't need systems engineering. Still, at scale, systems engineering can prevent costly consequences when many lives are at stake. The first flights were an experiment, but the Link Trainer cultivated responsibility. As Wilbur Wright himself put it, one can fly without motors but not without knowledge and skill. The trainers afforded far more nuanced instructions on performing in the total environment than simply flying by the seat of your pants. The blue boxes and the instructors who wielded them provided critical systems engineering support for what had been missing in aviation practice. Like the Wright brothers, Ed Link had no academic bona fides. He was drawn to challenges throughout his lifetime, observed Ralph Flexman, an aviation engineer who knew Ed Link for three decades. "While he did not go about looking for problems, when he came upon a situation or condition that produced unacceptable consequences, it always attracted his attention," Flexman noted. "Ed never worried about the solution to a problem until he thoroughly understood the problem in all its dimensions."

As noted earlier, standard technical engineering practice may not always allow for a deeper understanding of problems and reflection on the consequences of specific solutions. But that's precisely the starting point for systems engineering. Amid the calls for multidisciplinary, interdisciplinary, crossdisciplinary, and trans-disciplinary work, sound systems engineering exceeds disciplines and, crucially, requires discipline. Such considerations abound in Ed Link's work. He engineered unusual and initially unintuitive devices for unique environments without any formal education in those areas. "By missing a university education, I failed to learn my own limitations," he once said. "It made it possible for me to do things that had never been done before." His brilliance was in liberating himself from specialization that allowed him to transfer concepts from musical space to aerospace and then to hydrospace, and finally return to where he started.

"Ed Link is the most persistent son of a gun I ever met," Art "Silver Bar" McKee, a noted treasure diver, put it in 1957. "Look at this boat," he said. "On the way here the radar breaks. Ed flies a man in from Miami and then does 75% of the work himself. Out of Nassau an icebox line broke, so Ed repairs the whole icebox. At Great Inagua a bearing or something in the automatic pilot froze. Ed goes below and turns it down on the lathe. He made a new belt for the air conditioning. He made parts for the water pump and the main generator." Ed Link's direct experiences in hostile environments underpinned his engineering accomplishments. The resulting designs defied distances and depths and recognized the overlooked relationships among technology, behavior, and policy early on. Civics, too, is a matter of building competencies in multiple mediums, with its versions of cloudy skies and vicious waves. Our civic competencies, capabilities, and character depend on how we engage with and engineer our wicked problems using the concepts of operations. And vitally, only through civics can we improve the recognition of people, professions, and places that our world depends on. Ed Link is an underknown protagonist in engineering. Similarly, long before Silicon Valley, there was the engineering hub of the Susquehanna Valley and, much earlier, the Indus Valley.

In Binghamton, Ed Link often blushed when people called him a genius. "No-o-o, no, I'm not," he told a reporter. "First, know what the problem is," Ed Link offered. "Anybody can do the same thing I do if they're willing to work at it and study it. . . . Most people don't want to spend that much time." On a different occasion, when asked how an invention works, Ed Link replied, "Define the problem." When the same journalist asked Ed Link about the age of specialization, he said he didn't regret it. Still, he warned against overspecialization that "can result in forgetting other people's problems." Ed Link didn't have a command of language—he spoke less in words and more with his eloquence in engineering. There was the testing of self in his approach, even stretching personal limits from one domain to another. Ed Link's systems design undeniably had a placid sense of simplicity, veracity, and gravity, applicable to solving, resolving, and dissolving hard, soft, and messy problems. "If you are going to build these things," he said, "you have to have the integrity to test them yourself."

✳

DOWN THE HALL FROM the Link Trainer in the Binghamton airport was a faint wall-mounted exhibit on the city's evolution "from the smokestacks to high-tech." It contained a copy of *E-J Worker's Review* from 1920, Thomas J. Watson Sr.'s THINK desk plaque, and three generations of IBM technologies from Endicott—the vacuum-tube, solid-state, and solid-logic circuits. A nearby display promoted spices and marinades, featuring spiedie sauce, a mainstay of the summer state fair and the city's hot-air balloon rally.

You'll find some offices and an observation room on the airport's second floor. The long linear concourse to the north end of the building went past the desolate departure lounge, a barren baggage claim, and an old marketing phone board for hotels and taxis. Pillar advertisements promoted the university, philharmonic, hospitals, casino resorts, and local restaurants. A stairwell led to a well-lit panoramic viewing deck studded with potted plants.

The dreamlike stillness in the room felt like the hermitage of a quiet genius. Ed Link's mural, 8 feet by 20, was the centerpiece. Across from it, a small memory collection featured fading pictures of player pianos, blue-box production, the Apollo program, the Perry-Link *Deep Diver*, the *Johnson-Sea-Link* submersible, and modern simulators. A long airport bench looked out the glass wall. The jet bridges were retracted; and the asphalt runways and taxiways, wet and illuminated. The weather vanes responded to the thick breeze as cumulus, nimbus, and stratus rapidly remixed in silent commotion.

For the Sanskrit poet Kālidāsa, the original Romantic centuries before Shelley, Coleridge, and Wordsworth, cloudy inexactness hinted at more than their expressions, perhaps a lover's longing. The departing weather front thinned out glittery gray cauliflowers; aimless cotton puffs were airborne against a salmon-hue backdrop. The twilight shades were sharp as the sun sank behind the hills. A clock ticked in the background as light retired to darkness. Everything seemed calm; nothing, constant.

EDWIN A. LINK
1904 — 1981

Acknowledgments

I have written a wicked book, and feel spotless as the
lamb. . . . It is a strange feeling—no hopefulness is in it,
no despair.

—*Herman Melville to Nathaniel Hawthorne, 1851*

T HIS BOOK BECAME WHAT IT DID WHILE I WAS PRETEND-
ing to write a different book over the past six years. Several
people supported me through this journey, and I owe them my six
degrees of gratitude.

Tango Hotel Alpha November Kilo Sierra to editor Brendan
Curry, the reliable copilot in my flights of fancy. Assistant editor
Caroline Adams brought her keen and caring sense to my high-g
manuscript maneuvers. She fastened the seat belt on bumpy sen-
tences and showed the exit sign for many more. Copyeditor Sarah
Johnson completed critical cross-checks, as Robert Byrne, Rebecca
Homiski, Will Scarlett, Steve Colca, Gabrielle Nugent, Lauren
Abbate, and others at W. W. Norton & Company created the ideal
cabin conditions for a straight-and-level journey. My agent Michelle
Tessler, ever diligent and reassuring at air-traffic control, ensured
the project was safe and orderly from before liftoff to after land-
ing. Udayan Mitra and colleagues at HarperCollins India, Andrew
Gordon at David Higham Associates, and the team at Andrew
Nurnberg Associates International provided valuable support for
this ultralong-haul flight.

It's my privilege and pleasure to work with gifted colleagues at
the National Academy of Engineering and collaborators across the
National Academies of Sciences, Engineering, and Medicine—an
incomparable free-trade zone of intellect. I thank John Anderson,
Al Romig, Don Winter, and Nadine Aubry for giving me vital lead-

ership lessons and latitude. My mentor and friend Norm Augustine is a fount of wit, wisdom, and worldly guidance: "To think outside the box, go ask someone who isn't in the box." Harvey Fineberg, Rita Colwell, Bruce Alberts, Paul Citron, Bill Rouse, George Poste, Richard Foster, Chuck Phelps, David Butler, Kevin Finneran, and Radka Nebesky are models of generosity with their time and counsel, as were Steve Merrill and M.S. Swaminathan.

Roger McCarthy smoke-tested, Zach Pirtle sanity-tested, and Jag Bhalla wind-tunnel-tested the drafts. Geeta Bhatt, Lindsay Ruel, Stephen Nichols, Tom Snitch, and Ryan Alimento bravely beta-tested the book. For early and enlightening explorations, I thank Rino Rappuoli, Alfred Spector, Eswaran Subrahmanian, Dan Sarewitz, Tracy Lieu, Vint Cerf, Asad Madni, David Edgerton, Don Franck, David Blockley, Georges Amar, Darryl Farber, Monty Alger, Susan Fitzpatrick, Michael Batty, Ethan Frizzell, Nahum Gershon, Patrick Godfrey, Jay Labov, Randy Nesse, Venky Narayanamurti, Ric Parker, Holly Prigerson, Hari Pulakkat, Hillary Sillitto, Daniel Sniezek, Chris Washburne, Tariq Durrani, Don Gause, Harry Boyte, Trygve Throntveit, Scott Peters, Rebecca Shamash, Timothy Simpson, Mike Johns, Bill Stead, Denis Cortese, Darcy Gentleman, Susan Blumenthal, Al Bunshaft, and Geoffrey West. And for early interests and readings: Andy Russell, Colleen Carow, Chinny Krishna, Farrell Fitzpatrick, Narayan Hariharan, Flora Ogilvie, Jeff Genung, Jennifer Clark, Nerur Ganesh, Nicole Skarke, Aparna Subramaniam, Varsha Subramaniam, Jocelyn Widmer, Rishaba Swaminathan, Kate Fletcher, Paul Kopalek, Hadassah Mativetsky, Sumeet Shetty, Andrew Shaner, and Janet Hunziker.

Binghamton—the community and the university—served as a knowledge base camp to unify my technical, business, and civic leanings. David Sloan Wilson and Tom Kelly have been inspiring "smart grids" of acumen and altruism since I was a graduate student. I thank Shelly Dionne, Doug Johnson, Andy Clark, trustees, and staff of the Link Foundation, and importantly the late Marilyn Link, for their encouragement. I applaud the vital service of Gerald

Smith, Brian Frey, Broome County Historical Society, Roberson Museum and Science Center, and TechWorks! in communicating Binghamton's illustrious history. My thanks to Frank Cardullo and other instructors of the flight-simulation course at Binghamton University organized in cooperation with the American Institute of Aeronautics and Astronautics: Olen Atkins, James Davis, Thomas Galloway, Valerie Gawron, David Gingras, Jeffrey Schroeder, James Takats, and Andreas Tolk. Beth Kilmarx provided significant guidance during my research in the Bartle Library archives, as did the special collections curator Diane Newman and the late oceanographer George Maul at the Florida Institute of Technology. I am grateful to Wanda Brogdon and Julianna Andrews for their extensive assistance at the Library of Congress.

Longtime friend Barb Oakley has been a considerate "first responder" to my ideas. Derrick Martin, Russell Harrison, Claudia Grossmann, Lauren Shern, Lauren Bartolozzi, Francis Amankwah, Darul West, Adam Winkleman, David Dierkesheide, Jeffrey Peake, Bruce Hecht, Roland Wright, Jim Hartel, Anil and Vijaya Sharma, Darlene Karamanos, Arthur Coucouvitis, Mohan Ramaswami, Margo Martin, Nong Louie, Swami Venkatraman, and Raja Subramaniam have endured and defragmented my thinking over the years.

And finally, my boundless gratitude is for my parents and family. I present this book in memory of my father-in-law Sreekrishna Ramaswami. A steady and sagacious guide, he encouraged and engaged with this book before and unlike anyone else did. My wife Ramya's trusted, talismanic presence through the world's treats and turbulences exemplifies philosopher Simone Weil's prudence: "Attention is the rarest and purest form of generosity." She is my life's Link Trainer.

And DJH, you remain my beacon.

Sources and Resources

Collections

The British Library, London, United Kingdom

Broome County Public Library, Binghamton, New York

Edwin A. and Marion Clayton Link Collections, Glenn G. Bartle
Library, Binghamton University, State University of New York,
Binghamton, New York

Edwin A. and Marion C. Link Special Collection, John H. Evans Library,
Florida Institute of Technology, Melbourne, Florida

Library of Congress, Washington, DC

Smithsonian National Air and Space Museum Library, Washington, DC

Notes

Preface: Clocks and Clouds

xiii **clocks and clouds:** Based on Karl Popper's "Of Clocks and Clouds: An Approach
to the Problem of Rationality and the Freedom of Man," in *Objective Knowledge:
An Evolutionary Approach* (Oxford University Press, 1979).

xiii **Shakespeare:** *Antony and Cleopatra*, 4.14.3–8.

xiii **"chief organ of sentiment":** From John Constable's 1821 letter; see Claude Barbre,
"Constable's Skies," *Journal of Religion and Health* 43, no. 4 (2004): 391.

xiii **"no two days are alike":** See Mary Jacobus, *Romantic Things: A Tree, a Rock, a
Cloud* (University of Chicago Press, 2012), 11.

xiii **"Look, I've not tamed":** Charles West Churchman, "Guest Editorial: Wicked
Problems," *Management Science* 14, no. 4 (1967): B141–42.

xiii **"spreadthink":** John Warfield, "Spreadthink: Explaining Ineffective Groups,"
Systems Research and Behavioral Research 12, no. 1 (1995): 5–14.

xiv **"the monster":** Norman R. Augustine, "Is Biomedical Research a Good Investment?," *Journal of Clinical Investigation* 124, no. 12 (2014): 5087–89; "Toward
an Engineering 3.0," *Bridge* 50, no. 4 (2020): 79–82.

xiv **"aerosol words":** A term from Mark Dodgson in the August 21, 2016, *Future
Tense* podcast with Antony Funnell for the Australian Broadcasting Corporation.

xv **more *pluribus* than *unum*:** Arthur M. Schlesinger Jr., "E Pluribus Unum?," in *The
Disuniting of America: Reflections on a Multicultural Society* (W. W. Norton,
1992), 125–48.

Prologue: Airplane Mode

1   **"I change, but I cannot die"**: From Percy Bysshe Shelley, "The Cloud" (1820), in *Shelley and His Poetry*, Edward William Edmunds (George G. Harrap & Company, 1911), 115.

1   **"There's no love in a carbon atom"**: Manlio De Domenico et al., *Complexity Explained* (2019).

1   **"Say, it's only a paper moon"**: Music by Harold Arlen, lyrics by Edgar Yipsel Harburg and Billy Rose (1932), popularized by Nat King Cole, Ella Fitzgerald, and Benny Goodman.

1   **50 British cadets**: Tom Killebrew, *The Royal Air Force in Texas: Training British Pilots in Terrell during World War II* (University of North Texas Press, 2003); Tom Killebrew, *The Royal Air Force in American Skies: The Seven British Flight Schools in the United States during World War II* (University of North Texas Press, 2015).

1   **blue book**: From *Notes for Your Guidance*, issued by the Royal Air Force, taken from Killebrew, *Training British Pilots*, 67; Davenport Steward, "As the English See Us," *Saturday Evening Post*, October 11, 1941.

2   **500 novice students**: Gilbert S. Guinn, "British Aircrew Training in the United States 1941–1945," *Air Power History* 42, no. 2 (1995): 11.

2   **President Franklin Roosevelt**: Mary Stuckey, *The Good Neighbor: Franklin D. Roosevelt and the Rhetoric of American Power* (Michigan State University Press, 2013).

2   **"arsenal of democracy"**: From Roosevelt's Fireside Chat, December 29, 1940, The American Presidency Project, University of California–Santa Barbara.

2   **Lend-Lease Act**: See R. G. D. Allen's "Mutual Aid between the U.S. and the British Empire, 1941–45," *Journal of the Royal Statistical Society* 109, no. 3 (1946): 243–77; Warren F. Kimball, *The Most Unsordid Act: Lend-Lease, 1939–1941* (Johns Hopkins University Press, 1969).

2   **coat of arms**: Guinn, "British Aircrew Training," 6.

3   **corn on the cob and grits**: Killebrew, *Training British Pilots*, 83.

3   **segregation**: Killebrew, 68–69; Guinn, "British Aircrew Training," 16.

3   **Supermarine Spitfire . . . and Hawker Typhoon**: Killebrew, 84.

3   **"dust bowl"**: Killebrew, 65.

4   **Stearman . . . "ground loop"**: Killebrew, 30–31.

4   **landed his Stearman**: Killebrew, 76.

4   **Calvin Coolidge**: Ellen Pawlikowski, "Surviving the Peace: Lessons Learned from the Aircraft Industry in the 1920s and 1930s" (master's thesis, National Defense University, 1994), 2.

4   **Ed Link's pilot trainer**: Edward Molloy and Ernest Walter Knott, *The Link Trainer* (Chemical Publishing Company/Doray Press, 1941).

6   **10 properties of wicked problems**: Horst Rittel and Melvin Webber, "Dilemmas in a General Theory of Planning," *Policy Sciences* 4, no. 2 (1973): 155–69. The properties are: (1) There is no definitive formulation of a wicked problem; (2) Wicked problems have no stopping rule; (3) Solutions to wicked problems are not true-or-false, but good-or-bad; (4) There is no immediate and no ultimate test of a solution to a wicked problem; (5) Every solution to a wicked problem is a "one-shot operation," because there is no opportunity to learn by trial and error; (6) Wicked problems do not have an enumerable (or an exhaustively describable) set

of potential solutions, nor is there a well-described set of permissible operations that may be incorporated into the plan; (7) Every wicked problem is essentially unique; (8) Every wicked problem can be considered to be a symptom of another problem; (9) The existence of a discrepancy representing a wicked problem can be explained in numerous ways; (10) The planner has no right to be wrong.

6   **"problem solvers":** See Thomas Siller, Gerry Johnson, and Russell Korte, "Broadening Engineering Identity: Moving beyond Problem Solving," in *Engineering and Philosophy: Reimagining Technology and Social Progress*, ed. Z. Pirtle, D. Tomblin, and G. Madhavan (Springer, 2021), 181–95.

8   satisficing: A term coined by the Nobel-winning Herbert Simon in the 1950s although first discussed in his *Administrative Behavior* (Macmillan, 1947). It was subsequently popularized by others including Ackoff.

8   **reframed out of existence:** Russell Ackoff, "Systems Thinking and Thinking Systems," *System Dynamics Review* 10, no. 2–3 (1994): 187.

9   **automobile stripped apart:** Ackoff, "Systems Thinking and Thinking Systems," 180.

9   **prework over rework:** Richard Beasley, Andy Nolan, and Andrew Pickard, "When 'Yes' Is the Wrong Answer," *INCOSE International Symposium* 24, no. 1 (2014): 938–52.

10   **Indus Valley:** Jagat Pati Joshi, *Harappan Architecture and Civil Engineering* (Rupa & Co., 2008); Gregory L. Possehl, *The Indus Civilization: A Contemporary Perspective* (AltaMira Press, 2002).

10   **Prominent successes:** Agatha C. Hughes and Thomas P. Hughes, *Systems, Experts, and Computers: The Systems Approach in Management and Engineering, World War II and After* (MIT Press, 2000).

11   **Systems engineering:** A part of this section was previously published in slightly different form as a *This View of Life* interview with David Sloan Wilson: "Systems Engineering as Cultural Group Selection: A Conversation with Guru Madhavan," March 29, 2018.

12   **Rubik's Cube:** Ernő Rubik, *Cubed: The Puzzle of Us All* (Flatiron Books, 2020), 73, 153, 166–67.

13   **narrow explanations:** The term "monocausotaxophilia" is from Ernst Pöppel, who described it as "the love of single causes that explain everything" in "The Neural Lyre: Poetic Meter, the Brain, and Time," *Poetry* 142, no. 5 (1983): 284. "Monotheorism" is from Jag Bhalla's "Rational or Misrational? Logically Pursuing Mad Goals," *Big Think*, January 28, 2015.

13   **"Without contraries":** William Blake, *The Marriage of Heaven and Hell* (John W. Luce & Company, 1906).

13   **contradictory proxies that work like opposable thumbs:** Roger Martin, *When More Is Not Better: Overcoming America's Obsession with Economic Efficiency* (Harvard Business School Press, 2020); *The Opposable Mind: How Successful Leaders Win through Integrative Thinking* (Harvard Business School Press, 2007).

14   **"In a dark time":** Theodore Roethke, *Collected Poems of Theodore Roethke* (Doubleday, 1966), 239.

15   **"perfect all-weather operations":** Erik Conway, *Blind Landings: Low-Visibility Operations in American Aviation, 1918–1958* (Johns Hopkins University Press, 2006), 4.

15   **"were dismissed simply as":** James H. "Jimmy" Doolittle and Carroll V. Glines, *I Could Never Be So Lucky Again: An Autobiography* (Schiffer Publishing, 1995), 129.

16   **preparing a meal:** Stephen Darcy Chiabotti, "The Glorified Link: Flight Simula-

tion and Reform in Air Force Undergraduate Pilot Training, 1967–1980" (PhD diss., Duke University, 1986), 3.

16 **half or more:** Reported in *New York Daily News*; Susan van Hoek and Marion Clayton Link, *From Sky to Sea: A Story of Edwin A. Link* (Best Publishing Company, 1993), 47.

16 **"An airway exists on the ground":** Quote by airmail pioneer Paul Henderson from Nick Komons, *Bonfires to Beacons: Federal Civil Aviation Policy under the Air Commerce Act, 1926–1938* (Federal Aviation Administration, 1978), 125.

16 **"doubtful that any piece":** Keith Matzinger, "That X*/#&** Link Trainer," *Aerospace Historian* 31, no. 2 (1984): 125.

16 **"passion for wings":** Robert Wohl, *A Passion for Wings: Aviation and the Western Imagination, 1908–1918* (Yale University Press, 1994).

17 **a camera is an instrument:** David C. King, *Dorothea Lange: Photographer of the People* (Routledge, 2014), 7.

17 **Civics:** Peter Levine, "What Does the Word Civic Mean?," *Peter Levine* (blog), December 11, 2019; Levine, "Civic Renewal in America," *Philosophy and Public Policy Quarterly* 26, no. 1/2 (2006): 3.

Chapter One: Wicked Efficiency

21 **"run by faith":** Elting Morison, *The War of Ideas: The United States Navy 1870–1890* (United States Air Force Academy, 1969), 5; Morison, *Men, Machines, and Modern Times* (MIT Press, 1966).

21 **"Every portion of the warship":** William J. Baxter, "Suggestions for Increasing the Efficiency of Our New Ships," *Proceedings of the United States Naval Institute* 21, no. 3 (1895): 439–48.

21 **The drills . . . *Army, Navy Journal* . . . cockroaches:** Morison, *War of Ideas*, 3.

22 **"miracle of imagination":** Morison, 4.

22 **Battle of Manila Bay:** Philip R. Alger, "Errors of Gun-Fire at Sea," *Proceedings of the United States Naval Institute* 26, no. 4 (1900): 575–92. Other accounts suggest there was one casualty in Dewey's side, but of heart attack.

22 **9,500 projectiles:** Alger, "Errors of Gun-Fire at Sea," 578.

23 **"unqualified success":** Elting Morison, *Admiral Sims and the Modern American Navy* (Houghton Mifflin Company, 1942), 78.

23 **One magazine:** Franklin Chester, "The Greatest Fighting Machines Afloat," *Munsey's Magazine* 24 (October 1900): 22.

23 **"*Kentucky* is not a battleship at all":** Morison, *Admiral Sims*, 80.

23 **"We can send":** William Sowden Sims, *The Victory at Sea* (Doubleday, 1921), 32.

24 **"dreadfully backwards in all useful subjects":** Related notes and quotes are from Percy Scott, *Fifty Years in the Royal Navy* (John Murray, 1919), 3, 26, 29, 31, 73, 82–84.

26 **six targets out of eight:** Morison, *Admiral Sims*, 86.

26 **"Admiralty remained immovable":** Scott, *Fifty Years*, 184.

26 **"Like all authors":** Scott, 185.

26 **Bureau of Ordnance:** Benjamin F. Armstrong, "Continuous-Aim Fire: Learning How to Shoot," *Naval History Magazine*, April 2015.

27 **Roosevelt elevated Sims:** Morison, *Admiral Sims*, 104.

27 **3,000 percent:** Morison, 145.

27 **"It could sink the whole German navy":** John Edward Moore, "H.M.S. Dreadnought: Myths and Realities," *Proceedings and Papers of the Georgia Association*

of Historians 5 (1984): 48, citing Davis Harris Willson, *A History of England*, 2nd ed. (Dryden Press, 1972), 703.

27 **The capital ship's superior:** Armstrong, "Continuous-Aim Fire."

28 **"We have, indeed, no choice":** Sims, "The Inherent Tactical Qualities of All-Big-Gun, One-Caliber Battleships of High Speed, Large Displacement, and Gun-Power," *U.S. Naval Institute Proceedings* 32, no. 4 (1906): 1337–66.

28 **"largest battleship":** William McBride, "Nineteenth-Century American Warships: The Pursuit of Exceptionalist Design," in *Re-Inventing the Ship: Science, Technology and the Maritime World, 1800–1918,* ed. D. Leggett and R. Dunn (Ashgate, 2012), 204.

28 **series of letters:** December 15, 1920, letter, in *The Future of Navies. Great Ships or—? Leading Articles Reprinted from the Times, with Letters from Admiral Sir Percy Scott and Others* (Times Publishing Company, 1921), 21.

28 **"just as the motor vehicle":** "Doom of the Dreadnought," *Kalgoorlie Miner,* August 8, 1914, 1.

28 **"efficiency animals":** Peter Kroes cited in Byron Newberry, "Efficiency Animals: Efficiency as an Engineering Value," in *Engineering Identities, Epistemologies and Values: Engineering Education and Practice in Context,* ed. S. H. Christensen et al. (Springer International Publishing, 2015), 2:199–214.

29 **"In a society of cannibals":** Billy Vaughn Koen, *Discussion of the Method: Conducting the Engineer's Approach to Problem Solving* (Oxford University Press, 2003), 19.

29 **"Engineers *do* economics":** Robert B. Ekelund and Robert F. Hébert, *Secret Origins of Modern Microeconomics: Dupuit and the Engineers* (University of Chicago Press, 1999), 39.

29 **state efficiency and social equality:** Ekelund and Hébert, *Secret Origins,* 29.

30 **technical efficiency and economic efficiency:** Based on Stefan Mann and Henry Wüstemann, "Efficiency and Utility: An Evolutionary Perspective," *International Journal of Social Economics* 37, no. 9 (2010): 676–85.

30 **Adam Smith:** Adam Smith, *An Inquiry into the Nature and Causes of the Wealth of Nations,* vol. 1 (Methuen & Co., 1776).

31 **old Indian parable:** Adapted from a spiritual tale told by sage Ramakrishna Paramahamsa (1836–86).

31 **Jeremy Bentham:** *An Introduction to the Principles of Morals and Legislation* (Pickering, 1823); in John Broome, "Utility," *Economics & Philosophy* 7, no. 1 (1991): 1–12.

31 **"not confined to one human being":** Ekelund and Hébert, *Secret Origins,* 68.

32 **"isolated engineers":** Bruno Belhoste and Konstantinos Chatzis, "From Technical Corps to Technocratic Power," *History and Technology* 23, no. 3 (2007): 221.

32 **"aid to private decisions"; "routine, almost automatic":** Theodore M. Porter, "Objectivity and Authority: How French Engineers Reduced Public Utility to Numbers," *Poetics Today* 12, no. 2 (1991): 254, 262.

32 **"a tribute to democracy":** Porter, "Objectivity and Authority," 262. For more on value and effects of quantification, see: Theodore Porter, *Trust in Numbers: The Pursuit of Objectivity in Science and Public Life* (Princeton University Press, 1996); Jerry Muller, *The Tyranny of Metrics* (Princeton University Press, 2018); James Vincent, *Beyond Measure: The Hidden History of Measurement from Cubits to Quantum Constants* (W. W. Norton, 2022); Alfred Spector et al., *Data Science in Context* (Cambridge University Press, 2023).

32   Claude-Louis Navier . . . Charles Joseph Minard: Ekelund and Hébert, *Secret Origins*, 74–80; Manuela Mosca, "Jules Dupuit, the French 'Ingénieurs Économistes' and the Société d'Economie Politique," in *Studies in the History of French Political Economy: From Bodin to Walras*, ed. G. Faccarello (Routledge, 1998), 254–83.

33   "cold, reserved, cutting": Robert B. Ekelund, "A Critical Evaluation of Jules Dupuit's Contributions to Economic Theory and Policy" (PhD diss., Louisiana State University and Agricultural & Mechanical College, 1967), 20.

34   summarize their motivations: Ekelund and Hébert, *Secret Origins*, 68.

34   substitution calculus: Bernard Grall, "From Maintaining Roads to Measuring Utility: Dupuit's Substitution Calculus (1842–1844)," in *The Works of Jules Dupuit: Engineer and Economist of the French XIXth Century*, ed. Jean-Pascal Simonin and François Vatin, trans. Chris Hinton (Edi-Gestion, 2016).

34   Dupuit knew: Ekelund and Hébert, *Secret Origins*; Ekelund, "A Critical Evaluation"; Robert B. Ekelund, "Jules Dupuit and the Early Theory of Marginal Cost Pricing," *Journal of Political Economy* 76, no. 3 (1968): 462–71; Robert B. Ekelund, "The Economist Dupuit on Theory, Institutions, and Policy: First of the Moderns?," *History of Political Economy* 32, no. 1 (2000): 1–38.

34   "clouds of philosophy": Ekelund and Hébert, *Secret Origins*, 7.

35   *consumer surplus . . . differential pricing*: Alfred Marshall is credited with the term "consumer surplus," but his work grew from Dupuit's and is used here in that spirit. See Cheng-chung Lai and Tai-kuang Ho, "The Marginal School in France," in *History of Economic Ideas in 20 Talks*, ed. C. Lai and T. Ho (Springer Nature Singapore, 2022), 81–88.

36   "The only real utility": Arsène Jules Étienne Juvénal Dupuit, "De la mesure de l'utilité des travaux publics," *Annales des ponts et chaussées* (1844), trans. R. H. Barback as "On the Measurement of the Utility of Public Works," *International Economic Papers* 2 (1952): 83–110; reprinted in *Readings in Welfare Economics*, ed. Kenneth J. Arrow and Tibor Scitovsky (Richard D. Irwin, 1969), 258.

36   "institutions, technology, and human nature": Ekelund and Hébert, *Secret Origins*, 3–7, 29.

36   Economic freedom: Ekelund, "A Critical Evaluation," 15–16.

37   "if a man married his maid": Daniel Hirschman and Elizabeth Popp Berman, "Do Economists Make Policies? On the Political Effects of Economics," *Socio-Economic Review* 12, no. 4 (2014): 799.

37   "As with all sciences": Mosca, "Jules Dupuit," 271; Dupuit from around 1860.

37   cats, cannonballs, or economics: James Hartle, "Theories of Everything and Hawking's Wave Function of the Universe," in *The Future of Theoretical Physics and Cosmology: Celebrating Stephen Hawking's Contributions to Physics*, ed. G. W. Gibbons, E. P. S. Shellard, and S. J. Rankin (Cambridge University Press, 2003), 38–39.

38   HIV/AIDS crisis of the 1980s: See Steven Epstein, *Impure Science: AIDS, Activism and the Politics of Knowledge* (University of Chicago Press, 1998).

38   "Health and illness": Daniel Carpenter, "Is Health Politics Different?," *Annual Review of Political Science* 15, no. 1 (2012): 287–311.

38   HALYs . . . QALYs . . . DALYs: QALYs were developed in the 1960s, and DALYs in the 1990s. See Marthe Gold, "HALYs and QALYs and DALYs, Oh My: Similarities and Differences in Summary Measures of Population Health," *Annual Review of Public Health* 23, no. 1 (2002): 115–34; Sarah Whitehead and Shehzad

Ali, "Health Outcomes in Economic Evaluation: The QALY and Utilities," *British Medical Bulletin* 96, no. 1 (2010): 5–21; Julie Lajoie, "Understanding the Measurement of Global Burden of Disease" (paper prepared for the National Collaborating Centre for Infectious Diseases, Winnipeg, Manitoba, 2013, revised February 2015); and Guru Madhavan and Charles Phelps, "Valuing Health: Evolution, Revolution, Resistance, and Reform," *Value in Health* 22, no. 5 (2019): 505–10.

38   **"feelings thermometer":** Different valuation techniques give rise to different uncertainties. I have kept it simple by denoting the subjective element of one's health state. Moreover, such techniques may not consider distributional effects spanning one's lifetime or multiple generations.

39   **discounting:** See David A. Katz and H. Gilbert Welch, "Discounting in Cost-Effectiveness Analysis of Healthcare Programmes," *PharmacoEconomics* 3, no. 5 (1993): 276–85.

40   **"rigor distortis":** From Jag Bhalla's January 17, 2018, *Big Think* perspective.

40   **greater collective good:** Whitehead and Ali, "Health Outcomes in Economic Evaluation."

40   **Thomas Aquinas:** Based on the Wikipedia entry of "just price," February 2023.

40   **medication's "perceived value":** Norman R. Augustine, Guru Madhavan, and Sharyl J. Nass, eds., *Making Medicines Affordable: A National Imperative* (National Academies Press, 2018).

41   **health insurance and the legal rights:** Charles Phelps and Guru Madhavan, "Patents and Drug Insurance: Clash of the Titans?," *Science Translational Medicine* 10, no. 467 (2018): 6902.

41   **prioritizing new and improved vaccines:** This section is heavily based on my decade-long work for the National Academies and the US Department of Health and Human Services and the European Union Malaria Fund. See National Academies' *Ranking Vaccines* series (2012–15) edited by Guru Madhavan et al., and Charles Phelps and Guru Madhavan's *Making Better Choices: Design, Decisions, and Democracy* (Oxford University Press, 2021).

42   **Ebola:** See David Quammen, *Ebola: The Natural and Human History of a Deadly Virus* (W. W. Norton, 2014); Mary H. Moran, "Missing Bodies and Secret Funerals: The Production of 'Safe and Dignified Burials' in the Liberian Ebola Crisis," *Anthropological Quarterly* 90, no. 2 (2017): 399–421; Padraig Lyons et al., "Engaging Religious Leaders to Promote Safe Burial Practices During the 2014–2016 Ebola Virus Disease Outbreak, Sierra Leone," *Bulletin of the World Health Organization* 99, no. 4 (2021): 271–79.

42   **vaccine candidates for Ebola:** See Charles Phelps, Guru Madhavan, Rino Rappuoli, Rita Colwell, and Harvey Fineberg, "Beyond Cost-Effectiveness: Using Systems Analysis for Infectious Disease Preparedness," *Vaccine* 35 (2017): A46–A49.

43   **The World Health Organization:** "One Year into the Ebola Epidemic: A Deadly, Tenacious and Unforgiving Virus," WHO Newsroom Spotlight, 2015.

43   **"backward" . . . offensive:** Moran, "Missing Bodies and Secret Funerals," 406–7.

43   **developed new customs:** From Steve Hayes, "Kissing the Banana Trunk: Will You Commit and Act in the Fight Against Ebola?," *Huffington Post*, December 6, 2017.

43   **performance measurements:** From V. F. Ridgway, "Dysfunctional Consequences of Performance Measurements," *Administrative Science Quarterly* 1, no. 2 (1956): 240–41.

43   **lifelong iconoclast:** Benyaurd Wygant, "Admiral Sims as I Knew Him," *United States Naval Institute Proceedings* (October 1951): 1091.

44   **"effing debate":** The term is used by hosts Ron Baker and Ed Kless in their podcast, *The Soul of Enterprise.*

44   **"imperative of engineering . . . holy grail":** From Eugene Ferguson, Carl Mitcham, and Stanley Carpenter, quoted in Newberry, "Efficiency Animals," 200.

44   **"efficient cause of man":** Tomas Spencer, "Of the Efficient Cause," chap. 7 in *The Art of Logick Delivered in the Precepts of Aristotle and Ramus* (printed by John Dawson for Nicholas Bourne, 1628).

44   **"In other words, efficiency":** Henry Mintzberg, "A Note on That Dirty Word 'Efficiency,'" *Interfaces* 12, no. 5 (1982): 102.

45   **Energy efficiency:** Tessa Dunlop, "Mind the Gap: A Social Sciences Review of Energy Efficiency," *Energy Research & Social Science* 56 (2019): 101216.

45   **As with *Ensatina*:** Byron Newberry, "The Dialectics of Engineering," in *Engineering Identities, Epistemologies and Values: Engineering Education and Practice in Context*, ed. S. H. Christensen et al. (Springer International Publishing, 2015), 2:13–15. The "ring species" example is based on evolutionist Robert Stebbins and colleagues' research, which was popularized by Richard Dawkins in *The Ancestor's Tale: A Pilgrimage to the Dawn of Evolution* (Houghton Mifflin, 2004).

46   **elastic-manufacturing:** Guru Madhavan, "Tinker, Taylor, Soldiering, Spy: Escaping Efficiency Traps," *Bridge* (2021): 95–96; based on Erica Fuchs and colleagues' December 18, 2020, *Issues in Science and Technology* article on pandemic response and data.

46   **Global supply networks . . . when most needed:** Madhavan, "Tinker, Taylor, Soldiering, Spy."

Refrain: Pitch

47   **Dole wanted to produce:** Details of the Dole Derby are based on Lesley Forden, *Glory Gamblers: The Story of the Dole Race* (Nottingham Press, 1986); Jason Ryan, *Race to Hawaii: The 1927 Dole Air Derby and the Thrilling First Flights That Opened the Pacific* (Chicago Review Press, 2018). See also Jane Eshleman Conant, "Death Dogged the Dolebirds: Pioneer Pacific Fliers Wrote Tragic Chapter in Air History," *San Francisco Call-Bulletin*, October 10, 1955; Robert Hegenberger, "'The Bird of Paradise': The Significance of the Hawaiian Flight of 1927," *Air Power History* 38, no. 2 (1991): 6–18; William G. "Burl" Burlingame, "The Dole Derby," *Honolulu Star-Bulletin*, December 29, 2003.

47   **"greatest air race" . . . Roman gladiators:** Ryan, *Race to Hawaii*, 217.

48   **eight contenders:** The ninth plane, *Spirit of Peoria*, was disqualified for insufficient fuel capacity.

48   **Colonel Pineapple:** Ryan, *Race to Hawaii*, 187.

48   **"When we get over to Honolulu":** Ryan, 216.

48   ***El Encanto* was ready:** "Descriptions of the Dole Derby Planes," *Aviation*, August 22, 1927, 414–15.

48   **"I would rather have":** Ryan, *Race to Hawaii*, 228.

49   **"There it is":** Forden, *Glory Gamblers*, 118.

49   **"Let's go home":** Forden, 104.

50   **"For God's sake":** Ryan, *Race to Hawaii*, 253.

50   **One poet wrote:** Addison N. Clark, "Into the West," *Western Flying Magazine*, 1927; Forden, *Glory Gamblers*, 137.

50   **Newspapers declared:** Quoted in Ryan, *Race to Hawaii*, 263.

51   **The winning Art Goebel:** Ryan, 170, 267.

51   **Jensen was on a job:** Ryan, 269.

51   **"America has found":** Scott A. Berg, *Lindbergh* (G. P. Putnam's Sons, 1998), 188.

51   **The Binghamton region:** William Lawyer, *Binghamton: Its Settlement, Growth and Development, and the Factors in Its History, 1800–1900* (Century Memorial Publishing Co., 1900); John H. VanGorden, *The Susquehanna Flows On* (Wilcox Press, 1966); J. B. Wilkinson, Tom Cawley, and John Hart, *The Annals of Binghamton of 1840: With an Appraisal, 1840–1967* (Broome County Historical Society and Old Onaquaga Historical Society, 1967); Ross McGuire and Nancy Grey Osterud, *Working Lives: Broome County, New York, 1800–1930* (Roberson Center for the Arts and Sciences, 1980); Gerald R. Smith, *Sweeping across America: Stories of Broome County Citizens in American History* (Keystone Digital Press, 2016).

51   **"form a cabin":** Lawyer, *Binghamton*, 5.

51   **Bingham:** See Margaret Brown, "William Bingham, Eighteenth Century Magnate," *Pennsylvania Magazine of History and Biography* 61, no. 4 (1937): 432–34.

51   **Erie Canal:** See Peter Bernstein, *Wedding of the Waters: The Erie Canal and the Making of a Great Nation* (W. W. Norton, 2005).

52   **Factory furnaces:** Mass production of cigars created an early wholesale factory model in Binghamton. Over 6,000 workers in 70 firms rolled over 150 million cigars.

52   **Parlor City:** Tom Cawley, "Binghamton's 'Parlor City' Nickname Explained," *Binghamton Press & Sun-Bulletin*, January 17, 1964.

52   **"clean, fair, and grand":** Ed Aswad and Suzanne Meredith, *Binghamton* (Images of America; Arcadia Publishing, 2001), 11.

52   **In 1898:** Arthur Reblitz and David Bowers, *Treasures of Mechanical Music* (Vestal Press, 1981); Reblitz, *Player Piano Servicing and Rebuilding* (Vestal Press, 1985); Reblitz and Bowers, *The Golden Age of Automatic Musical Instruments: Remarkable Music Machines and Their Stories* (Mechanical Music Press, 2001). See also Arthur Loesser, *Men, Women and Pianos: A Social History* (Simon & Schuster, 1954; repr. Dover, 1990); James R. Gaines, *The Lives of the Piano* (Holt, Rinehart and Winston, 1981); Stuart Isacoff, *A Natural History of the Piano: The Instrument, the Music, the Musicians—from Mozart to Modern Jazz and Everything in Between* (Knopf, 2011).

52   **"Should one applaud?":** Brian Dolan, *Inventing Entertainment: The Player Piano and the Origins of an American Musical Industry* (Rowman & Littlefield, 2009), 34.

52   **85 percent:** Reblitz, *Player Piano Servicing*, 1.

52   **popularity:** Gaines, *Lives of the Piano*; David Suisman, *Selling Sounds: The Commercial Revolution in American Music* (Harvard University Press, 2012).

52   **"every touch in technique":** Dolan, *Inventing Entertainment*, 46.

52   **one Niagara Falls company:** David Bowers, *Put Another Nickel In* (Vestal Press, 1966), 27.

53   **Jacques de Vaucanson:** Dolan, *Inventing Entertainment*, 40.

54   **alienated artists:** Dolan, 33.

54   **"menace of mechanical music":** John Philip Sousa, "The Menace of Mechanical Music," *Appleton's Magazine* 8, no. 3 (September 1906): 279.

54   "Automatic instruments were always": Reblitz and Bowers, *Golden Age*, 5.

54   "Chronic mechanitis": Dolan, *Inventing Entertainment*, 35.

54   "endless rolls": For technical details on the Automatic rolls see Charles H. Hamilton and George R. Thayer's 1908 US patent, "Web-controlling mechanism for self-playing instruments," US937933A.

55   "distinctly a delight": Reblitz and Bowers, *Treasures of Mechanical Music*, 408.

55   High-end clubs: See Kerry Segrave, *Jukeboxes: An American Social History* (McFarland, 2002).

55   "most pleasant surprise": C. Sharpe Minor to George Link, May 28, 1925.

56   "architectural symphony": Carl Bronson, "Artist Scores Triumph on Link Organ," February 28, 1928 (from Link Collection news clipping, unknown publication).

56   "winged gospel": Joseph Corn, *The Winged Gospel: America's Romance with Aviation, 1900–1950* (Oxford University Press, 1983).

56   "air-mindedness": Corn, *Winged Gospel*, vii, 136.

56   Picasso: "The Scallop Shell: 'Notre Avenir est dans l'Air,'" The Collection: Modern and Contemporary Art, Metropolitan Museum of Art.

56   "Take possession of the air": Wohl, *Passion for Wings*, 154.

56   "air is an extremely dangerous": Winston Churchill, "Adventures in the Air," *Cosmopolitan*, June 24, 1924.

56   "Goggles, gloves": Peter Pigott, *Brace for Impact: Air Crashes and Aviation Safety* (Dundurn, 2016), 21, 34, 38–39.

57   internal sensations: In one incident, a pilot realized that his aircraft was out of control and, worse, flying upside down only when his pocket watch fell out and hit his face. See Timothy P. Schultz, *The Problem with Pilots: How Physicians, Engineers, and Airpower Enthusiasts Redefined Flight* (Johns Hopkins University Press, 2018), 45.

57   Union Pacific Railroad: T. A. Heppenheimer, *Turbulent Skies: The History of Commercial Aviation* (John Wiley & Sons, 1995), 8.

57   Rand McNally: Maurer Maurer, *Aviation in the U.S. Army, 1919–1939* (United States Air Force, 1987), 36.

57   "with a few drops of homing pigeon": Heppenheimer, *Turbulent Skies*, 9.

57   enemy territories: See Conway, *Blind Landings*, 13.

57   survey of "sky roadways": Henry Lehrer, *Flying the Beam: Navigating the Early US Airmail Airways, 1917–1941* (Purdue University Press, 2014), 38.

58   colorful commentary: Bruce Etyinge and Rex Uden, *Landing Field Guide and Pilot's Log Book* (Etyinge & Uden, 1920), 53, 56.

58   helpful hints: Etyinge and Uden, *Landing Field Guide*, 12.

58   "lighted airway": Pigott, *Brace for Impact*, 38–39; Lehrer, *Flying the Beam*, 82.

60   east to west: Todd La Porte, "The United States Air Traffic Control System: Increasing Reliability in the Midst of Rapid Growth" (working paper, University of California, Berkeley, 1988), 2; Heppenheimer, *Turbulent Skies*, 11.

60   "Sometimes you couldn't": Flint Whitlock and Terry L. Barnhart, *Capt. Jepp and the Little Black Book: How Barnstormer and Aviation Pioneer Elrey B. Jeppesen Made the Skies Safer for Everyone* (Savage Press, 2007), 116.

60   "He knew, for example": Whitlock and Barnhart, *Capt. Jepp*, 110.

60   400,000 people: La Porte, "Air Traffic Control System," 2.

60   "When Undertaking Very Hard Work": Lehrer, *Flying the Beam*, 82–83; Chris Forsyth and Daegan Miller, "Keep Direction by Good Methods," *Places Journal*, February 2021.

61 "Unfortunately, it wasn't": van Hoek and Link, *From Sky to Sea*, 11.

61 "just dumb enough to be a genius": Richard Whitmire, "Ed Link: I'm Not a Genius," *Binghamton Press & Sun-Bulletin*, August 15, 1976.

61 nomadic barnstormers: Paul O'Neil et al., *Barnstormers & Speed Kings* (Time-Life Books, 1981), 26.

61 "a bunch of parts": Description of Curtiss JN-4H Jenny from Smithsonian National Postal Museum, *Airmail in America*.

62 "That's a hell of a way": Lloyd Kelly and Robert Parke, *The Pilot Maker* (Grosset & Dunlap, 1970), 20.

62 Wilbur Wright trained: Rebecca Hancock Cameron, *Training to Fly: Military Flight Training, 1907–1945* (Air Force History and Museums Program, 1999), 34–37.

63 "kiwi-trainer": Max Baarspul, "A Review of Flight Simulation Techniques," *Progress in Aerospace Sciences* 27, no. 1 (1990): 1–120; Michael Moroney and William Lilienthal, "Human Factors in Simulation and Training: An Overview," in *Human Factors in Simulation and Training*, ed. D. A. Vincenzi et al. (CRC Press, 2009), 18.

64 Ruggles . . . fitness to fly: Cameron, *Training to Fly*, 265–67.

64 Ocker and Crane's conception: William C. Ocker and Carl J. Crane, *Blind Flight in Theory and Practice* (The Naylor Company, 1932).

64 two interlinked problems: Conway, *Blind Landings*, 12.

65 "And that was one of the things": Ed Link, interview by Wanda Wood, September 18, 1978, Broome County Oral History Project, Binghamton University Special Collections, 4–5.

65 Alexander Calder: Jed Perl, *Calder: The Conquest of Time; The Early Years: 1898–1940* (Knopf, 2017), 507.

Chapter Two: Wicked Vagueness

67 Diana Zhang: Based on related case laws and T. Leigh Anenson, "Great Expectations: The Role of the Consumer in Determining Defective Air Bag Design," *Tort Trial & Insurance Practice Law Journal* 38, no. 3 (2002): 963–98.

67 "partial ignorance"; "social experiment"; "ongoing success": The terms are from Mike W. Martin and Roland Schinzinger, *Introduction to Engineering Ethics*, 2nd ed. (McGraw-Hill, 2010), 78–80; Jameson Wetmore, "Engineering with Uncertainty: Monitoring Air Bag Performance," *Science and Engineering Ethics* 14, no. 2 (2008): 201–18.

68 let's consider vagueness: See R. M. Sainsbury, "Concepts without Boundaries," in *Vagueness: A Reader*, ed. Rosanna Keefe and Peter Smith (MIT Press, 1996); Nicholas J. J. Smith, *Vagueness and Degrees of Truth* (Oxford University Press, 2008); Kees van Deemter, *Not Exactly: In Praise of Vagueness* (Oxford University Press, 2012).

68 bald people: The *phalakros* paradox is discussed in Timothy Williamson, *Vagueness* (Routledge, 1994); Anna Mahtani, "The Instability of Vague Terms," *Philosophical Quarterly (1950–)* 54, no. 217 (2004): 570–76.

68 Where does one system end: Recall the tiny differences in the *Ensatina* salamanders and try to make sense of overlapping language. Is vagueness, like the ring species, an evolutionary flaw or feature? See "What Is a Species?" in van Deemter, *Not Exactly*, 20–30.

68   **paradox of *mañana*:** Dorothy Edgington, "Vagueness by Degrees," in Keefe and Smith, *Vagueness: A Reader*, 294–315.

69   **gradual change:** van Deemter, *Not Exactly*, 15.

69   **"a grid, a net" . . . "magnetic poles":** Mark Sainsbury, "Is There Higher-Order Vagueness?," *Philosophical Quarterly (1950–)* 41, no. 163 (1991): 167–82.

69   **"defeat any definition":** H. G. Wells, *First and Last Things: A Confession of Faith and a Rule of Life* (G. P. Putnam's Sons, 1908), 26–27.

70   ***passive* restraints to *supplemental* restraints:** Jameson Wetmore, "Delegating to the Automobile: Experimenting with Automotive Restraints in the 1970s," *Technology and Culture* 56, no. 2 (2015): 440–63, 458.

70   **airbags reduced fatalities:** National Highway Traffic Safety Administration, *Fatality Reduction by Air Bags: Analyses of Accident Data through Early 1996* (technical report DOT HS 808 470, Department of Transportation, August 1996), v, 1, 2, 23; L. A. Wallis and I. Greaves, "Injuries Associated with Airbag Deployment," *Emergency Medicine Journal* 19, no. 6 (2002): 490; Robert E. Antosia, Robert A. Partridge, and Alamjit S. Virk, "Air Bag Safety," *Annals of Emergency Medicine* 25, no. 6 (1995): 794–98.

70   **In 1996:** The Alexandra Greer incident is discussed in an Associated Press account, "Air-Bag Concerns Heightened after Child Is Decapitated in Fender-Bender," *Los Angeles Times*, November 28, 1996.

70   **Bill Clinton:** "President Clinton's Weekly Radio Address," CNN, December 28, 1996.

71   **seat belt compliance:** In 1997, Australia, Canada, and Germany reported over 90 percent seat belt use. Warren Brown and David B. Ottaway, "Small Victims of a Flawed Safety Device," *Washington Post*, June 2, 1997.

71   **Reports from early 1997 . . . 15 miles per hour:** Cited in Wetmore, "Engineering with Uncertainty," 210.

72   **"a social solution":** Wetmore, 215.

72   **three-point message:** Wetmore, 211.

72   **Between 1994 and 2000:** Wetmore, 214.

72   **two interlinked questions . . . "decide how much salt":** Ezra Hauer, "Safety in Geometric Design Standards. II: Rift, Roots and Reform," in *Proceedings of the Second International Symposium on Highway Geometric Design, June 14–17, 2000*, ed. R. Krammes and W. Brilon (Mainz, Germany, 2000), 24–25.

73   **"what a committee of":** Ezra Hauer, "An Exemplum and Its Road Safety Morals," *Accident Analysis & Prevention* 94 (2016): 168–79.

73   **"unpremeditated level of safety":** Hauer, "Safety in Geometric Design," 24.

75   **"I easily remember":** William B. Rouse, *Beyond Quick Fixes: Addressing the Complexity and Uncertainties of Contemporary Society* (Oxford University Press, 2023), chap. 5.

75   **"Who speaks of vagueness":** Bertrand Russell, "Vagueness," *Australasian Journal of Psychology and Philosophy (June 1923):* 84–92; also collected in Keefe and Smith, *Vagueness: A Reader*.

76   **new phenomena:** Philip Anderson, "More Is Different," *Science* 177, no. 4047 (1972): 393–96; Steven Strogatz et al., eds., "Fifty Years of 'More Is Different,'" special issue, *Nature Reviews Physics* 4, no. 8 (2022): 508–10.

76   **abstraction and aggregation:** Jens Rasmussen, "The Role of Hierarchical Knowledge Representation in Decisionmaking and System Management," *IEEE Transactions on Systems, Man, and Cybernetics* 15, no. 2: 234–43.

76   **Building-information modeling:** Jimmy Abualdenien and André Borrmann, "Vagueness Visualization in Building Models across Different Design Stages," *Advanced Engineering Informatics* 45 (2020): 101107.

78   **"worship of hard numbers":** John C. Bogle, *Don't Count on It! Reflections on Investment Illusions, Capitalism, "Mutual" Funds, Indexing, Entrepreneurship, Idealism, and Heroes* (John Wiley & Sons, 2011), 15.

78   **pleasurable clarity:** C. Thi Nguyen, "The Seductions of Clarity," *Royal Institute of Philosophy Supplements* 89 (2021): 227–55.

78   **"indicator culture":** Sally Engle Merry, *The Seductions of Quantification: Measuring Human Rights, Gender Violence, and Sex Trafficking* (University of Chicago Press, 2016).

79   **"cosmology episodes":** Karl Weick, interview with Diane Coutu, "Sense and Reliability," *Harvard Business Review*, April 2003.

79   **over the subsequent week:** See *The Guardian* reporting on April 16, 2010 (by Graeme Wearden), April 24, 2010 (by Jamie Doward and Cal Flyn), and April 25, 2011 (by Ian Sample).

80   **John Cleese:** Robert Booth, "Non Flying Circus," *Guardian*, April 16, 2010.

80   **civilians stranded:** Thomas Birtchnell and Monika Büscher, "Stranded: An Eruption of Disruption," *Mobilities* 6, no. 1 (2011): 1–9.

80   **Eyjafjallajökull:** Readings included Ágúst Gunnar Gylfason et al., *The 2010 Eyjafjallajökull Eruption, Iceland,* ed. Barði Þorkelsson (report to ICAO, 2012), 13–15, 48–49, 53–56, 63–65, 101–102, 117–119; Guðrún Nína Petersen, "A Short Meteorological Overview of the Eyjafjallajökull Eruption 14 April–23 May 2010," *Weather* 65, no. 8 (2010): 203–7.

80   **particulate matter:** Hanne Krage Carlsen et al., "A Survey of Early Health Effects of the Eyjafjallajökull 2010 Eruption in Iceland: A Population-Based Study," *BMJ Open* 2, no. 2 (2012): e000343.

80   **"I will show you fear":** T. S. Eliot's "The Waste Land" verse is from Peter Brooker, "Fear in a Handful of Dust: Aviation and the Icelandic Volcano," *Significance* 7, no. 3 (2010): 112–15.

81   **"worst peacetime aviation crisis":** Graeme Wearden, "Ash Cloud Costing Airlines £130m a Day," *Guardian*, April 16, 2010; Alan Cowell and Nicola Clark, "Britain Opens Its Airspace as Travel Crisis Begins to Ebb," *New York Times*, April 20, 2010; Alok Jha, "How Icelandic Volcano Issued Warnings Months before Its Eruption," *Guardian*, November 17, 2010.

81   **$1.7 billion in losses:** "Airlines Lost More Than $1.7 Billion by Tuesday: IATA," Reuters, April 21, 2010.

81   **furloughed:** Nick Wadhams, "Iceland Volcano: Kenya's Farmers Losing $1.3m a Day in Flights Chaos," *Guardian,* April 18, 2010.

81   **Amy Donovan and Clive Oppenheimer:** Amy Donovan and Clive Oppenheimer, "The 2010 Eyjafjallajökull Eruption and the Reconstruction of Geography," *Geographical Journal* 177, no. 1 (2011): 4–11; Donovan and Oppenheimer, "Volcanoes on Borders: A Scientific and (Geo)Political Challenge," *Bulletin of Volcanology* 81, no. 5 (2019): 31.

81   **Laki:** Alexandra Witze and Jeff Kanipe, *Island on Fire* (Pegasus Books, 2015), 17–18, 21–24.

81   **"was of a permanent nature":** Benjamin Franklin, "Meteorological Imaginations and Conjectures," May 1784, National Archives.

81   **the French Revolution:** Greg Neale, "How an Icelandic Volcano Helped Spark the French Revolution," *Guardian*, April 15, 2010.

82   **"masterpiece of understatement":** Macarthur Job, *Air Crash* (Aerospace Publications, 1992), 2:106.

82   **destroyed a $100 million aircraft . . . melting temperature:** Peter Sammonds, Bill McGuire, and Stephen Edwards, *Volcanic Hazard from Iceland: Analysis and Implications of the Eyjafjallajökull Eruption* (UCL Institute for Risk and Disaster Reduction, University College London, 2010).

83   **convening and coordination:** UK Civil Aviation Authority web pages; Padhraic Kelleher, *Report of Proceedings, Volcanic Ash International Teleconferences, 17–23 April 2010*; background notes from Brian Collins.

84   **"winning a golden ticket":** Press Association, "First Passengers to Land at Heathrow Describe Relief," *Guardian*, April 20, 2010.

84   **"how liberating it is":** "Humbled by a Volcano, We Can Only Sit in Wonder," *Guardian*, April 17, 2020.

84   **"like going up the stairs":** Kristine McKenna, "Art Carny," *Los Angeles Times*, January 27, 1991.

84   **"language *goes on holiday*":** Ludwig Wittgenstein, *Philosophical Investigations*, 4th ed., ed. P. M. S. Hacker and Joachim Schulte, trans. G. E. M. Anscombe, P. M. S. Hacker, and Joachim Schulte (Blackwell Publishing, 2009; 1st ed. 1953), 23e.

84   **"efficiency loss":** Barton Lipman, "Why Is Language Vague?," Department of Economics, Boston University, 2009. Also discussed in Matthew James Green and Kees van Deemter, "The Elusive Benefits of Vagueness: Evidence from Experiments," in *Vagueness and Rationality in Language Use and Cognition*, ed. R. Dietz (Springer International Publishing, 2019), 63–86.

84   **"certainty can be fatal":** Harry Beckwith, *The Invisible Touch: The Four Keys to Modern Marketing* (Warner Books, 2000), 3.

85   **"computing with words":** Lotfi Zadeh, "Fuzzy Logic = Computing with Words," *IEEE Transactions on Fuzzy Systems* 4, no. 2 (1996): 103–11.

85   **Alan Greenspan:** Greenspan's 1987 quote is in several sources, including, for example, Petra Maria Geraats, "The Mystique of Central Bank Speak" (Cambridge Working Paper in Economics, 2005), and Douglas R. Holmes, "How the Fed Learned to Talk," *New York Times*, February 1, 2014.

85   **weight-loss programs:** S. Himanshu Mishra, Arul Mishra, and Baba Shiv, "In Praise of Vagueness: Malleability of Vague Information as a Performance Booster," *Psychological Science* 22, no. 6 (2011): 733–38.

86   **"stand a good chance":** See J. L. Austin, *Sense and Sensibilia*, reconstructed by G. J. Warnock (repr., Oxford University Press, 1979), 125; and *How to Do Things with Words*, 2nd ed. (Clarendon Press, 1975).

86   **scenario-planning exercise:** Claire Witham et al., "Practising an Explosive Eruption in Iceland: Outcomes from a European Exercise," *Journal of Applied Volcanology* 9, no. 1 (2020): 1; Larry Mastin et al., "Progress in Protecting Air Travel from Volcanic Ash Clouds," *Bulletin of Volcanology* 84, no. 1 (2021).

86   **House of Lords:** Collins's response to question 236 from Lord Willetts in the Evidence Session No. 24 of the House of Lords, Risk Assessment and Risk Planning Committee, May 26, 2021.

Refrain: Roll

88 **"great value as means"**: Ed Link, Combination Training Device for Student Aviators and Entertainment Apparatus, US patent 1,825,462, September 29, 1931.

89 **"physical movements and sensations"**: Chihyung Jeon, "The Virtual Flier: The Link Trainer, Flight Simulation, and Pilot Identity," *Technology and Culture* 56, no. 1 (2015): 30.

90 didn't merely mimic: Jeon, "The Virtual Flier," 36.

91 Ford Model T: van Hoek and Link, *From Sky to Sea*, 48.

91 **"Just a quarter"**: Max Hill, "Once Two-Bit Carnival Ride, Link Trainer Now Most Vital in Turning Out U.S. Fighters," *Binghamton Press*, February 27, 1943.

91 **"Marion made"**: Ed Link, interview by Wanda Wood, 9

91 desperate Ed Link: van Hoek and Link, *From Sky to Sea*, 39–40.

91 **"I put a venturi"**: van Hoek and Link, 43.

91 flying billboards: Kelly and Parke, *Pilot Maker*, 47.

92 called themselves **"airlines"**: John Correll, "The Air Mail Fiasco," *Air Force Magazine*, March 1, 2008.

92 equivalent weight of airmail: Correll, "Air Mail Fiasco," quoting Oliver E. Allen.

92 **"crazy quilt of routes"**: Conway, *Blind Landings*, 31.

92 6,500 forced landings: From the Smithsonian National Postal Museum exhibit, *Airmail in America*.

92 wet their fingers: Cameron, *Training to Fly*, 22.

92 **When pilots launched**: Even commercial flight-training schools relocated to ensure all-year flying. Cameron, *Training to Fly*, 22.

92 two types of weather-related accidents: Conway, *Blind Landings*, 19–20.

93 **"Aircraft that sit"**: Conway, 4.

93 **"legalized murder"**: Correll, "Air Mail Fiasco."

93 Billy Mitchell: See John Lancaster, *The Great Air Race: Glory, Tragedy, and the Dawn of American Aviation* (Liveright, 2022).

94 Commonwealth: van Hoek and Link, *From Sky to Sea*, 48.

94 pivotal place: In "Flying Training: The American Advantage in the Battle for Air Superiority against the Luftwaffe," Kenneth P. Werrell notes that the United States dramatically ramped up pilot training "at an astounding rate, from a mere 225 men who pinned on pilot wings in the last half of 1939, to 2,500 in the last quarter of 1941, rising to a peak of 29,000 in the second quarter of 1944." *Air Power History* 61, no. 1 (2014): 36.

94 **"shadows of imagination"**: Samuel T. Coleridge, *Biographia Literaria; or, Biographical Sketches of My Literary Life and Opinions* (Leavitt, Lord & Company, 1834), 175.

94 regimented real-time editing: Lex Parrish, *Space-Flight Simulation Technology* (Howard W. Sams, 1969), 16–17.

95 **Some resisted learning**: "Link Nickelodeon Rebuild & Interview with Edwin Link Jr, November 6, 1965" (source unknown; Link Collection).

95 **"The turns did happen"**: Conway, *Blind Landings*, 24.

95 This new cognitive approach: Schultz, *Problem with Pilots*, 58.

95 Lacking a visible horizon: Based on Assen "Jerry" Jordanoff, *Through the Overcast: The Art of Instrument Flying* (Funk & Wagnalls, 1938), 305–6; Jimmy Doolittle and Carroll Glines, *I Could Never Be So Lucky Again: An Autobiography* (Schiffer Publishing, 1995; originally published in 1991 by Bantam), 132–35.

96   **They sought to identify:** *Solving the Problem of Fog Flying: A Record of the Activities of the Fund's Full Flight Laboratory to Date* (Daniel Guggenheim Fund for the Promotion of Aeronautics, 1929), 5.

96   **Jimmy Doolittle:** Doolittle and Glines, *I Could Never Be*; "Jimmy Doolittle," Home of Heroes, accessed July 2023.

96   **Schneider Trophy:** Jerry Murland, *The Schneider Trophy Air Races: The Development of Flight from 1909 to the Spitfire* (Pen & Sword Books, 2021).

96   **Doolittle topped the speed record:** Richard P. Hallion, "Pioneer of Flight: Doolittle as Aviation Technologist," *Air Power History* 40, no. 4 (1993): 9.

96   **Harry Guggenheim borrowed him:** Doolittle and Glines, *I Could Never Be*, 127.

97   **five linked problems:** *Solving the Problem of Fog Flying*, 4.

97   **disperse the fog:** *Solving the Problem of Fog Flying*, 44.

97   **Consolidated NY-2 biplane:** See *Equipment Used in Experiments to Solve the Problem of Fog Flying: A Record of the Instruments and Experience of the Fund's Full Flight Laboratory* (Daniel Guggenheim Fund for the Promotion of Aeronautics, 1930), 6.

97   **Guggenheim encouraged Doolittle:** Richard Hallion, *Legacy of Flight: The Guggenheim Contribution to American Aviation* (University of Washington Press, 1977).

97   **"However, despite all":** Carroll Glines, *Jimmy Doolittle: Daredevil Aviator and Scientist* (Macmillian, 1972), 90.

98   **Bronson Murray Cutting:** Key references were related newspaper articles published on May 7, 1935, in the *New York Times*, *Washington Post*, *Baltimore Sun*, and *Chicago Daily Tribune*; Nick A. Komons, *The Cutting Air Crash: A Case Study in Early Federal Aviation Policy* (US Federal Aviation Administration, 1984); George E. Hopkins, *Flying the Line: The First Half Century of the Air Line Pilots Association* (Air Line Pilots Association, 1982).

99   **"unintentional collision":** Komons, *Cutting Air Crash*, 32.

100   **over 40 instruments:** E. J. Lovesey, "Information Overload in the Cockpit," *IEE Colloquium on Information Overload* (Institution of Electrical Engineers, 1995); R. Schroer, "Cockpit Instruments (A Century of Powered Flight: 1903–2003)," *IEEE Aerospace and Electronic Systems Magazine* 18, no. 7 (2003): 13–18.

100   **"instrument explosion":** E. J. Lovesey, "The Instrument Explosion—a Study of Aircraft Cockpit Instruments," *Applied Ergonomics* 8, no. 1 (1977): 23–30.

100   **"The day of the throttle jockey":** David Mindell, *Digital Apollo: Human and Machine in Spaceflight* (MIT Press, 2008); quoted in Schultz, *Problem with Pilots*, 220.

100   **"A modern pilot":** Schultz, 122.

101   **To realistically reflect:** Tom Cawley, "Link Trainer Secrets Are a Saga of Science but Can't Be Told Now," *Binghamton Press*, June 20, 1942.

101   **flew more like an airplane:** June 8, 1945, news clipping, Link Collection.

101   **"Don't make the mistake":** Jordanoff, *Through the Overcast*, 314–15.

101   **women's opportunities in aviation:** Carole Briggs, *At the Controls: Women in Aviation* (Lerner Publications, 1991).

101   **Jacqueline Cochran . . . Amelia Earhart:** Janene G. Leonhirth, "They Also Flew: Women Aviators in Tennessee, 1922–1950" (master's thesis, Middle Tennessee State University, 1990), 3–7.

102   **"They are simply thoroughly":** Amelia Earhart, *The Fun of It: Random Records*

of My Own Flying and of Women in Aviation (Harcourt Brace and Company, 1932), 162.

102 *Friendship*: See Amelia Earhart, *20 Hrs. 40 Min.: Our Flight in the* Friendship (G. P. Putnam's Sons, 1928).

102 **Powder Puff Derby:** Gene Nora Jessen, *The Powder Puff Derby of 1929: The First All Women's Transcontinental Air Race* (Sourcebooks, 2002).

102 **In one distressing instance:** Corn, *Winged Gospel*, 79.

103 **"Sweetie":** Quoted letters from Judy Litoff and David C. Smith, "American Women in a World at War," *OAH Magazine of History* 6, no. 3 (2002): 9–10.

103 **Black women:** Paul Louis Dawson, "Luis De Florez and the Special Devices Division" (PhD diss., George Washington University, 2005), 204.

104 **WASP:** WAFS and WFTD were organized in September 1942 and incorporated into WASP in August 1943. See Sarah Byrn Rickman, *WASP of the Ferry Command: Women Pilots, Uncommon Deeds* (University of North Texas Press, 2016); Katherine Sharp Landdeck, *The Women with Silver Wings: The Inspiring True Story of the Women Airforce Service Pilots of World War II* (Crown, 2020); and related information on the websites of the Airforce Historical Support Division and National WWII Museum.

104 **"The gods must envy me":** Letter from Marion Stegeman, April 24, 1943; Litoff and Smith, 11.

104 **Link Trainer instructors:** Corn, *Winged Gospel*, 89.

104 **Eugene Jacques Bullard:** Jim Haskins, *Black Eagles: African Americans in Aviation* (Scholastic, 1995), 5–9.

105 **Bessie Coleman:** Haskins, *Black Eagles*, 17–43; Doris L. Rich, *Queen Bess: Daredevil Aviator* (Smithsonian Institution Press, 1993); Flint Whitlock, "Racial Discrimination against Pilots: An Historical Perspective," in *Ethical Issues in Aviation*, ed. E. Hoppe (Ashgate, 2011), 137–44.

105 **Hubert Fauntleroy Julian:** David Shaftel, "The Black Eagle of Harlem," *Air & Space*, December 2008.

105 **William Jenifer Powell:** Haskins, *Black Eagles*, 50–66.

106 **"While originally it"** . . . **Banning and Allen:** Julia Lauria-Blum, "James H. Banning and the Flying Hoboes Transcontinental Flight," *Metropolitan Airport News*, May 13, 2020.

106 **"with the gracefulness of a bird":** Banning article in the *Pittsburgh Courier*, December 17, 1932; reprinted in Joseph J. Corn, ed., *Into the Blue: American Writing on Aviation and Spaceflight* (Library of America, 2011), 78–81.

106 **Tuskegee:** See Charles E. Francis, *The Tuskegee Airmen: The Men Who Changed a Nation*, 5th ed., ed. Adolph Caso (Branden Books, 2008; first published 1955); Charles W. Dryden, *A-Train: Memoirs of a Tuskegee Airman* (University of Alabama Press, 1997); Samuel L. Broadnax, *Blue Skies, Black Wings: African American Pioneers of Aviation* (University of Nebraska Press, 2008).

106 **"I always heard":** Quote as reported in David Stout, "Charles Anderson Dies at 89; Trainer of Tuskegee Airmen," *New York Times*, April 17, 1996.

106 **"Keep us flying!"** . . . **"Your letters and gifts":** "The Tuskegee Airmen: Training and Stateside Experiences," website for the Tuskegee Airmen National Historical Site, National Park Service, accessed July 2023.

108 **dead reckoning:** H. H. Shufeldt and G. D. Dunlap, *Piloting and Dead Reckoning*, 4th ed., ed. Bruce Allan Bauer (Naval Institute Press, 1999).

108 **higher altitudes:** Philip Van Horn Weems, *Air Navigation* (McGraw-Hill, 1943), 355.

108 **almanac:** Boyden Sparkes, "Teaching 'Lindy' Navigation," *Popular Science Monthly*, August 1928, 53–54, 108.

108 **presenting geography:** Edwin A. Link, "Helping Hand," *Connecting Link* 4, no. 1 (January 1947).

Chapter Three: Wicked Vulnerability

111 **July 1918:** My chief readings about the molasses disaster were Stephen Puleo, *Dark Tide: The Great Boston Molasses Flood of 1919* (Beacon Press, 2019); John Mason's account in *Yankee Magazine*, January 1965, 52–53, 109–11, posted on "Eric Postpischil's Molasses Disaster Pages," Eric Postpischil's Domain, October 30, 2021; and articles in the *Boston Globe* and *Boston Post* in 1919. More recent analyses were in *Los Angeles Times* (September 1995), *History* (February 2014, January 2019), *Boston Globe* (January 2015), *New York Times* and *New Scientist* (November 2016), and *Mental Floss* and WBUR News (January 2019).

111 **Isaac Gonzalez:** Puleo, *Dark Tide*, 24–25.

111 **"essential ingredient in":** Puleo, 48.

111 **the 1760s:** See Gilman Ostrander, "The Colonial Molasses Trade," *Agricultural History* 30, no. 2 (1956): 81–82.

112 **"very tolerable drink":** William Penn's quote from Gilman Ostrander, *The Molasses Trade of the Thirteen Colonies* (University of California, 1948), 6.

112 **"sturdy, sound, and ready to use":** Puleo, *Dark Tide*, 17.

114 **"like dinosaurs in a tar pit":** Susan Doll and David Morrow, "The Great Boston Molasses Flood: Newspaper Coverage and the Shaping of an Extraordinary Disaster," in *Ordinary Reactions to Extraordinary Events*, ed. R. Browne and A. Neal (Bowling Green State University Popular Press, 2001), 198.

114 **stresses far exceeded:** Ronald A. Mayville, "The Great Boston Molasses Tank Failure of 1919," *Civil + Structural Engineer*, September 1, 2014.

114 **tank's safety factor:** Puleo, *Dark Tide*, 207.

114 **zigzag damage pattern:** Earl Parker, *Brittle Behavior of Engineering Structures* (John Wiley & Sons, 1957), 255–56.

114 **Liberty ships:** Parker, *Brittle Behavior*, 270–73.

115 **242 storage tank . . . "caused by human errors" . . . Sounder engineering:** James I. Chang and Cheng-Chung Lin, "A Study of Storage Tank Accidents," *Journal of Loss Prevention in the Process Industries* 19, no. 1 (2006): 51, 58.

115 **"There's no disaster":** Richard Bach, *One* (Dell, 1988), 121.

115 **willing to accept:** William W. Lowrance, *Of Acceptable Risk: Science and the Determination of Safety* (William Laufmann, 1976).

116 **Guy Tozzoli:** Details were reconstructed from James Glanz and Eric Lipton, *City in the Sky: The Rise and Fall of the World Trade Center* (Holt, 2003) and from various obituaries.

117 **Steel: 200,000 tons . . . National Parks:** Glanz and Lipton, *City in the Sky*, 177.

117 **"It's going to take"; "I don't care":** Glanz and Lipton, 239.

118 **critical utilities:** Thomas O'Rourke, Arthur Lembo, and Linda Nozick, "Lessons Learned from the World Trade Center Disaster about Critical Utility Systems," in *Beyond September 11th: An Account of Post-Disaster Research* (University of Colorado, 2003), 269–90.

118 **AT&T mobilized:** Mitchell L. Moss and Anthony Townsend, "Response, Restoration, and Recovery: September 11 and New York City's Digital Networks," *Crisis Communications: Lessons from September 11*, ed. A. M. Noll (Rowman & Littlefield, 2003), 57.

119 **"complex maladaptive systems":** For discussion on differences between "CAS1" and "CAS2" systems from both evolutionary and engineering perspectives see David Sloan Wilson and Guru Madhavan, "Complex Maladaptive Systems," *Bridge* 50, no. 4 (Winter 2020): 61–63.

119 **linked ensemble models:** Some of this discussion appeared in my article "Do-It-Yourself Pandemic: It's Time for Accountability in Models," *Issues in Science and Technology*, July 1, 2020.

119 **analyses reported:** William Pitts, Kathryn Butler, and Valentine Junker, *Federal Building and Fire Safety Investigation of the World Trade Center Disaster: Visual Evidence, Damage Estimates, and Timeline Analysis* (Report No. NIST NCSTAR 1-5A; National Institute of Standards and Technology, 2005).

120 **fire-resistance rating:** See NIST, *Federal Building and Fire Safety Investigation of the World Trade Center Disaster: Final Report of the National Construction Safety Team on the Collapse of the World Trade Center Towers* (NIST NCSTAR 1, 2005); Shyam Sunder, "20 Years Later: NIST's World Trade Center Investigation and Its Legacy," August 18, 2021; Kathryn Butler, "Putting Together the Big Picture for the World Trade Center Disaster Investigation," *Taking Measure: Just a Standard Blog*, August 25, 2021, NIST.

122 **"paradox of robustness":** Steven Frank, "Maladaptation and the Paradox of Robustness in Evolution," *PLoS One* 2, no. 10 (2007): e1021.

122 **"I'd ride the elevator":** Undated World Trade Centers Association video tribute to Guy Tozzoli, courtesy of a YouTube video, uploaded May 10, 2013.

122 **Funeral Oration:** Thomas Hobbes of Malmesbury, "The History of the Grecian War (The Second Book of the History of Thucydides)," in *The English Works of Thomas Hobbes*, ed. W. Molesworth (John Bohn, 1843), 8:198.

123 **"What If Hurricane Ivan"** Shirley Laska, *Sociological Inquiry* 78, no. 2 (2008): 174–78, originally in *Natural Hazards Observer* 24, no. 2 (November 2004).

123 **barreled with an energy equivalent:** Ivor van Heerden and Mike Bryan, *The Storm: What Went Wrong and Why during Hurricane Katrina* (Viking, 2006), 13.

123 **migration four times:** Joanne M. Nigg, John Barnshaw, and Manuel R. Torres, "Hurricane Katrina and the Flooding of New Orleans: Emergent Issues in Sheltering and Temporary Housing," *Annals of the American Academy of Political and Social Science* 604 (2006): 113–28.

123 **Katrina damaged:** See the executive summary and overview section in IPET, *Performance Evaluation of the New Orleans and Southeast Louisiana Hurricane Protection System: Final Report of the Interagency Performance Evaluation Task Force*, vol. 1 (June 2009).

123 **New Orleans was established:** J. David Rogers, "Development of the New Orleans Flood Protection System Prior to Hurricane Katrina," *Journal of Geotechnical and Geoenvironmental Engineering* 134, no. 5 (2008): 602–17; C. R. Kolb and R. T. Saucier, "Engineering Geology of New Orleans," in *Geology under Cities*, ed. R. F. Legget (Geological Society of America, 1982), 95–118.

123 **1722 Great Hurricane:** Douglas Brinkley, *The Great Deluge: Hurricane Katrina, New Orleans, and the Mississippi Gulf Coast* (Harper Perennial, 2006), 5–7.

124 **Louisiana Purchase . . . 1844 flood:** George Pabis, "Delaying the Deluge: The

Engineering Debate over Flood Control on the Lower Mississippi River, 1846–1861," *Journal of Southern History* 64, no. 3 (1998): 425–26.

124 **"curse to the soil":** C. Vann Woodward, *Origins of the New South 1877–1913* (Louisiana State University Press, 1970), 180; quoted in John Dean Davis, "Levees, Slavery, and Maintenance," *Technology's Stories* (August 2018).

124 **"the first object of the settler":** Andrew Atkinson Humphrey, *Report upon the Physics and Hydraulics of the Mississippi River* (Government Printing Office, 1867), 80.

124 **Humphreys:** Martin Reuss, "Andrew A. Humphreys and the Development of Hydraulic Engineering: Politics and Technology in the Army Corps of Engineers, 1850–1950," *Technology and Culture* 26, no. 1 (1985): 2–3.

125 **third of the federal budget:** Charles A. Camillo, *Divine Providence: The 2011 Flood in the Mississippi River and Tributaries Project* (Mississippi River Commission, 2012), 17.

125 **over 80 percent:** Zachary Pirtle, "How the Models of Engineering Tell the Truth," in *Philosophy and Engineering: An Emerging Agenda*, ed. I. Poel and D. Goldberg (Springer Netherlands, 2010).

125 **"The system did not perform":** IPET, *Performance Evaluation*, 1:2, 1:127.

125 **"The promise or perception":** IPET, 1:8.

126 **"toys that save millions":** Bill Addis, " 'Toys That Save Millions'—A History of Using Physical Models in Structural Design," *Structural Engineer*, 91, no. 4 (2013): 12–27.

127 **failure of I-walls:** Thomas L. Brandon, Stephen G. Wright, and J. Michael Duncan, "Analysis of the Stability of I-Walls with Gaps between the I-Wall and the Levee Fill," *Journal of Geotechnical and Geoenvironmental Engineering* 134, no. 5 (2008): 692–700.

127 **"hydraulic short circuit":** J. Michael Duncan, Thomas L. Brandon, Stephen G. Wright, and Noah Vroman, "Stability of I-Walls in New Orleans during Hurricane Katrina," *Journal of Geotechnical and Geoenvironmental Engineering* 134, no. 5 (2008): 681.

127 **"Our truth is the intersection":** Richard Levins, "The Strategy of Model Building in Population Biology," *American Scientist* 54, no. 4 (1966): 423.

127 **paragraph-length story:** See Jorge Luis Borges's "On Exactitude in Science" in *Collected Fictions*, trans. Andrew Hurley (Penguin Books, 1999), 325.

127 **"a handy fiction":** Neil Gaiman, *Neverwhere* (Harper Perennial, 2003), 8–9.

128 **"organized ignorance":** See Scott Frickel and M. Bess Vincent's 2007 paper "Hurricane Katrina, Contamination, and the Unintended Organization of Ignorance" in *Technology in Society* 29, no. 2: 181–88.

128 **undamageable bridges:** Irma Richter, ed., *The Notebooks of Leonardo da Vinci* (Oxford University Press, 1952), 294–95.

128 **wounded soldier:** Piers Blaikie et al., *At Risk: Natural Hazards, People's Vulnerability and Disasters* (Routledge, 1994); P. M. Kelly and W. N. Adger, "Theory and Practice in Assessing Vulnerability to Climate Change and Facilitating Adaptation," *Climatic Change* 47, no. 4 (2000): 325–52.

128 **" 'situations' which people":** Benjamin Wisner et al., *At Risk: Natural Hazards, People's Vulnerability and Disasters*, 2nd ed. (Routledge, 2004), 14.

129 **"viewing vulnerability":** Karen O'Brien et al., "What's in a Word? Conflicting Interpretations of Vulnerability in Climate Change Research" (CICERO Working Paper 2004:04, Center for International Climate and Environmental Research, Oslo, Norway), 3.

129 **engaging with vulnerabilities:** Anique Hommels, Jessica Mesman, and Wiebe E. Bijker, eds., *Vulnerability in Technological Cultures: New Directions in Research and Governance* (MIT Press, 2014); Katherine T. Fox-Glassman and Elke U. Weber, "What Makes Risk Acceptable? Revisiting the 1978 Psychological Dimensions of Perceptions of Technological Risks," *Journal of Mathematical Psychology* 75 (2016): 157–69.

129 **United States and the Netherlands:** Wiebe Bijker, "American and Dutch Coastal Engineering: Differences in Risk Conception and Differences in Technological Culture," *Social Studies of Science* 37, no. 1 (2007): 143–51.

130 **except to suffer local damage:** Scott Gabriel Knowles, *The Disaster Experts: Mastering Risk in Modern America* (University of Pennsylvania Press, 2011), 6–7.

130 **if tigers escaped a zoo:** Chauncey Starr, "Risk Management, Assessment, and Acceptability," *Risk Analysis* 5, no. 2 (1985): 97–98.

132 **"boring apocalypses":** Hin-Yan Liu et al., "Governing Boring Apocalypses: A New Typology of Existential Vulnerabilities and Exposures for Existential Risk Research," *Futures* 102 (2018): 6–19.

Refrain: Yaw

133 **Morton and Watson:** Taken from National Transportation Safety Board's 1968 Aircraft Accident Report: Delta Air Lines, Inc. DC-8, N802E, Kenner, Louisiana, March 30, 1967 (Adopted: December 20, 1967), Washington, DC: Department of Transportation. Related syndicated news reports were consulted from the *Lubbock Avalanche-Journal* (March 31, 1967) and the former *Argus Fremont* (April 1, 1967).

136 **Eastern Air Lines:** From Civil Aeronautics Board's Aviation Accident Report: Eastern Air Lines Flight 304, 7.

136 **"designated mysteries":** See, for example, Adriane Quinlan, "50 Years after Eastern Air Lines Flight 304 Crashed into Lake Pontchartrain Leaving No Survivors, Something Still Remains," *Times-Picayune*, February 25, 2014.

137 **"nothing but a mechanical spy":** Pigott, *Brace for Impact*, 101.

137 **David Warren:** Based on Marcus Williamson's July 31, 2010, obituary in the *Independent* and Roger Connor's presentation "The Black Box: Creating Resiliency in Air Transport for 77 Years: But, Where Did It Come From?," 2014 IEEE/AIAA 33rd Digital Avionics Systems Conference.

137 **best place . . . "never saw an airplane":** Scott M. Fisher, "Father of the Black Box," HistoryNet, June 29, 2017.

138 **"indestructible machine":** Quotes and related background from Greg Siegel, "Technologies of Accident: Forensic Media, Crash Analysis, and the Redefinition of Progress" (PhD diss., University of North Carolina at Chapel Hill, 2005), 73, 76–80.

138 **withstand impact shocks:** Greg Siegel, *Forensic Media: Reconstructing Accidents in Accelerated Modernity* (Duke University Press, 2014), 82.

138 **"live truthfully under imaginary circumstances":** Larry Silverberg, *The Sanford Meisner Approach: An Actor's Workbook—Workbook One* (Smith and Kraus, 1994), 9.

138 **analog processors:** Masaaki Hirooka, *Innovation Dynamism and Economic Growth: A Nonlinear Perspective* (Edward Elgar Publishing, 2006), 245–46.

139 **the simulator:** John Killick, "Proxy Flight," *Pegasus*, January 1954, 8.

139 **VAMP:** Kelly and Parke, *Pilot Maker*, 136–37.

139 **"simulation syndrome"**: From the title of James Der Derian's "The Simulation Syndrome: From War Games to Game Wars," *Social Text* 24 (1990): 187–92.

139 **specialized simulators**: P. W. Caro, "Flight Crew Training Technology: A Review," NASA Ames Research Center, prepared by Seville Training Systems, 1984.

140 **"In the culture of simulation"**: Sherry Turkle, *Life on the Screen* (Simon & Schuster, 1995), 24.

141 **"Microsoft Flight Simulator"**: For more on Bruce Artwick and the philosophy behind the product design, see Preston Lerner, "Pilot Program," *Air & Space Quarterly*, March 22, 2023.

142 **forms of fidelity**: Don Harris, *Human Performance on the Flight Deck* (Ashgate, 2011), 126–33.

142 **g-forces as experienced**: Kelly and Parke, *Pilot Maker*, 70.

142 **"lack of vertigo"**: Jeon, "The Virtual Flier," 46.

143 **"The pilots become so preoccupied"**: Philip Klass, "Link Simulator Boosts B-47 Potential," *Aviation Week* 56, no. 24 (June 16, 1952).

143 **"pucker factor"**: Quote from Robert W. Weight, "Flight Simulators in the RAAF: A Bold Step Forward or Back to the Future?," *Wings* (Spring 2013): 36; Peter Hobbins, "Emulating the 'Pucker Factor': Faith, Fidelity and Flight Simulation in Australia, 1936–58," *Journal of Transport History* 44, no. 1 (2022): 3–26.

143 **United Airlines and Continental**: Andy Pasztor and Susan Carey, "United Continental Pilots Split on Training Simulators," *Wall Street Journal*, June 20, 2011.

144 **"Off we go into the wild blue yonder"**: From the official song of the US Air Force, often simply referred to as "Wild Blue Yonder."

145 **"gamify" the experience**: Timothy Lenoir, "All but War Is Simulation: The Military-Entertainment Complex," *Configurations* 8, no. 3 (2000): 295.

145 **"Not that I wish to"**: Ed Link to Margaret Weems, April 10, 1944.

146 **Alex Seiden**: National Research Council, *Modeling and Simulation: Linking Entertainment and Defense* (National Academies Press, 1997), 168–69.

146 **Mountains cannot resemble . . . operating environment**: See, for example, Patrick Crogan, *Gameplay Mode: War, Simulation, and Technoculture* (University of Minnesota Press, 2011); and Ralph Norman Haber, "Flight Simulation," *Scientific American* 255, no. 1 (1986): 96–103.

147 **their latest software**: For an in-depth review see David Allerton, *Flight Simulation Software: Design, Development and Testing* (John Wiley & Sons, 2023).

148 **"dromoscopic vision"; "The world flown over"**: Quotes from Crogan, *Gameplay Mode*, 37, 41. Based on Paul Virilio, *Negative Horizon*, trans. Michael Degener (Continuum, 1989), 105–19.

149 **visual detail**: Ivan Sutherland, "The Ultimate Display," *Proceedings of the International Federation for Information Processing Congress* 65, no. 2 (1965): 506–8.

149 **"been adding one thing"**: From Ed Link's letter to C. S. Jones, October 25, 1941.

150 **"must abandon the notion"**: From Eduardo Salas, Clint A Bowers, and Lori Rhodenizer, "It Is Not How Much You Have but How You Use It: Toward a Rational Use of Simulation to Support Aviation Training," *International Journal of Aviation Psychology* 8, no. 3 (1998): 197–208.

150 **"I sort of lost interest"**: From Mary Beth Herzog, *Vero Beach Press Journal*, May 1, 1977.

150 **"I started as a grease monkey"**: From Clarence E. Lovejoy, *New York Times*, February 23, 1956.

Chapter Four: Wicked Safety

153 **the Great Lafayette:** See Derek Tait, *The Great Illusionists* (Pen & Sword Books, 2018); Arthur Setterington, *The Life and Times of the Great Lafayette* (Abraxas Publications, 1991). Related coverage in *Fife Free Press, & Kirkcaldy Guardian* (1911), *Scotsman* (July 15, 2013, by David McLean), *Edinburgh Reporter* (May 3, 2011).

154 **theater fires:** Based on William Paul Gerhard, "The Safety of Theatre Audiences and of the Stage Personnel against Danger from Fire and Panic," *American Architect*, October 21 and 28, 1899 (reprinted from the publications of the British Fire Prevention Committee), 7–10.

154 **Magic numbers:** See Margaret Law and Paula Beever, "Magic Numbers and Golden Rules," *Fire Safety Science* 4 (1994): 79–84.

156 **"flame thrower":** From a research profile of Bisby: *Nature*, November 23, 2020, 694.

156 **municipal fire service:** Resources include Stephen Dando-Collins, *The Great Fire of Rome: The Fall of the Emperor Nero and His City* (Da Capo Press, 2010); Joseph J. Walsh, *The Great Fire of Rome: Life and Death in the Ancient City* (Johns Hopkins University Press, 2019); Anthony A. Barrett, *Rome Is Burning: Nero and the Fire That Ended a Dynasty* (Princeton University Press, 2020).

157 **"code speak":** See Angus Law and Graham Spinardi, "Performing Expertise in Building Regulation: 'Codespeak' and Fire Safety Experts," *Minerva* 59, no. 4 (2021): 515–38.

157 **fire-safety design practices:** See Graham Spinardi, Luke Bisby, and Jose Torero, "A Review of Sociological Issues in Fire Safety Regulation," *Fire Technology* 53, no. 3 (2017): 1011–37; Graham Spinardi, "Fire Safety Regulation: Prescription, Performance, and Professionalism," *Fire Safety Journal* 80 (2016): 83–88.

157 **"Don't learn safety by accident":** From the 1951 International Brotherhood of Electrical Workers' motto.

158 **Fukushima:** See Evan Osnos, "The Fallout," *New Yorker*, October 10, 2011; David Lochbaum et al., *Fukushima: The Story of a Nuclear Disaster* (New Press, 2014); Yoichi Funabashi, *Meltdown: Inside the Fukushima Nuclear Crisis* (Brookings Institution Press, 2021).

158 **vulnerable to tsunamis:** In Japan, hundreds of stone tablets or "tsunami stones," some from the 14th century, have provided stark warnings: "Do not build your homes below this point!" See Martin Fackler, "Tsunami Warnings, Written in Stone," *New York Times*, April 20, 2011, and Danny Lewis, *Smithsonian Magazine*, "These Century-Old Stone 'Tsunami Stones' Dot Japan's Coastline," August 31, 2015.

158 **Japanese mindset:** Quote from *The National Diet of Japan: The Official Report of the Fukushima Nuclear Accident Independent Investigation Commission*, Executive Summary, Message from the Chairman (2012), 9.

159 **"adequate protection":** Bill Ostendorff and Kimberly Sexton, "Adequate Protection after the Fukushima Daiichi Accident: A Constant in a World of Change," *Nuclear Law Bulletin* 91, no. 1 (2013): 23–41.

160 **quirky geographic split:** This case study appeared in Guru Madhavan, John L. Anderson, and Alton D. Romig Jr., "Engineering for Inevitable Surprises," *PNAS Nexus* 1, no. 1 (2022): pgac014; with related references of accounts in the *Scientific American* blog (March 25, 2011), *Wall Street Journal* (March 26, 2011), and

IEEE Spectrum (April 6, 2011); Bill Canis, "The Motor Vehicle Supply Chain: Effects of the Japanese Earthquake and Tsunami," Congressional Research Service, May 23, 2011; Vasco Carvalho et al., "Supply Chain Disruptions: Evidence from the Great East Japan Earthquake," Columbia Business School Research Paper 17-5 (2016).

160   **Boeing didn't know:** Christopher Tang et al., "Managing New Product Development and Supply Chain Risks: The Boeing 787 Case," *Supply Chain Forum* 10, no. 2 (2009): 74–86.

160   **Airbus also grappled:** Nicola Clark, "The Airbus Saga," *New York Times*, December 11, 2006.

161   **seafood consumption:** Dalnim Lee et al., "Factors Associated with the Risk Perception and Purchase Decisions of Fukushima-Related Food in South Korea," *PLoS One* 12, no. 11 (2017): e0187655.

161   **"How many of you have":** From Ostendorff's talk, "Lessons for the U.S. from the Japanese Nuclear Crisis," at Advancing U.S. Resilience to a Nuclear Catastrophe, Center for Biosecurity, University of Pittsburgh Medical Center conference held on May 19, 2011, in Washington, DC.

161   **14,000 bird strikes:** Cecilia Nilsson et al., "Bird Strikes at Commercial Airports Explained by Citizen Science and Weather Radar Data," *Journal of Applied Ecology* 58, no. 10 (2021): 2029–39.

161   **natural threat:** John Downer, "When the Chick Hits the Fan: Representativeness and Reproducibility in Technological Tests," *Social Studies of Science* 37, no. 1 (2007): 10.

162   **bird species . . . mammal species:** Richard A. Dolbeer, Michael J. Begier, Phyllis R. Miller, John R. Weller, and Amy L. Anderson, *Wildlife Strikes to Civil Aircraft in the United States, 1990–2020*, Serial Report Number 27 (Federal Aviation Administration, 2021).

162   **"onus, therefore":** Downer, "When the Chick Hits," 11.

162   **engines are reliable enough:** Downer, 20.

162   **114,000 calendar years:** Downer, 233, with a reference to John Rushby's 1993 report *Formal Methods and the Certification of Critical Systems* (Technical Report CSL93-7).

163   **Reliability . . . Tories:** John Downer, "The Aviation Paradox: Why We Can 'Know' Jetliners but Not Reactors," *Minerva* 55, no. 2 (2017): 231, 235, 238.

163   **controlling every detail:** Paul E. Bierly and J. C. Spender, "Culture and High Reliability Organizations: The Case of the Nuclear Submarine," *Journal of Management* 21, no. 4 (1995): 639–56.

164   **"fail-safe" solutions:** A classic paper by Stanley Kaplan and B. John Garrick, "On the Quantitative Definition of Risk," *Risk Analysis* 1, no. 1 (1981): 11–27.

164   **fail-safe system . . . safe-to-fail system:** See Jack Ahern, "From Fail-Safe to Safe-to-Fail: Sustainability and Resilience in the New Urban World," *Landscape and Urban Planning* 100, no. 4 (2011): 341–43; Mikhail Chester, Thaddeus R. Miller, and Tischa A. Muñoz-Erickson, "Rethinking Infrastructure in an Era of Unprecedented Weather Events," *Issues in Science and Technology* (Winter 2018); Mikhail Chester, B. Shane Underwood, and Constantine Samaras, "Keeping Infrastructure Reliable under Climate Uncertainty," *Nature Climate Change* 10, no. 6 (2020): 488–90.

164   **drivers have insurance:** Jakub Kubrynski, *DevSkiller Tech Blog*, April 4, 2018.

164   **much-quoted editorial:** *Wall Street Journal*, May 14, 1952.

164   **"You may share it with others":** Statement of Hyman G. Rickover, Radiation Safety and Regulation: Hearings before the Joint Committee on Atomic Energy, 87th Congress, June 1961, p. 366.

165   **"inverted miracle":** Paul Virilio, *Politics of the Very Worst*, trans. Michael Cavaliere, ed. Sylvère Lotringer (Semiotext(e), 1999), 89.

165   **"atrophy of vigilance":** George Busenberg, "The Evolution of Vigilance: Disasters, Sentinels and Policy Change," *Environmental Politics* 8, no. 4 (1999): 90–109.

165   **"normalization of deviance":** Diane Vaughan, *The Challenger Launch Decision: Risky Technology, Culture, and Deviance at NASA* (University of Chicago Press, 1996).

165   **"we provoke the hazards":** Barry Turner, "The Sociology of Safety," *Engineering Safety*, ed. D. I. Blockley (McGraw-Hill, 1992), 189.

165   **Swiss cheese:** See James Reason, *Human Error* (Cambridge University Press, 1991).

166   **"normal accidents":** See Charles Perrow, *Normal Accidents: Living with High Risk Technologies* (Basic Books, 1984); Chris Clearfield and András Tilcsik, *Meltdown: Why Our Systems Fail and What We Can Do about It* (Penguin Press, 2018).

166   **complex as a nuclear reactor; "high-reliability organization":** See Gene Rochlin, Todd R. La Porte, and Karlene H. Roberts, "The Self-Designing High-Reliability Organization: Aircraft Carrier Flight Operations at Sea," *Naval War College Review* 40, no. 4 (1987): 76–92; Karlene H. Roberts, "Managing High Reliability Organizations," *California Management Review* 32, no. 4 (1990): 101–14; Todd R. LaPorte and Paula M. Consolini, "Working in Practice but Not in Theory: Theoretical Challenges of 'High-Reliability Organizations,'" *Journal of Public Administration Research and Theory* 1, no. 1 (1991): 19–48.

166   **some organizations:** Nick Pidgeon, "Complex Organizational Failures: Culture, High Reliability, and the Lessons from Fukushima," *Bridge* 42, no. 3 (2012): 18–22; Nick Pidgeon and Mike O'Leary, "Man-Made Disasters: Why Technology and Organizations (Sometimes) Fail," *Safety Science* 34, no. 1–3 (2000): 15–30.

166   **"mindful organizing" . . . "error free":** Karl E. Weick and Kathleen M. Sutcliffe, *Managing the Unexpected: Sustained Performance in a Complex World*, 3rd ed. (Jossey-Bass, 2015), 12, 18, 21–22. See also Nassim Nicholas Taleb, *Antifragile: Things That Gain from Disorder* (Random House, 2012).

167   **In engineering, safety:** My view is influenced by Nancy Leveson; see Leveson, *Engineering a Safer World: Systems Thinking Applied to Safety* (MIT Press, 2011) and *Safeware: System Safety and Computers* (Addison-Wesley, 1995).

167   **both can coexist:** Leveson, *Engineering a Safer World*, 7–9; see also Nancy Leveson et al., "Moving Beyond Normal Accidents and High Reliability Organizations: A Systems Approach to Safety in Complex Systems," *Organization Studies* 30, no. 2–3 (2009), 227–49.

167   **ferry system:** Nancy Leveson and John P. Thomas, *STPA Handbook* (Massachusetts Institute of Technology, March 2018), 9.

168   **"Productivity and safety":** Nancy Leveson, *Safety III: A Systems Approach to Safety and Resilience* (Massachusetts Institute of Technology, July 1, 2020).

168   **Normal-accident theory . . . High-reliability organizations:** Leveson et al., "Moving Beyond Normal Accidents," 227–30.

169   **A process isn't a hard solution; "In engineering, like flying":** Nathan J. Slegers, Ronald T. Kadish, Gary E. Payton, John Thomas, Michael D. Griffin, and Dan Dumbacher, "Learning from Failure in Systems Engineering: A Panel Discussion," *Systems Engineering* 15, no. 1 (2012): 74–82.

169   **PowerPoint:** Edward R. Tufte, *The Cognitive Style of PowerPoint: Pitching Out Corrupts Within* (Graphics Press, 2006); James Thomas, *McDreeamie-Musings* (blog), April 15, 2019.

169   **"a social instrument":** Ian Parker, "Absolute PowerPoint," *New Yorker*, May 28, 2001.

170   **"The foam debris":** Nancy Leveson et al., "Effectively Addressing NASA's Organizational and Safety Culture: Insights from Systems Safety and Engineering Systems," presented at the MIT Engineering Systems Division Symposium, March 2004, 4–5, from the August 2003 Columbia Accident Investigation Board report, 195.

170   **one in five Americans:** Leveson, *Safety III*, 17–18.

170   **"blame is the enemy of safety":** Leveson, *Engineering a Safer World*, 56.

171   **"There's nowhere to hide"; "It's become the meaning":** Svetlana Alexievich, *Voices from Chernobyl*, trans. Keith Gessen (Dalkey Archive Press, 2005), 45, 219.

171   **isn't some linguistic quibble:** Steve Rayner and Robin Cantor, "How Fair Is Safe Enough? The Cultural Approach to Societal Technology Choice," *Risk Analysis* 7, no. 1 (1987): 5.

171   **definers—not the defined:** Toni Morrison, *Beloved* (Knopf, 1987), 190.

171   **"psychedelic" dreamworld:** Jerome Lederer, "Ideal Safety System for Accident Prevention," *Journal of Air Law and Commerce* 34, no. 3 (1968): 336–42.

172   **Moulton; "Engineers, however":** Lederer, "Ideal Safety System," 339.

172   **"ultrasafe" systems:** René Amalberti, "The Paradoxes of Almost Totally Safe Transportation Systems," *Safety Science* 37, no. 2 (2001): 109–26.

173   **"natural or ecological safety":** Amalberti, "Paradoxes," 118.

173   **many different forms of expertise:** Law and Spinardi, "Performing Expertise in Building Regulation."

174   **international meeting of fire-safety experts:** Bisby's lecture was at the International Congress Fire Safety & Science at the Institute for Safety in Arnhem, the Netherlands, on November 14 and 15, 2018.

Refrain: Surge

175   **White House Cabinet Room:** From John F. Kennedy Library, President's Office Files, Presidential Recordings Collection, Tape 63A, November 21, 1962, from Presidential Recordings, Miller Center, University of Virginia. I encountered this in Erinn Catherine McComb's dissertation, "Why Can't a Woman Fly? NASA and the Cult of Masculinity, 1958–1972" (Mississippi State University, 2012), 274.

175   **political symbolism:** See Walter A. McDougall, *The Heavens and the Earth: A Political History of the Space Age* (Basic, 1985); Teasel Muir-Harmony, *Operation Moonglow: A Political History of Project Apollo* (Basic, 2020).

178   **A crucial hurdle:** See John McLeod, "Manned Spacecraft Simulation," *Proceedings of the May 21–23, 1963, Spring Joint Computer Conference* (Detroit, Michigan, 1963), 401–9.

178   **ENIAC; "spurring rumors":** Wikipedia, "ENIAC," August 2023; Gregory Farrington, "ENIAC: Birth of the Information Age," *Popular Science*, March 1996, 74.

178   **$2 million each:** The computers were IBM 9600/9604 family. Mark-I and Mark-II could barely compute the equations of motions at the requisite speed.

178   **instructions per second:** "Sampling frequency" specifies performing at least

twice as quickly as the natural frequency of the simulated process. Abbey's team needed to execute the equations 10 times faster to produce real-time simulations of the variables. See Link Capabilities Statement, April 1963, courtesy of TechWorks!

179   **systems integration:** Manfred von Ehrenfried, *Apollo Mission Control: The Making of a National Historic Landmark* (Springer International Publishing, 2018), 130.

179   **human impulse:** Based on Stanley H. Goldstein, "Astronaut Training: An Administrative History of Projects Mercury, Gemini and Apollo" (master's thesis, University of Colorado, 1984), 3–4; and Goldstein's *Reaching for the Stars: The Story of Astronaut Training and the Lunar Landing* (Praeger, 1987).

180   **time in simulators:** See C. H. Woodling et al., "Apollo Experience Report: Simulation of Manned Space Flight for Crew Training" (NASA Technical Note D-7112, 1973), 1–3.

180   **"The final simulation":** Gene Kranz, *Failure Is Not an Option: Mission Control from Mercury to Apollo 13 and Beyond* (Simon & Schuster, 2000).

180   **nearly 700 switches:** Hamish Lindsay, *Tracking Apollo to the Moon* (Springer-Verlag London, 2001), 206–7.

180   **sophisticated visual experience:** Based on Paul Ceruzzi, *Beyond the Limits: Flight Enters the Computer Age* (MIT Press, 1989), 170; James Tomayko, *Computers Take Flight: A History of NASA's Pioneering Digital Fly-by-Wire Project* (NASA History Office, 2000); Joe Dahm, *Evening Press* (Binghamton), November 24, 1964, 17, 32.

181   **10 tons devoted:** Kelly and Parke, *Pilot Maker*, 160.

181   **"Apollo gave new":** Chiabotti, "The Glorified Link," 61, 63.

181   **14-hour days:** See David M. Harland's "Preparations," in *The First Men on the Moon: The Story of Apollo 11* (Springer Praxis Books, 2007).

182   **When the *Eagle* landed:** *Apollo 11* Technical Air-to-Ground Voice Transmission (GOSS NET 1), Tape 70, no. 24, p. 377.

182   **"normal design" . . . "radical design":** Walter Vincenti, "Engineering Knowledge, Type of Design, and Level of Hierarchy: Further Thoughts About What Engineers Know . . . ," in *Technological Development and Science in the Industrial Age*, ed. P. Kroes and M. Bakker (Kluwer Academic Publishers, 1992), 17–34.

183   **weight-loss project:** Courtney Brooks, James M. Grimwood, and Loyd S. Swenson, *Chariots for Apollo: The NASA History of Manned Lunar Spacecraft to 1969* (NASA, 1979), 172–75.

184   **"organization, education, and training":** Jean-Jacques Servan-Schreiber, quoted in Stephen Johnson, *The Secret of Apollo: Systems Management in American and European Space Programs* (Johns Hopkins University Press, 2002), 5.

184   **success of such projects:** Stephen Johnson, "Philosophical Observations and Applications in Systems and Aerospace Engineering," in *Engineering and Philosophy: Reimagining Technology and Social Progress*, ed. Z. Pirtle, D. Tomblin, and G. Madhavan (Springer, 2021), 94.

184   **"We can lick gravity":** Often attributed to Wernher Von Braun.

184   **"Spacecraft that":** Johnson, *Secret of Apollo*, 4.

184   **six-step procedure:** Brooks, Grimwood, and Swenson, *Chariots for Apollo*, 169.

185   **"insurance for technical success":** Johnson, *Secret of Apollo*, 223.

185   **"Military officers":** Johnson, 17.

185 "We were lucky all the time": Vincent Davis, "NASA Retiree Reflects on Apollo 13 Crisis," *San Antonio Express-News*, April 10, 2010.

185 "no training rhythm to it": Francis E. "Frank" Hughes, interview by Rebecca Wright, Houston, Texas, September 10, 2013, Edited Oral History Transcript, NASA Johnson Space Center Oral History Project.

186 "Houston, we've had a problem": *Apollo 13* Technical Air-to-Ground Voice Transcription, NASA Apollo Spacecraft Program Office, Manned Spacecraft Center, Houston, Texas, April 1970.

187 The Link simulator: Based on "Deliverance from Disaster," *Connecting Link* (Summer 1970): 2–5.

187 Apollo 13 needed: See Jim Lovell and Jeffrey Kluger, *Apollo 13* (Mariner Books, 2006); Fred Haise and Bill Moore, *Never Panic Early: An Apollo 13 Astronaut's Journey* (Smithsonian Books, 2022).

187 "successful failure"; "unsung heroes": Frank Hughes's foreword for Brian Woycechowsky, *Lunar Module Moon-Referenced Equations of Motion* (Center for Technology & Innovation, 2021), vii.

188 "Makes you feel kind of creepy": Kurt Vonnegut, *Player Piano* (1952; Dial Press, 1999), 32.

188 operational preservation: John Chilvers, "Curatorial Lessons from Other Operational Preservationists: Towards a Methodology for Computer Conservation," *Proceedings of Making IT Work* (British Computer Society, National Museum of Computing, 2017), 44–53.

189 Whirlpool: See Gerald Smith, "How a Household Name Got Its Start in Binghamton," *Binghamton Press & Sun-Bulletin*, June 16, 2016.

190 military flight simulators: Dave Peters (cofounder of Diamond Visionics) and Richard Mecklenborg also significantly contributed to simulation through the wide-angled collimated window display technology, with Mecklenborg also advancing the dome display for the British vertical take-off aircraft.

191 Bennett wrote a monograph: *Visualizing Software: A Graphical Notation for Analysis, Design, and Discussion* (Marcel Dekker, 1992), 3–6.

192 "Another flaw": Kurt Vonnegut, *Hocus Pocus* (1990; Berkley, 1997), 238.

192 "Any sufficiently": Debbie Chachra's June 14, 2017, tweet.

Chapter Five: Wicked Maintenance

195 "telescoped a half century's": From Peter Maust, "Preventing 'Those Terrible Disasters': Steamboat Accidents and Congressional Policy, 1824–1860" (PhD diss., Cornell University, 2012), 3; sourced from Richard C. Wade, *The Urban Frontier* (University of Chicago Press, 1959), 70.

195 Senator Daniel Webster . . . Benton: From John G. Burke, "Bursting Boilers and the Federal Power," *Technology and Culture* 7, no. 1 (1966): 12.

196 "mine of gunpowder": James T. Lloyd, *Lloyd's Steamboat Directory, and Disasters on the Western Waters* (James T. Lloyd & Co., 1856), 90.

196 fragments of the ship deck: Henry Howe, *Historical Collections of Ohio* (Derby, Bradley, & Co., 1848), 223–24.

196 investigative committee: See *Report of the Committee Appointed by the Citizens of Cincinnati, April 26, 1838, to Enquire into the Causes of the Explosion of the Moselle and to Suggest Such Preventive Measures as May Be Best Calculated to*

Guard Hereafter against Such Occurrences (Alexander Flash; Looker & Ramsay, Printers, 1838), 28.

197 **"inflate the ambition":** Quoted in Jeff Suess, "Steamboat Explosion Led to Federal Regulations," *Cincinnati Enquirer*, March 14, 2018.

197 **Between 1807 and 1853:** Maust, "Preventing 'Those Terrible Disasters,'" 4; source, John K. Brown, *Limbs on the Levee: Steamboat Explosions and the Origins of Federal Public Welfare Regulation, 1817–1852* (International Steamboat Society, 1989), 13.

197 **One member of Congress:** Maust, 4.

197 **1845 court decision:** *Spencer v. Campbell*, 9 Watts & Serg. 32 (1845), described in Burke, "Bursting Boilers," 17–18.

198 **Congress passed:** Maust, "Preventing 'Those Terrible Disasters,'" 14; source, Forest G. Hill, "Formative Relations of American Enterprise, Government, and Science," *Political Science Quarterly* 75, no. 3 (September 1960): 416.

198 **"a wise and frugal":** From Thomas Jefferson's First Inaugural Address, March 4, 1801.

198 **Samuel Langhorne Clemens:** See Sherry Nord Marron's "Mark Twain's Work as a Steamboat Pilot Earned His Pen Name," *Los Angeles Times*, April 28, 2017.

198 **"lure of technology":** See Steven Usselman, "The Lure of Technology and the Appeal of Order: Railroad Safety Regulation in Nineteenth Century America," *Business and Economic History* 21 (1992): 290–99.

199 **degraded them:** Maust, "Preventing 'Those Terrible Disasters,'" 10.

199 **poor repair:** See Rev. Chester D. Berry, *Loss of the Sultana and Reminiscences of Survivors* (Darius D. Thorp, Printer and Binder, 1892), 25; Hugh Berryman et al., "The Ill-Fated Passenger Steamer Sultana: An Inland Maritime Mass Disaster of Unparalleled Magnitude," *Journal of Forensic Science* 33, no. 3 (1988): 842–50; Gene Eric Salecker, *Disaster on the Mississippi: The Sultana Explosion, April 27, 1865* (Naval Institute Press, 2015).

199 **innovation . . . "sales pitch":** Lee Vinsel and Andrew L. Russell, *The Innovation Delusion: How Our Obsession with the New Has Disrupted the Work That Matters Most* (Currency, 2020), 11.

199 **"the engineer learns most on the scrapheap":** Claude A. Claremont, *Spanning Space* (Sir Isaac Pitman & Sons, 1939), 13.

200 **"Infrastructure is all about maintenance":** Quoted in Andrew L. Russell and Lee Vinsel, "After Innovation, Turn to Maintenance," *Technology and Culture* 59, no. 1 (2018): 17; sourced from "AHR Conversation: Historical Perspectives on the Circulation of Information," *American Historical Review* 116, no. 5 (2011): 1409.

200 **"slow disaster . . . disaster multiplier":** Scott Knowles, "Deferred Maintenance: The American Disaster Multiplier," *Technology's Stories* 4, no. 1 (2016): 1, 2, 7.

201 **"Crime, intelligence":** Victor Hugo, *Les Misérables*, trans. Isabel Hapgood, in *The Works of Victor Hugo: Les Misérables* (Kelmscott Society Publishers, 1887), 108.

201 **"grow to the size":** Sarah Griffiths, "How to Vanquish the Fatberg Menace," *Engineering & Technology* (July 2022): 52–55.

201 **wastewater treatment:** See Martin V. Melosi, *The Sanitary City: Urban Infrastructure in America from Colonial Times to the Present* (Johns Hopkins University Press, 1999); Lina Zeldovich, *The Other Dark Matter: The Science and Business of Turning Waste into Wealth and Health* (University of Chicago Press,

2021); Chelsea Wald, *Pipe Dreams: The Urgent Global Quest to Transform the Toilet* (Avid Reader Press, 2021).

202   **material constitution:** David Edgerton, *The Shock of the Old: Technology and Global History since 1900* (Profile Books 2019), ix.

204   **"Oh God, it's hard":** Elardo also said that "the workers at New York's treatment plants are outright amazing in their ability to maintain the integrity of our services to the public with these challenges."

204   **"A history of sanitation":** John Joseph Cosgrove, *History of Sanitation* (Standard Sanitary Mfg. Co., 1909), 2.

204   **remote-controlled submersibles:** Griffiths, "How to Vanquish."

204   **Mohenjo-Daro:** See Harold Farnsworth Gray, "Sewerage in Ancient and Mediaeval Times," *Sewage Works Journal* 12, no. 5 (1940): 939–46; Cedric Webster, "The Sewers of Mohenjo-Daro," *Water Pollution Control Federation Journal* 34, no. 2 (1962): 116–23; M. Jansen, "Water Supply and Sewage Disposal at Mohenjo-Daro," *World Archaeology* 21, no. 2 (1989): 177–92; Vernon L. Scarborough, "Water Management Adaptations in Nonindustrial Complex Societies: An Archaeological Perspective," *Archaeological Method and Theory* 3 (1991): 101–54. The discussion on Mohenjo-Daro also appeared in my essay "Creative Intolerance," *Issues in Science and Technology* 39, no. 4 (2023): 31–33.

205   **Around 2600 BCE . . . fourth century BCE:** From Steven J. Burian and Findlay G. Edwards, "Historical Perspectives of Urban Drainage," in *Global Solutions for Urban Drainage* and presented at the Ninth International Conference on Urban Drainage, Portland, Oregon (2002); Petr Hlavinek, "New/Old Ways for Storm Water—Learning from the History," presented at the Fifth International Water History Association Conference, Tampere, Finland (2007); Sanna-Leena Rautanen et al., "Sanitation, Water and Health," *Environment and History* 16, no. 2 (2010): 173–94.

206   **"large enough to allow":** Wikipedia, "Cloāca Maxima," July 2023.

206   **Japan:** Jayant S. Joshi and Rajesh Tewari, "Public Health and Sanitation in the Nineteenth Century Japan," *Proceedings of the Indian History Congress* 64 (2003): 1259–71; Susan B. Hanley, "Urban Sanitation in Preindustrial Japan," *Journal of Interdisciplinary History* 18, no. 1 (1987): 1–26.

206   **Public Health Act:** Ian Morley, "City Chaos, Contagion, Chadwick, and Social Justice," *Yale Journal of Biology and Medicine* 80, no. 2 (2007): 61–72.

207   **Ellis Chesbrough:** See Louis Cain, "Raising and Watering a City: Ellis Sylvester Chesbrough and Chicago's First Sanitation System," *Technology and Culture* 13, no. 3 (1972): 353–72; Benjamin Sells, *The Tunnel under the Lake: The Engineering Marvel That Saved Chicago* (Northwestern University Press, 2017).

207   **poor sanitation:** See Edwin Chadwick, *The Present and General Condition of Sanitary Science: An Address* (James Meldrum, 1889), 4.

207   **since 1840:** Annabel Ferriman, "BMJ Readers Choose Sanitation as Greatest Medical Advance since 1840," *British Medical Journal* 334 (2007): 111.

207   **"rendezvous of all exhaustions":** Hugo, *Les Misérables*, 88.

208   **Engineers, much like nurses:** From Janna van Grunsven et al., "How Engineers Can Care from a Distance: Promoting Moral Sensitivity in Engineering Ethics Education," in *Thinking through Science and Technology: Philosophy, Religion, and Politics in an Engineered World*, ed. G. Miller, H. M. Jerónimo, and Q. Zhu (Rowman & Littlefield International, 2023), 141–64.

208   **emotional sensitivity:** Sabine Roeser, "Emotional Engineers: Toward Morally Responsible Design," *Science and Engineering Ethics* 18, no. 1 (2012): 103–15.

208   **"broken world thinking":** Steven Jackson, "Rethinking Repair," in *Media Technologies: Essays on Communication, Materiality, and Society*, ed. T. Gillespie, P. Boczkowski, and K. Foot (MIT Press, 2014), 221–39.

208   **electricity and eyeglasses:** Mark Thomas Young, "Now You See It (Now You Don't): Users, Maintainers and the Invisibility of Infrastructure," in *Technology and the City: Towards a Philosophy of Urban Technologies*, ed. M. Nagenborg, T. Stone, M. González Woge, and P. E. Vermaas (Springer International Publishing, 2021), 101–19.

210   **trillion dollars in assets . . . 10 times:** David LaShell, "How the New York MTA Manages $1 Trillion in Assets," *WhereNext Magazine*, August 28, 2018.

210   **maintenance consumes:** See Robert L. Glass, "Frequently Forgotten Fundamental Facts about Software Engineering," *IEEE Software* 18, no. 3 (May/June 2001): 112.

210   **multibillion-dollar capital programs:** From Clifton Hood, *722 Miles: The Building of the Subways and How They Transformed New York* (Simon & Schuster, 1993), 13.

211   **"megaprojects":** From Bent Flyvbjerg, "What You Should Know about Megaprojects and Why: An Overview," *Project Management Journal* 45, no. 2 (2014): 6–19. See related discussion on engineering considerations in Guru Madhavan et al., "Delivering Effectively on Large Engineering Projects," *PNAS Nexus* 2, no. 9 (2023): pgad281.

211   **West Coast Main Line:** See UK National Audit Office, *The Modernisation of the West Coast Main Line: Report by the Comptroller and Auditor General* (HC 22 Session 2006–2007; The Stationery Office, 2006); in International Council on Systems Engineering Transportation Working Group, *Systems Engineering in Transportation Projects: A Library of Case Studies* (Issue 8.0), November 19, 2016.

212   **Suez Canal . . . over 200 percent:** Flyvbjerg, "What You Should Know," 10.

212   **New York City:** See Hood, *722 Miles*, 76–77.

213   **"Success doesn't depend":** From the New York Transit Museum; Vivian Heller, *The City Beneath Us: Building the New York Subway* (W. W. Norton, 2004), 20.

213   **"city-born, city-bred"; He came from:** From Henry Petroski, "Engineering: William Barclay Parsons," *American Scientist* 96, no. 4 (2008): 280–83; "William Barclay Parsons," *New York Times*, May 10, 1932; R. S. Weston, "William Barclay Parsons (1859–1932)," *Proceedings of the American Academy of Arts and Sciences* 68, no. 13 (1933): 655–58.

213   **"Reverend Parsons":** Hood, *722 Miles*, 77.

213   **"We have the worst transit problem":** New York Transit Museum; and Heller, *City Beneath Us*, 20.

214   **One politician . . . Another contractor:** Quotes from Stefan Höhne, *Riding the New York Subway: The Invention of the Modern Passenger* (MIT Press, 2021), 21.

214   **"Parsons's Ditch":** Tom Malcolm, "The Renaissance Man of New York's Subways: William Barclay Parsons, Transportation Engineer Extraordinaire," *TR News*, January/February 2006, 10.

214   **object for the subject . . . subject for the object:** Höhne, *Riding the New York*

Subway, 5, quoting Marx's *Grundrisse: Foundations of the Critique of Political Economy*.

214 **"social media"; unruly commuters:** Höhne, *Riding the New York Subway*, 6.

214 **"mega-machine":** Höhne, 17–18.

214 **"spiritual side":** Quote from Richard G. Weingardt, "William Barclay Parsons Jr," *Leadership and Management in Engineering* 7, no. 2 (2007): 84–88.

215 **"The perfect design":** Richard P. Rumelt, *The Crux: How Leaders Become Strategists* (Public Affairs, 2022), 127.

216 **"The sourball of every revolution":** From Mierle Laderman Ukeles's proposal for her 1969 exhibition *Care* ("Manifesto!" as she highlighted it for "maintenance art"), in Patricia C. Phillips, *Mierle Laderman Ukeles: Maintenance Art* (Queens Museum and DelMonico Books/Prestel, 2016).

216 **"Actual heroism":** David Foster Wallace, *The Pale King: An Unfinished Novel* (Little, Brown, 2011), 231.

216 **"perhaps the quintessential":** Ken Alder, *Engineering the Revolution: Arms and Enlightenment in France, 1763–1815* (Princeton University Press, 1997), 15.

217 **"dangerously volatile relationship":** Iain Sinclair, "Negative Equivalent," *London Review of Books*, January 19, 2023.

217 **Consecration ceremonies:** The act of sprinkling holy waters (*abhishekam*) from a vessel (*kumbha*) is called, in Tamil, *Kumbhabhishekam*.

217 **"Conservation, repair and maintenance":** Naman Ahuja, "Make Do and Mend," *Caravan*, July 31, 2021.

218 **"civil self" and a "consuming self":** Joe Macleod, *Endineering: Designing Consumption Lifecycles That End as Well as They Begin* (AndEnd, 2021), 33.

219 **Endineering; "Although it is not anti-business":** Macleod, *Endineering*, 121–22.

219 **Four engines:** See Edgerton, *Shock of the Old*, 87–90.

219 **Stratofortress:** The B-52 upgrade details are from articles published on Defense One (website), by Tara Copp (May 28, 2021); "The War Zone" (of the website The Drive), by Tyler Rogoway (October 10, 2021) and Joseph Trevithick (August 27, 2018; September 24 and December 22, 2021); and the *Financial Times* by Sylvia Pfeifer (September 27, 2021).

219 **"attritable" designs:** John Colombi et al., "Attritable Design Trades: Reliability and Cost Implications for Unmanned Aircraft," *Proceedings of the 2017 Annual IEEE International Systems Conference (SysCon)* (Montreal, Canada, 2017).

220 **operators:** Steve Shapin, "What Else Is New?," *New Yorker*, May 7, 2007.

220 **"tends to evaporate":** Michael North, *Novelty: A History of the New* (University of Chicago Press, 2013), 3.

220 **daily grind:** Guru Madhavan, "The Grind Challenges," *Issues in Science and Technology* 38, no. 4 (Summer 2022): 17–19.

Refrain: Sway

221 **"part fish and part fowl":** Coles Phinizy, "The Missing Link," *Sports Illustrated*, July 15, 1957, 27–29.

221 **Linkanoe:** Kelly and Parke, *Pilot Maker*, 87–88.

222 **"water and air":** Whitmire, "Ed Link: I'm not a Genius."

222 **Looe Key:** van Hoek and Link, *From Sky to Sea*, 84–86.

222 **"I took up golf":** Phinizy, "The Missing Link."

222 **"A beautiful sea garden":** van Hoek and Link, *From Sky to Sea*, 87.

223 **"When we made those"**: Marion Clayton Link, *Sea Diver: A Quest for History under the Sea* (University of Miami Press, 1964), 3.

223 **"I was able to swim"**: Marion Clayton Link, *Sea Diver*, 26.

223 **"I found out"**: Quote from Kelly and Parke, *Pilot Maker*, 90.

224 **Columbus's first landfall**: Ed Link, "My Flights over Columbus' Routes, Dictated aboard 'Sea Diver,'" transcribed April 20, 1955, 1, 6–7; Link, "Big Argument about His First Landing," *Life*, October 12, 1959; Florida Institute of Technology Link Special Collection, and based on Ed and Marion Link, *A New Theory on Columbus' Voyage through the Bahamas* (Smithsonian Institution, 1958).

224 **"sailor's imagination"**: "Columbus Landing Spot Disputed," *New York Times*, March 10, 1958, 1, 31.

224 **"a place of spectacular"**: van Hoek and Link, *From Sky to Sea*, 139.

224 **"It took guts"**: van Hoek and Link, 148.

225 **British cannon**: van Hoek and Link, 159.

225 **World Heritage Site**: UNESCO, with acknowledgment to Ed Link, noted there's "no national or regional comparison for Port Royal as it is the only authentic sunken city in the Western Hemisphere."

225 *Sea Diver II*: Marion Clayton Link, *Windows in the Sea* (Smithsonian Institution Press, 1973).

225 **explore Caesarea**: Charles T. Fritsch and Immanuel Ben-Dor, "The Link Expedition to Israel, 1960," *Biblical Archaeologist* 24, no. 2 (1961): 50–59.

226 **Apollonius of Tyana**: Charles T. Fritsch with Glanville Downey et al., eds., *Studies in the History of Caesarea Maritima* (Scholars Press, 1975), 7.

226 **engineering spectacle**: Jerry Handte, "Links Hope to Find Ancient Religious Tablets on Expedition to Site of Caesarea," *Sunday Press* (Binghamton), February 21, 1960.

226 **Book of Jonah**: From Jonah 1:5 (New King James Version), quoted in Fritsch and Ben-Dor, "Link Expedition to Israel," 53.

226 **size of specific structures**: Ed Link's undated manuscript (likely circa 1960), "Survey Trip to Israel," Florida Institute of Technology Link Special Collection, 2.

227 **relics to the Israeli government**: Eliav Simon, "Scientific Ship Searches Israeli Harbor for Lost City Herod Built," *Milwaukee Journal*, July 26, 1960.

227 **underwater living**: Based on James Goff, "Aerospace Concepts Applied to Deep Submergence Vehicle Simulation," *SAE Transactions* 76 (1968): 1239–43.

227 **Bushnell**: Brenda Milkofskym, "David Bushnell and His Revolutionary Submarine," ConnecticutHistory.org, September 6, 2019.

229 **Jacques Cousteau**: See Jacques Cousteau and Frederic Duma, *The Silent World* (Harper, 1953); Brad Matesen, *Jacques Cousteau: The Sea King* (Pantheon, 2009).

229 **"big meal of cabbage"**: Quote from Peter Fairley, "Caroline, The Goat, Paves the Way," *London Evening Standard*, October 20, 1962.

229 **"Man-in-Sea"**: A key reference was Ed Link, "Our Man-in-Sea Project," *National Geographic*, May 1963.

229 **prospectus to the US Navy**: Based on the Man-in-Sea program prospectus from General Precision, Inc., February 14, 1963, and as described by Andrew Clark, "The Link Legacy," *Marine Technology Society Journal* 49, no. 6 (2015): 43–44.

230 **"We cannot afford to neglect"**: From Ed Link, draft untitled report on deep submergence, September 6, 1963, Florida Institute of Technology Link Special Collection, 6.

230 **USS *Thresher***: See F. N. Spiess and A. E. Maxwell, "Search for the 'Thresher,'" *Science* 145, no. 3630 (1964): 349–55.

230   "should not be viewed" . . . "good engineering": Francis Duncan, *Rickover and the Nuclear Navy: The Discipline of Technology* (Naval Institute Press, 1990), 85.

231   "it could be possible": Link, draft untitled report, 6.

231   "We should not have lost": Joe MacInnis, *Breathing Underwater: The Quest to Live in the Sea* (Viking Canada, 2004), 17.

232   the SPID: Based on articles in the *Miami News, San Francisco Chronicle*, and *Binghamton Press* from early July 1964.

232   "I have always": Quoted in MacInnis, *Breathing Underwater*, 74; Robert Sténuit, *The Deepest Days: A Remarkable Odyssey of Undersea Adventure and the Longest, Deepest Dive Ever Made* (Hodder and Stoughton, 1966).

232   "Our mission was accomplished": Ed Link, "Outpost under the Ocean," *National Geographic*, April 1965, 532.

233   maximum maneuverability: Goff, "Aerospace Concepts Applied."

233   "Living in the sea": Wallace Mitchell, "Leave Living in the Sea to Fish, Says Scientist," *Honolulu Advertiser*, July 13, 1967.

233   "Volkswagens of the deep": Ed Link, "Working Deep in the Sea," in *The World in 1984*, ed. N. Calder (Penguin Books, 1964), 1:103–5.

234   Great Stirrup Cay: Tom Huser, *Palm Beach Post-Times*, March 10, 1968.

234   highly maneuverable submersible: See R. Frank Busby, *Manned Submersibles* (Office of the Oceanographer of the Navy, 1976).

235   *Johnson-Sea-Link*: Based on Timothy Askew, "Submersibles for Science: Johnson-Sea-Link I & II," in *Proceedings of Oceans 1984* (Marine Technology Society and IEEE Ocean Engineering Society, Washington, DC, September 1984), 612–16.

235   June 18, 1973: June 19–20, 1973, articles in the *Evening Press* and *Sun-Bulletin* (Binghamton), and in the *Floridian* (by Don North), January 11, 1976; and Clark, "The Link Legacy."

236   "At this moment": MacInnis, *Breathing Underwater*, 219.

236   "this loss need never": van Hoek and Link, *From Sky to Sea*, 306.

236   "work blind": Tom Cawley, *Evening Press* (Binghamton), April 3, 1975.

236   "It made us realize": Mary Beth Herzog, *Vero Beach Press Journal*, May 1, 1977.

236   "I'm demanding": Annie Laurie Morgan, unpublished draft profile of Marilyn Link, 1977.

237   "Our seas are": Tom Cawley, "Edwin Link Starts New Career at 71," *Binghamton Press*, 1975 [undated in author's copy].

237   "haven't taken"; "There is ignorance at all levels": Barry Holtzclaw, *Sun-Bulletin* (Binghamton), July 26, 1971.

238   *Johnson-Sea-Link*s helped locate: Timothy Askew's 1987 conference paper, "Johnson-Sea-Link Submersibles' Role in the Challenger Recovery," *Proceedings of Oceans '87* (Marine Technology Society and IEEE Ocean Engineering Society, Nova Scotia), 1225–28.

238   "Were it not for CORD": Clark, "The Link Legacy," 52.

238   "In the offshore business": Clark, 50; General Precision, Inc., "The *Man in Sea* Program," prospectus prepared for US Navy, February 14, 1963.

239   "Few major fields": Clark, 45.

239   *Titanic*: See Joe MacInnis, *Titanic in a New Light* (Thomasson-Grant, 1992).

Chapter Six: Wicked Resilience

241 **Henry VI; Columbus:** Michael Castagna, *Shipworms and Other Marine Borers* (Fishery Leaflet 505), US Bureau of Commercial Fisheries, June 1961.

241 **"endless frontier":** Charles F. Carroll, "Wooden Ships and American Forests," *Journal of Forest History* 25, no. 4 (1981): 213.

242 **supply networks:** Robert Greenhalgh Albion, *Forests and Sea Power: The Timber Problem of the Royal Navy, 1652–1862* (Harvard University Press, 1926), 4.

242 **engineers as much as admirals:** Albion, *Forests and Sea Power*, 6.

242 **Thomas Tredgold:** Key biographical readings included L. G. Booth, "Thomas Tredgold (1788–1829): Some Aspects of His Work," *Transactions of the Newcomen Society* 51, no. 1 (1979): 57–94, and "Thomas Tredgold (1788–1829): Some Further Aspects of His Life and Work," *Transactions of the Newcomen Society* 69, no. 1 (1997): 237–47.

242 **cast iron; faith in the wood:** Booth, "Some Aspects of His Work," 65.

242 **short paper:** See Thomas Tredgold's "On the Transverse Strength and Resilience of Timber," *Philosophical Magazine* 51, no. 239 (1818): 214–16.

242 **"in the power of resisting":** Tredgold, "On the Transverse Strength," 216.

242 **Scottish nobleman:** See Peter R. Shergold, Ian Inkster, and J. Lowe, "Civil Engineering and the Admiralty: Thomas Tredgold, Edward Deas Thomson and Early Steam Navigation, 1827–1828," *Great Circle* 4, no. 1 (1982): 41.

243 **Bhagavad Gītā:** In chapter 2 (verses 50–70), available in accessible translations such as Eknath Eswaran's 2007 second edition from Nilgiri Press and Laurie Patton's 2008 Penguin Classics edition.

243 **"In a storm":** The chosen phrasing is from Deepak Chopra, "The Gift of Resilience," SFGate.com, October 10, 2022. See also Robert Jordan, "The oak fought the wind and was broken, the willow bent when it must and survived," in *The Fires of Heaven* (TOR Fantasy, 1993), 617.

243 **the word "resilience":** Quotes from David E. Alexander, "Resilience and Disaster Risk Reduction: An Etymological Journey," *Natural Hazards and Earth System Sciences* 13, no. 11 (2013): 2707–9.

243 **"the last man who":** After the title of Andrew Robinson's biography of Thomas Young (Oneworld Publications, 2006).

244 **"a species of strength":** Thomas Tredgold, *Elementary Principles of Carpentry*, 2nd ed. (1820; E. L. Carey and A. Hart, 1837), 73.

244 **"Trees planted in the worst":** Quoted in Tredgold, "On the Transverse Strength," 214.

244 **engineers of civilizations:** Based on Tredgold, *Elementary Principles of Carpentry*, viii.

244 **self-reinforcing "concrete":** See Norbert J. Delatte, "Lessons from Roman Cement and Concrete," *Journal of Professional Issues in Engineering Education and Practice* 127, no. 3 (2001): 109–15; Marie D. Jackson et al., "Mechanical Resilience and Cementitious Processes in Imperial Roman Architectural Mortar," *Proceedings of the National Academy of Sciences* 111, no. 52 (2014): 18484; Zahra Ahmad, "Why Modern Mortar Crumbles, but Roman Concrete Lasts Millennia," ScienceShots, *Science*, July 3, 2017.

244 **"mobile cell sites and mobile command centers":** Braden R. Allenby and Mikhail Chester, "Learning from Engineers," *Issues in Science and Technology*, April 21, 2020.

244 **"a lumberman's carnival":** Quote from Richard Tucker, "The World Wars and

Globalization of Timber Cutting," in *Natural Enemy, Natural Ally: Toward an Environmental History of Warfare*, ed. R. Tucker and E. Russell (Oregon State University Press, 2004), 126.

244 **"a great reservoir":** Quote from Tucker, "World Wars and Globalization," 127.

245 **"the quantitative":** C. S. Holling, "Resilience and Stability of Ecological Systems," *Annual Review of Ecology and Systematics* 4, no. 1 (1973): 1.

245 **engineering resilience and ecological resilience:** C. S. Holling, "Engineering Resilience Versus Ecological Resilience," in *Engineering within Ecological Constraints*, ed. P. Schulze (National Academy Press, 1996), 31–44.

245 **persistence, cohesion, and adaptation:** See David D. Woods, "Four Concepts for Resilience and the Implications for the Future of Resilience Engineering," *Reliability Engineering & System Safety* 141 (2015): 5–9; Erik Hollnagel, "Resilience Engineering and the Built Environment," *Building Research & Information* 42, no. 2 (2014): 221–28.

245 **as in a clear lake:** Marco A. Janssen and John M. Anderies, "Robustness Trade-Offs in Social-Ecological Systems," *International Journal of the Commons* 1, no. 1 (2007): 46.

245 **"is essentially static":** Holling, "Resilience and Stability," 2.

246 **"panarchy":** Lance H. Gunderson and C. S. Holling, eds., *Panarchy: Understanding Transformations in Human and Natural Systems* (Island Press, 2002).

246 **"Just as the 'price'":** Ron Westrum, "Resilience and Restlessness," in *Resilience Engineering Perspectives, Remaining Sensitive to the Possibility of Failure*, ed. E. Hollnagel, C. P. Nemeth, and S. Dekker (Ashgate, 2008), 1:1.

246 **In the prologue:** Tredgold, *Elementary Principles of Carpentry*, ix.

246 **"Art of directing":** Booth, "Some Aspects of His Work," 61.

246 **"All good art is political!":** Quote by Morrison featured in Kevin Nance, "The Spirit and the Strength: A Profile of Toni Morrison," *Poets & Writers* (November/December 2008). Emphasis in the original.

247 **"Each self-important creature":** Quote from Booth, "Some Aspects of His Work," 62.

247 **one of his encyclopedias:** Booth, "Further Aspects of His Life," 238.

248 **"is that of defining problems":** Rittel and Webber, "Dilemmas in a General Theory," 159.

249 **Robustness makes sense:** Based on Janssen and Anderies, "Robustness Trade-Offs," 43–45.

250 **adapting to their circumstance:** See Mark D. Seery et al., "Whatever Does Not Kill Us: Cumulative Lifetime Adversity, Vulnerability, and Resilience," *Journal of Personality and Social Psychology* 99, no. 6 (2010): 1025–41.

250 **a study of individuals:** Quotes from Sneha Shankar et al., "'I Haven't Given Up and I'm Not Gonna': A Phenomenographic Exploration of Resilience among Individuals Experiencing Homelessness," *Qualitative Health Research* 29, no. 13 (2019): 1850–61.

250 **"My fear is being found on the street":** Quoted in Jack Tsai et al., "Is Homelessness a Traumatic Event? Results from the 2019–2020 National Health and Resilience in Veterans Study," *Depression and Anxiety* 37, no. 11 (2020): 1137–45.

252 **by nearly one-half:** National Academies, *Permanent Supportive Housing* (National Academies Press, 2018), 22 (Box 2-1).

252 **working model:** Tricia C. Bailey, "Eliminating Veterans' Homelessness: Connecting a Deserving Population with the Right Resources," MITRE, September 2012.

See also Steve Vogel (December 26, 2011), Josh Hick (November 21, 2013), and Emily Wax-Thibodeaux (October 2, 2014) in the *Washington Post*.

253   **boating metaphor:** C. S. Holling, Lance H. Gunderson, and Donald Ludwig, "In Quest of a Theory of Adaptive Change," in Gunderson and Holling, *Panarchy*, 16–17 (Box 1-2).

253   **"Have we vanquished an enemy":** George Herbert Leigh Mallory, "Mont Blanc from the Col Du Géant by the Eastern Buttress of Mont Maudit," in *Alpine Journal: A Record of Mountain Adventure and Scientific Observation by Members of the Alpine Club*, ed. G. Yeld (Longmans, Green and Co., 1918), 162.

255   **"antiproperty":** James Carse, *Finite and Infinite Games: A Vision of Life as Play and Possibility* (Simon & Schuster, 1985), 131.

255   **In December 2015 . . . befouled liquid:** CNN, BBC, Reuters, and WSJ reported the incidents.

255   **Garbage City:** Chethan Kumar, "Bengaluru Chokes on Its Own Waste," *Times of India*, November 12, 2017.

255   **waste was dumped:** Eswaran Subrahmanian, Yoram Reich, and Sruthi Krishnan, *We Are Not Users: Dialogues, Diversity, and Design* (MIT Press, 2020), 87–88.

256   **fifth of the total waste:** Cited in Sunil Kumar et al., "Challenges and Opportunities Associated with Waste Management in India," *Royal Society Open Science* 4, no. 3 (2017): 160764. See also Assa Doron and Robin Jeffrey, *Waste of a Nation: Garbage and Growth in India* (Harvard University Press, 2018).

256   **"create more informed citizens":** Subrahmanian, Reich, and Krishnan, *We Are Not Users*, 91–95.

256   **trilingual:** In Kannada language, the game is called Kaasu Kasa, and in Tamil, Kaasu-Kuppai.

256   **"in a state of constant becoming":** Subrahmanian, Reich, and Krishnan, *We Are Not Users*, 87.

257   **A game . . . fidelity to realism:** See John R. Raser, *Simulation and Society: An Exploration of Scientific Gaming* (Allyn & Bacon, 1969); Richard D. Duke and Jac L. A. Geurts, *Policy Games for Strategic Management* (Dutch University Press with Purdue University Press and Rozenberg Publishers, 2004); Louise Sauvé et al., "Distinguishing between Games and Simulations: A Systematic Review," *Journal of Educational Technology & Society* 10, no. 3 (2007): 247–56; Sebastian Deterding, *The Gameful World: Approaches, Issues, Applications* (MIT Press, 2015); Paul T. Grogan and Sebastiaan A. Meijer, "Gaming Methods in Engineering Systems Research," *Systems Engineering* 20, no. 6 (2017): 542–52.

257   **sponge:** See Sebastiaan Meijer, "The Power of Sponges: Comparing High-Tech and Low-Tech Gaming for Innovation," *Simulation & Gaming* 46, no. 5 (2015): 512–35.

258   **US Marine Corps:** See Larry Sutton, "High tech versus low tech training," in *Eighth International Wildland Firefighter Safety Summit: 10 Years Later*, ed. B. W. Butler and M. E. Alexander (International Association of Wildland Fire, 2005), 1–9.

258   **India and China:** Kumar et al., "Challenges and Opportunities."

258   **trash on India's streets:** Noah M. Sachs, "Garbage Everywhere: What Refuse in India's Streets Reveals about America's Hidden Trash Problem," *Atlantic*, June 20, 2014.

258   **fatality rates:** Heather Horn, "The Secret World of 'Garbagemen,'" *Atlantic*, April 1, 2013.

259   **total dry biomass:** Emily Elhacham et al., "Global Human-Made Mass Exceeds All Living Biomass," *Nature* 588, no. 7838 (2020): 442–44.

259   **"safe operating space":** Linn Persson et al., "Outside the Safe Operating Space of the Planetary Boundary for Novel Entities," *Environmental Science & Technology*, 56, no. 3 (2022): 1510–21.

259   **one-quarter of the world's municipal solid waste:** Ed Cook and Costas Velis, *Global Review on Safer End of Engineered Life* (Royal Academy of Engineering and the Lloyd's Register Foundation, 2020); Ruth Boumphrey's perspective on open burning of waste appears in *The Engineer*, September 14, 2021.

259   **"Stare into dumpsters":** John Hoffman, *The Art and Science of Dumpster Diving* (Loompanics Unlimited, 1993), 130.

259   **"unpleasant but marginal problem":** Justin McGuirk's quotes and related ideas are from "The Waste Age," *Aeon*, January 4, 2022.

260   **function of resilience:** Eloise Taysom and Nathan Crilly, "Resilience in Sociotechnical Systems: The Perspectives of Multiple Stakeholders," *She Ji: The Journal of Design, Economics, and Innovation* 3, no. 3 (2017): 165–82.

260   **"a whole cloud of philosophy":** Wittgenstein, *Philosophical Investigations*, 233e.

260   **"labyrinth of paths":** Wittgenstein, 88e.

261   **"mountains long conceived":** John Muir, *Travels in Alaska* (Houghton Mifflin, 1915), 67.

261   **Krishna:** Based on Lewis Hyde, *Trickster Makes This World: Mischief, Myth, and Art* (Farrar, Strauss and Giroux, 1998), 71–72.

261   **weren't *im*moral but *a*moral; "mythic embodiment":** Hyde, *Trickster Makes This World*, 7, 10.

261   **like the tricksters:** See Melinda Harm Benson and Robin Kundis Craig, *The End of Sustainability* (University Press of Kansas, 2017).

262   **"makes resilience a dirty word":** David Oliver, "When 'Resilience' Becomes a Dirty Word," *British Medical Journal* 358 (2017): j3604.

262   **cultivate the pearl grains:** Siambabala Bernard Manyena, "The Concept of Resilience Revisited," *Disasters* 30, no. 4 (2006): 434–50.

262   **"ilities":** James R. Enos, "Achieving Resiliency in Major Defense Programs through Nonfunctional Attributes," *Systems Engineering* 22, no. 5 (2019): 389–400.

263   **"People are trapped":** James Baldwin, "Stranger in the Village," *Harper's Magazine*, October 1953, 43.

263   **many things in one:** Ben Anderson, "What Kind of Thing Is Resilience?," *Politics* 35, no. 1 (2015): 60–66.

263   **S-shaped curve:** Jonas Salk, "The Survival of the Wisest," *Phi Delta Kappan* 56, no. 10 (1975): 667–69. This discussion also appears in Guru Madhavan et al., "Engineering for Inevitable Surprises."

Refrain: Heave

264   **"probably received":** Moroney and Lilienthal, "Human Factors in Simulation and Training," 4.

264   **early simulators were games:** Based on Pat Harrigan and Matthew Kirschenbaum, eds., *Zones of Control: Perspectives on Wargaming* (MIT Press, 2016); Charles Homans, "War Games: A Short History," *Foreign Policy*, August 31, 2011.

265 "approximation of war"; "was always just an approximation": Jon Peterson, "A Game out of All Proportions: How a Hobby Miniaturized War," in Harrigan and Kirschenbaum, *Zones of Control*, 4, 23.

265 **Dungeons & Dragons:** See Jon Peterson, *Game Wizards: The Epic Battle for Dungeons & Dragons* (MIT Press, 2021); Ben Rigg, *Slaying the Dragon: A Secret History of Dungeons & Dragons* (St. Martin's Press, 2022).

265 **gaming the unthinkable:** See Sharon Ghamari-Tabrizi, "Simulating the Unthinkable: Gaming Future War in the 1950s and 1960s," *Social Studies of Science* 30, no. 2 (2000): 163–223.

265 **Jack Thorpe:** Jack Thorpe, "Future Views: Aircrew Training 1980–2000," Bolling Air Force Base, Washington, DC, September 15, 1987. See Thorpe's related retrospective, "Trends in Modeling, Simulation, & Gaming: Personal Observations about the Past Thirty Years and Speculation About the Next Ten," paper presented at the 2010 Interservice/Industry Training, Simulation, and Education Conference (I/ITSEC).

266 **high contextual complexity:** Lenoir, "All but War Is Simulation," 311–12.

266 **"Alice's looking glass":** Thorpe, "Future Views," 5.

266 **"Instead of communicating":** Thorpe, 8–9.

267 **"Fast, approximate, and cheap":** Thorpe, "Trends in Modeling," 33.

267 **Pentagon and Hollywood:** Based on Sharon Ghamari-Tabrizi, "The Convergence of the Pentagon and Hollywood," in *Memory Bytes: History, Technology, and Digital Culture*, ed. L. Rabinovitz and A. Geil (Duke University Press, 2004).

267 **"flying carpet":** From Project ODIN (1990–91), described in Thorpe, "Trends in Modeling," 11. Related reading included Duncan Miller and Jack Thorpe, "SIM-NET: The Advent of Simulator Networking," *Proceedings of the IEEE* 83, no. 8:1114–23; see also Crogan, *Gameplay Mode*, and Sharon Weinberger, *The Imagineers of War* (Knopf, 2017).

267 **"distant intimacy":** John Williams, "Distant Intimacy: Space, Drones, and Just War," *Ethics & International Affairs* 29, no. 1 (2015): 93–110.

268 **"reorientation is paired with":** Crogan, *Gameplay Mode*, 57.

268 **"War is the province of uncertainty":** Quoted in John Rhea, "Planet Simnet," *Air Force*, August 1989, 64.

268 **"What's the difference between fighting":** Quote by Richard Lindheim, executive vice president of Paramount Digital Entertainment, from Ghamari-Tabrizi, "Convergence of the Pentagon and Hollywood," 165.

268 **"light as a feather":** Walt Whitman, *Democratic Vistas and Other Papers* (Walter Scott, 1888), 71.

269 **hourly training costs:** The US Air Force estimate is Dehmel-type simulator for Boeing's B-50 Superfortress, from Hobbins, "Emulating the 'Pucker Factor.'"

269 **organ factory:** Barry Holtzclaw, *Sun-Bulletin* (Binghamton), July 26, 1971; "The Story of the Link Orchestral Organ," Roberson Center for the Arts and Sciences, possibly 1968; Billy Nale, "The Renaissance of the Organ," *Music Magazine* (late 1960s [undated in author's copy]).

269 **"I've had a fun career":** Tom Cawley, *Binghamton Press*, September 30, 1964.

270 **"Don't you ever believe":** "Innovators," *Business in New York State*, November/December 1968, 6–8.

270 **Telford:** Samuel Smiles, *Lives of the Engineers: History of Roads* (John Murray, 1874), 206.

270 **EJ:** William Inglis, *George F. Johnson and His Industrial Democracy* (Hunting-

ton, 1935); Kristina Wilcox, "Factory Town in Transition: A Community's Reaction to Change" (PhD diss., Georgetown University, 2015).

270   **"nifty shoes":** Ed Aswad and Suzanne Meredith, *Endicott-Johnson* (Images of America; Arcadia Publishing, 2003), 14.

270   **"except ordinary decency":** Richard Sherwood Saul, "An American Entrepreneur: George F. Johnson" (PhD diss., Syracuse University, 1966), 295.

271   **"negotiated loyalty":** Gerald Zahavi, "Negotiated Loyalty: Welfare Capitalism and the Shoeworkers of Endicott Johnson, 1920–1940," *Journal of American History* 70, no. 3 (1983): 602–20.

271   **"efficiency expenses"; "veritable beehive of industry":** Zahavi, *Workers, Managers, and Welfare Capitalism: The Shoeworkers and Tanners of Endicott Johnson, 1890–1950* (University of Illinois Press, 1988), 38, 6.

272   **"EJ Homes"; "There can be no security":** Mark Simonson, "Shoe Firm Put Workers' Housing First in Endicott," *Daily Star* (Oneonta), May 31, 2008.

272   **"godfather of sports":** Zahavi, *Workers, Managers, and Welfare Capitalism*, 51.

272   **"workingman's advocate" and "the working people":** Zahavi, 51.

272   **"My father's family":** Wilcox, "Factory Town in Transition," 34.

273   **Thomas J. Watson Sr.:** Thomas Graham Belden and Marva Robins Belden, *The Lengthening Shadow: The Life of Thomas J. Watson* (Little, Brown, 1962), 157, 166, 167, 194.

273   **"No errors":** McGuire and Osterud, *Working Lives*, 81. A similar commitment to accuracy proved decisive in the 1890 census, conducted in just six months with engineer Herman Hollerith's punch-card tabulator, with ancestral roots in the Jacquard loom and player pianos.

273   tabulating machines: Belden and Belden, *Lengthening Shadow*, 162–63; related useful background in Lars Heide, *Punched-Card Systems and the Early Information Explosion, 1880–1945* (Johns Hopkins University Press, 2009).

273   **With the motto THINK:** Belden and Belden, *Lengthening Shadow*, 158.

274   **"ever onward IBM":** Belden and Belden, 135–36.

274   religious tones: Belden and Belden, 127.

274   **IBM newspaper mused:** Belden and Belden, 128–29. Quote from Elbert Green Hubbard, *National Poultry, Butter & Egg Bulletin*, October 1923, 13.

274   **"He always shaved"** . . . **"people caring about":** Quotations in this paragraph from Belden and Belden.

275   **"We don't have to simulate":** Link Aviation Devices, Inc., Inter-Departmental Correspondence, Subject: "The Future of the Company," November 8, 1944, 1, Link Collection.

275   **"Many English workers":** *Binghamton Press* article, May 1949 [undated in author's copy].

276   **Ronald Reagan:** From September 12, 1984, remarks at Reagan-Bush rally in Endicott, New York, Ty Cobb Field at Union-Endicott High School, Ronald Reagan Presidential Library and Museum.

276   **2012 well-being survey:** Gallup-Healthways Well-Being Index; Binghamton followed McAllen-Edinburg-Mission, Texas. See related reporting by Angela Haupt in *US News & World Report*, March 13, 2012.

276   **"They were confident":** Wilcox, "Factory Town in Transition," 8.

276   1,300 workers: Online IBM archives.

277   **"biggest eyesores":** Jeff Platsky, "Wanted: Developers for IBM Country Club, Assistance Provided," *Binghamton Press & Sun-Bulletin*, June 20, 2019.

277 **"Remember: the past won't"**: Joseph Brodsky, "San Pietro," in *A Part of Speech*, trans. Barry Rubin (Farrar, Straus and Giroux, 1980), 147.

277 **"We kind of grew up"**: Katie Sullivan Borrelli, "Endicott Johnson Shoe Co.'s Former Employees Say Company Was 'Like Family,'" *Binghamton Press & Sun-Bulletin*, October 16, 2018.

279 **"walked up and down"**: Ronald Capalaces, *When All the Men Were Gone: World War II and the Home Front; One Boy's Journey through the War Years* (Lazarus Publishing, 2010), 25.

279 **"the past jostling the present"**: Quote from Julia Van Haaften, *Berenice Abbott: A Life in Photography* (W. W. Norton, 2018), 196.

281 **aerodynamic Art Deco bus station:** The Streamline Moderne style, the website Treasures of the Tier notes, was "intended to depict aerodynamics and a sense of speed, the design is attributed to Louisville architect William S. Arrasmith, who designed over sixty moderne Greyhound terminals in his career, of which only a half-dozen exist today." See Frank Wrenick and Elaine V. Wrenick, *The Streamline Era Greyhound Terminals: The Architecture of W. S. Arrasmith* (McFarland, 2007).

281 **Armando Dellasanta:** Little Venice's Manhattan and Binghamton Rooms feature the impressionist's energetic paintings of the cities that gave him the names "Binghamton's Van Gogh" and "urban Monet."

Epilogue: Civicware

285 **"The descent beckons"**: "The Descent," in *The Collected Poems of William Carlos Williams*, vol. 2, *1939–1962*, ed. Christopher MacGowan (New Directions Books, 1991), 2:245.

285 **"The many fictional inner journeys"**: "Dire Cartographies: The Roads to Ustopia," in *In Other Worlds: SF and the Human Imagination* (Doubleday, 2011), 70.

285 **"Let's start from the beginning"**: Directed by Alfred Hitchcock, screenplay by John Michael Hayes (Paramount Pictures, 1954).

286 **"depth technology"**: Judith Roof, "Depth Technologies," in *Technospaces: Inside the New Media*, ed. Sally R. Munt (Continuum, 2001), 21–22.

286 **"It exists for only"**: Michael Kelly, "David Gergen, Master of the Game," *New York Times Magazine*, October 31, 1993.

286 **"Suit yourself"**: Edward Abbey, *Desert Solitaire: A Season in the Wilderness* (1968; repr., University of Arizona Press, 1988), 37.

287 **hypocognition:** Based on Kaidi Wu and David Dunning, "Hypocognition: Making Sense of the Landscape Beyond One's Conceptual Reach," *Review of General Psychology* 22, no. 1 (2018): 25–35.

289 **Prehistoric stonework:** Aldo Faisal et al., "The Manipulative Complexity of Lower Paleolithic Stone Toolmaking," *PLOS One* 5, no. 11 (2010): 13718.

289 **aircraft's altitude:** The risk of substituting aircraft's altitude for other metrics is based on a conversation with Dennis Snower, who makes a similar point in his April 2020 *Global Solutions Journal* article with Katharina Lima de Miranda, "Recoupling Economic and Social Prosperity," 20.

290 **"piecemeal" social engineering:** Karl Popper, "The Poverty of Historicism, II. A Criticism of Historicist Methods," *Economica* 11, no. 43 (1944): 119–37.

291 **evolutionary approach:** Selected resources include Peter J. Richerson and Robert Boyd, *Not By Genes Alone* (University of Chicago Press, 2005); Joseph Hen-

rich, *The Secret of Our Success* (Princeton University Press, 2015); Kevin Leland, *Darwin's Unfinished Symphony* (Princeton University Press, 2017); Michael Muthukrishna, Michael Doebeli, Maciej Chudek, and Joseph Henrich, "The Cultural Brain Hypothesis," *PLoS Computational Biology* 14, no. 11 (2018): e1006504; Michele Gelfand, *Rule Makers, Rule Breakers* (Scribner, 2018); David Sloan Wilson, *This View of Life* (Pantheon, 2019); and David Sloan Wilson, Guru Madhavan, Michele Gelfand, and Rita R. Colwell, "Multilevel Cultural Evolution: From New Theory to Practical Applications," *Proceedings of the National Academy of Sciences* 120, no. 16 (2023): e2218222120.

291 **environmental degradation:** Russell Ackoff, *Redesigning the Future: A Systems Approach to Societal Problems* (Wiley, 1974), 173–74.

292 **"Engineers in Texas":** Darshan Karwat, "Creating a New Moral Imagination for Engineering," *Issues in Science and Technology*, June 2, 2022.

292 **projectification:** Philip Scranton, "Projects as a Focus for Historical Analysis: Surveying the Landscape," *History and Technology* 30, no. 4 (2014): 354; Yvonne-Gabriele Schoper et al., "Projectification in Western Economies: A Comparative Study of Germany, Norway and Iceland," *International Journal of Project Management* 36, no. 1 (2018): 71–82; Anders Jensen et al., "The Projectification of Everything: Projects as a Human Condition," *Project Management Journal* 47, no. 3 (2016): 21–34.

293 **One current standard:** Information on technical standard ISO 42020 (2019) and heuristics is courtesy of discussion at the Royal Academy of Engineering Insight Session on a Model for Transdisciplinary Engineering, November 23, 2021, with Peter Brook, Mike Pennotti, and David Rousseau. The 2009 book (and third edition of) *The Art of Systems Architecting* (CRC Press) by Mark W. Maier and Eberhardt Rechtin lists over 180 design heuristics.

293 **"cognitive violations and biases":** Additional examples are "endowment effect," "choice-supportive bias," "loss aversion," "selective perception," "availability heuristic," "clustering illusion," "fundamental attribution error," "illusory correlation," and "just-world phenomenon," chosen from hundreds more in behavioral research.

293 **"does not provide":** Carl Mitcham, "The True Grand Challenge for Engineering: Self-Knowledge," *Issues in Science and Technology* 31, no. 1 (Fall 2014).

293 **public welfare considerations:** See Erin Cech's 2013 paper "Culture of Disengagement in Engineering Education?," *Science, Technology, & Human Values* 39, no. 1: 42–72; and her chapter, "The (Mis)Framing of Social Justice: Why Ideologies of Depoliticization and Meritocracy Hinder Engineers' Ability to Think about Social Injustices," in *Engineering Education for Social Justice: Critical Explorations and Opportunities*, ed. J. Lucena (Springer Netherlands, 2013), 67–84.

294 **"They not only didn't know the answers":** Erin Cech, interview with Jacqui Thornton, *Nature*, November 4, 2022.

294 **"technical optimum"; "the religion of engineers"; "detect by the means":** Friedrich A. Hayek, *The Counter-Revolution of Science: Studies on the Abuse of Reason* (Free Press, 1952), 96, 102.

294 **"demosclerosis":** From Jonathan Rauch in a September 5, 1992, essay for the *National Journal*.

295 **"Great events":** Quote from William Gass, "Paul Valéry: The Later Poems and Prose," *New York Times*, August 27, 1972.

295 **exalted status:** Corn, *Winged Gospel*, 26.

296   **"Due recognition"**: Charles Taylor, "The Politics of Recognition," in *Multiculturalism: Examining the Politics of Recognition*, ed. A. Gutmann (Princeton University Press, 1994), 25–26.

296   **fly without motors**: Wilbur Wright's May 13, 1900, letter to Octave Chanute, Library of Congress Manuscript Division, Octave Chanute Papers.

297   **"While he did not go"**: "Edwin A. Link, Prepared by Ralph E. Flexman," April 5, 1985, unpublished.

297   **"By missing a university education"**: MacInnis, *Breathing Underwater*, 20.

297   **"Ed Link is the most"**: Phinizy, "The Missing Link," 27–29.

298   **"No-o-o, no, I'm not"**: Whitmire, "Ed Link: I'm Not a Genius."

298   **"Define the problem" . . . "can result in"**: 1979 interview in the *Sun-Bulletin* (Binghamton), September 8, 1981.

299   **Kālidāsa**: The lyrical Sanskrit gem "Meghaduta," or "cloud messenger," from the fourth to fifth century CE, has been translated in Thomas Clark's 1882 book *Meghaduta: The Cloud Messenger* (Trubner and Co.). See also Arthur W. Ryder's *Kalidasa: Translations of Shakunthala and Other Works* (J. M. Dent & Sons/E. P. Dutton).

Index

9/11, 116–20, 122, 244
99th Pursuit Squadron, 106–7

Abbey, Gerald Gene, 175–77, 178–79,
    190–91
Abbott, Henry, 124
abstraction, 191
accidents, 166–69. *See also various
    accidents*
Achaemenid Persian Empire, 205
Ackoff, Russell, 7, 8, 9, 291
acrobatic showmen, 61–62, 64–65, 90–91,
    105
aerial advertising, 91
aeroneurosis, 137
aerosol words, xiv
Affordable Care Act of 2010, 76–77
Air Mail Act, 91–92
Air Navigation (Weems), 108
air shows, 61–62, 90–91, 105. *See also*
    barnstormers
air travel, 10–11
airbags, 67, 69–71
airmail pilots
    Air Mail Act, 91–92
    air mapping, 57–58
    Army Air Corps, 93
    lighted airways, 58–59
    Lindbergh as, 47, 93
    navigation routes, 92
    road maps, 57
    standardized directions and, 59–60
Airport (film), 139
air-traffic control simulator, 74
Aldrin, Buzz, 182
Alexievich, Svetlana, 171
Alger, Philip, 22–23
Allen, Thomas, 106
Aloha, 49, 50, 108
Amalberti, René, 172–73
anarchists, 112
Anderson, Charles Alfred, 106
animals, 161–62
Antoinette Flying School, 63

Apollo 1, 179
Apollo 11, 180, 183–84
Apollo 13, 186–87
Apollo configuration management,
    184–85
Apollo simulator program, 177, 179, 180–
    81, 185–86, 187, 189
aquanauts, 229
Aquinas, Thomas, 40
Armstrong, Neil Alden, 179, 182
Army Air Corps, 93
Army Corps of Engineers, 125
The Art of Logic (Spencer), 44
ash concentration, 81, 82–84
Atomic Energy Act, 159
AT&T, 244
authority, 158
Automatic Musical Company, 54
automatic surveillance, 267
autopilot, 100–101
aviation
    acrobatic showmen, 61–62, 64–65,
        90–91, 105
    airplanes as symbol, 56
    air-traffic control simulator, 74
    animals and, 161–62
    autopilot, 100–101
    beacons, 58–59, 60, 99
    Black people, 103–7
    blind flight, 64, 94–96, 97–98
    celestial sightings, 108–9
    checklist duties, 98
    contact flying, 15, 57–58, 92–93
    Delta Flight 9877, 133–36
    Eastern Air Lines Flight 304, 136
    end of systems, 219
    as entertainment, 88, 90–91, 145–46
    flight-data recorders, 137–38
    flying in bad weather, 15, 50–51, 92–93,
        96–99, 109
    gliders, 96
    ground training, 16, 62–64
    health of pilots, 137
    instruments, 100–101

aviation (*continued*)
  motion trainers, 63–64, 142–43
  natigation skills, 108
  No. 1 British Flying Training Schools,
    2–4
  rotational axes, 11
  safety protocols, 170, 171–72
  sextants, 107
  speeds per hour, 3
  translational axes, 11
  vertigo, 142
  volcanic eruptions, 79–81, 82–84
  women, 5, 101–4, 105
  See also airmail pilots; flight contests;
    flight simulation; flight training;
    Link Trainers; Wright, Orville
    and Wilbur

B-52 Stratofortress bomber, 269
Bacon, Francis, 243
balance, 95–96
baldness, 68
Baltimore, 22
BANANA (Build Absolutely Nothing
  Anywhere Near Anyone), 215
Banning, James, 106
Barbarite, Gabby, 237–38
barnstormers, 61–62, 64–65, 90–91, 105
baselines, 6
bathyscaphe, 228, 230
bathysphere, 227
Bazalgette, Joseph, 207
beacons, 58–59, 60, 99
Bengaluru, India, 255, 256–57, 258, 260
Bennett, Bill, 190–92
Bentham, Jeremy, 31
Benton, Thomas Hart, 195–96
Bessie Coleman Aero Club, 106
Bhagavad Gītā, 243
Bijker, Wiebe, 129–30
Billing, Eardley, 63
Bingham, William, 51
Binghamton, NY, 51–52, 55, 189, 270–74,
  275–82, 285–86, 298–99
Bird of Paradise, 47
birds, 161–62
Bisby, Luke, 156–57, 173–74
black box recorders, 137–38
Black people, 3, 103–7
Blake, William, 13
blind flight, 64, 94–96, 97–98

Blind Flight in Theory and Practice (Ocker
  and Crane), 64
bodies, 121
Boeing, 160
Boeing Stratotanker, 180
boilers. *See* steam power
Bonaparte, Napoleon, 33
book overview, xiv–xv
Borges, Jorge Luis, 127
Boston, 111–16
bridges, 211
Britain, 1–3, 206–7, 241–42. *See also*
  Royal Navy
British cadets, 1–4
British Flying Training Schools, 2–4
Building-information modeling (BIM), 76
Bullard, Eugene Jacques, 104–5
Bureau of Ordnance, 26
burials, 43
Bushnell, David and Ezra, 227

Cabled Observation and Rescue Device
  (CORD), 236, 238
Caesarea, Israel, 226–27
Calder, Alexander, 65
cancer, 121
Capalaces, Ronald, 279
capitalism, 196–97
car accidents, 131
Cardullo, Frank, 89
carousel, 280
Carpenter, Daniel, 38
Carse, James, 255
Cech, Erin, 293–94
Celestial Navigation Trainer, 107, 108–9,
  110, 145
celestial sightings, 108–9
Center for Technology & Innovation, 188
centrifuge models, 126–27
Chadwick, Edwin, 206–7
chairs, 69
Challenger space shuttle, 170, 238
Chaplin, Sydney, 61–62
chaturanga, 264
checklist duties, 98
Chenango Canal, 52
Chernobyl, 171
Chesbrough, Ellis, 207
Church, Ellen, 102
Churchill, Winston, 56
Churchman, Charles West, xiii

circuses, 91
civic engineering, 10
civics
    overview, xv, 291, 298
    vs. consuming self, 218
    disengagement in engineering, 293–95
    laurel hazard, 296
    as unstable, 17–18, 288–89
Civil Aeronautics Act, 99–100
Civil Aeronautics Authority, 100
Claremont, Claude, 199–200
Clark, Andrew, 238–39
Clausewitz, Carl von, 267–68
Clayton, Marion, 90, 91
Cleese, John, 80
Clinton, Bill, 70
Cloāca Maxima, 206
clock metaphor, xiii–xiv
cloud metaphor, xiii–xiv
Cochran, Jacqueline, 101–2
cognitive fidelity, 142
Coleman, Bessie, 105
Coleridge, Samuel Taylor, 243
collaborative planning, 86–87
Collins, Brian, 79, 83, 84, 86–87
Collins, Michael, 182
Columbia space shuttle, 169–70
Columbus, Christopher, 223–24, 241
communal code of conduct, xiv
competition, 196–97
complex adaptive systems, 118–19
computers, 138–39, 150, 178, 188
Con Edison, 121
concept of operations, xiv
Connor, Roger, 89, 90
ConOps (concepts of operation), 11,
    263, 287–88. *See also* efficiency;
    maintenance; resilience; vagueness;
    vulnerability
Consolidated NY-2 biplane, 97
Constable, John, xiii
consumer surplus, 35–36
consuming self, 218
contact flying, 15, 57–58, 92–93
Continental Airlines, 143
Conway, Erik, 15, 92–93, 95
Coolidge, Calvin, 4
Copeland, Royal, 99
Corn, Joseph, 55, 56, 102–3
cosmology episodes, 79
cost-benefit calculations, 36

Cousteau, Jacques, 229
Covid-19 pandemic, 46, 86–87
cows, 57
Crane, Carl, 64, 95–96
Curtiss, Glenn, 62–63
Curtiss Jennys, 61, 64
Cutting, Bronson Murray, 98–99

da Vinci, Leonardo, 128
Dallas Spirit, 49, 50
Davis, William, 49–50
Debain, Alexandre-François, 53
decompression chambers, 228–29, 231–32,
    240
Deep Diver, 233
Deep Submergence Systems Review
    Group, 231
The Deepest Days (Sténuit), 232
Deepsea Challenger, 239
Defense Advanced Research Projects
    Agency (DARPA), 265–66
Delta Flight 9877, 133–36
democracy, xv
Dewey, George, 21–22
Diamond Visionics, 146–49
diet programs, 85–86
differential pricing, 35–36
disability-adjusted life years (DALYs),
    38–40, 42
diving, 222–25
Dole, James, 46
Dole Derby, 46–51, 108
Doolittle, Jimmy, 15, 96–98
Doran, Mildred, 49, 50
Douglas DC-2 aircraft, 98–99
Douglas DC-8 aircraft, 133–36
Downer, John, 162–63
Dreadnought, 27
Dreams (Kurosawa), 170
Drebbel, Cornelis, 227
driving routes, 33
Dungeons & Dragons (D&D), 265
Dupuit, Jules, 33–36, 41

Eagle, 182
Earhart, Amelia, 102, 108
From the Earth to the Moon (Verne), 179–80
Eastern Air Lines Flight 304, 136
Ebola, 41–43
economics, 29–31, 33–42
Edgerton, David, 202

Edgington, Dorothy, 69
Edinburgh Empire, 154, 156, 157
education, 71–72
Edwin A. Link Field, 285
efficiency
    overview, 11, 13–14
    of defense companies, 189–90
    economic, 30, 34, 38, 39–41
    vs. efficacy, 44
    energy, 44–45
    of government, 189–90
    hard, 30, 34, 45
    messy, 37, 40, 44, 45
    multicritera approach, 42–43
    optimizing, 28–29
    other ConOps and, 46
    vs. safety, 197
    social benefits and, 29–30
    soft, 29, 30, 44, 45
    technical vs. economic, 30
    as vague, 84–85
    and vulnerabilities, 46
    as vulnerability, 160
    war ships, 21–28
    wicked, 45
    See also ConOps
Egyptians, 205
Ekelund, Robert, 36
El Encanto, 47, 48
Elardo, Pam, 200–201, 202–3, 208
electricity, 160
The Elementary Principles of Carpentry
    (Tredgold), 243, 246
emosclerosis, 294
end of systems, 218–20
Endicott-Johnson Shoes (EJ), 270–73,
    276–77, 279
energy efficicency, 44–45
engineering
    overview, xv
    accountability, 125
    as care providers, 208
    civic, 10
    vs. civics, 18
    civics engagement and, xv
    code of conduct, 172
    as cultural choice, 11–12
    disruption and, 200
    economics and, 29–30, 34
    education, 293
    expertise, 174

heroism, 216–17
    inequality and, 293–94
    as male-dominated, 203
    mathematics and, 32–33, 34
    morality, 208
    normal, 182
    optimization, 28–29
    partial ignorance operations, 67–68
    performance-based design, 174
    problem-sovling and, 6–7
    projects and, 292
    radical, 182–83
    salamanders and, 45
    sharing burden of, 290
    social, 290–94
    systems engineering (*see* systems
        engineering/thinking)
    technical efficiency, 30
    and technology, 291
    testing, 130
    trade-offs, 28, 31
    user testing, 67–68
    vagueness and, 67–68, 69, 71–73
ENIAC, 178
Ensatina salamanders, 45
Erwin, Bill, 49, 50
Essay Concerning Human Understanding
    (Locke), 207–8
ethics, 39, 217, 267, 268
Etruria, 206
Eubulides of Miletus, 68
evacuations, 154
Excellent, 24
expendability, 219–20
expertise, 174, 287
Eyjafjallajökull volcano, 79–81, 82,
    83–84
Eytinge, Bruce, 58

fail-safe systems, 164
failure, 69, 168, 169, 170, 174
fatbergs, 201
Federal Aviation Administration, 10–11
femininity, 101–2
finite element modeling, 126, 127
fires, 173–74, 179
Fire-safety codes, 154–55, 174
first-degree price discrimination, 35
Fisher, Jackie, 27
FIT Aviation, 140–42, 143–44, 149
Flexman, Ralph, 297

flight contests, 47–50, 96, 102, 108
flight simulation
   overview, 139–44
   Abbey and, 177
   braking pedals, 168
   for combat, 265–66
   as cost-effective, 268–69
   details of Link's, 138–39
   Diamind Visionics, 146–49
   FIT Aviation, 140–42, 143–45
   gamified, 145–47
   laparoscopies, 147
   level of detail, 149–50
   VAMP, 139
   See also Apollo simulator program;
      Link Trainers; low-tech similua-
      tion; space flight simulation
flight training
   Black Americans rejections, 105
   British Air Force celestial, 109
   Celestial Navigation Trainer, 107, 108–
      9, *110*, 145
   Delta Flight 9877, 133–36
   ground-based, 17, 62–64
   history of, 63–64
   Link's flying school, 88–89
   Link's training, 61–62
   motion training, 63–64, 142–43
   vs. simulation, 150
   Wright training, 62–63
   See also Apollo simulator program;
      flight simulation; Link Trainers;
      low-tech similuation; Mercury
      simulator
flight-data recorders, 137–38
Florida Atlantic University, 238–39
Flying Hoboes, 106
Flyvbjerg, Bent, 211
focus groups, 74–75
fog, 96–97, 98–99
Ford Motor Company, 271
France, 184
Frank, Steve, 121–22
Freedom 7, 177
French Air Service, 64
Friendship, 102
Fukushima, 158–59, 160
Fulton, Robert, 227
future neglect, 215
"Future Views" (Thorpe), 265–66
fuzzy logic, 85

Galunggung volcano, 82
games, 256–58, 260, 264–65
garbage, 254–60
GAT-1, 192–93
Gawron, Valerie, 108–9
Gdovin, David, 146–49
Glazner, Chris, 247–49, 251, 252–53, 254
gliders, 96
global supply networks, 160
Goddard, Norman, 46, 48
Goebel, Art, 49–50, 51
Golden Eagle, 49, 50, 108
Goldstein, Stanley, 179
Gonzalez, Isaac, 111, 112, 116
Gordon, Louis, 102
grasscutter, 64
Great East Japan Earthquake, 158–59
Great Fire of 1666, 155
Great Fire of London, 155
Great Lafayette (Sigmund Neuberger),
   153–54, 157–58
Greater Binghamton Airport, 285–86,
   298–99
Greenspan, Alan, 85
Grenfell Tower fire, 173–74
ground-based training, 17, 62–64. *See also*
   Link Trainers
Grumman, 183
Guggenheim, Daniel and Harry, 96–97
Gulf War, 267–68
Gunderson, Lance, 246, 261
gunnery on ships, 21–28

Haise, Fred, 186–87
Handshake Ritual (Ukeles), 216
Harbor Branch Oceanographic Institute,
   237–38, 239
Hartle, James, 37
Hauer, Ezra, 72–73
Hawkins, Kenneth, 48
health, 37–38, 137
health financing, 38–42, 76–77
health-adjusted life years (HALYs), 38–40,
   42
heroism, 216–17
high-reliability organization, 166–69
HIV/AIDS, 38
Hobson, 164
Hodgson, Marion Stegeman, 104
Holling, Crawford "Buzz," 244–45, 261
homelessness, 248–53, 254

Hoover, Herbert, 92
housework, 36–37
housing, 271–72
Hughes, Frank, 185–86
Hugo, Victor, 201, 207
human behavior/performance, 74
Humphreys, Andrew, 124
Hunt, John, 178
Hurricane Ivan, 122–23
Hurricane Katrina, 123, 125–27, 128
Hurricane Rita, 125
hypocognition, 287

IBM, 272–75, 276–77
Iceland, 86
ideologies, 8
illness, 121
India, 204–5, 217–18, 255–57, 260
indicators, 78
innovation, 184, 192, 199–200, 216–17,
    219–20
instruments, 100–101. *See also* blind flight
Interagency Performance Evaluation Task-
    force (IPET), 125–27
Irving, Livingston "Lone Eagle," 49

Jackson, Andrew, 195
Jacquard, Joseph Marie, 53
James, Jim, 191
Japan, 94, 158–59, 160, 206
Jefferson, Thomas, 227
Jensen, Martin, 49, 50, 51
Jeppesen, Elrey Borge, 59–60
Johnson, George F., 270–73, 280
Johnson, Seward, Sr., 235
Johnson, Stephen, 184, 185
Johnson-Sea-Links, 233–35, 238
Jones, Charles "Casey," 94
    about, 14
Jordanoff, Assen, 101
Josephus, Flavius, 226–27
Julian, Hubert Fauntleroy, 105
Jurassic Park (film), 146

Karwat, Darshan, 291–92
Katla volcano, 86
Kelly, Tom, 275, 277–78, 280
Kelsey, Benjamin, 97
Kennedy, John F., 175–76
Kentucky, 23
kettles and tea, xiii–xiv

Killebrew, Tom, 3
Kilmer, Sylvester Andral, 280
kiwi-trainer, 63
Knowles, Scott, 200
Koen, Billy Vaughn, 29
Kranz, Gene, 180
kriegsspiele, 264
Krishna, 243, 261
Ku Klux Klan, 271
Kurosawa, Akira, 170

Lafayette, (Great), 153–54, 157–58
Laki volcano, 81–82
Landing Field Guide and Pilot's Log Book
    (Eytinge), 58
language, 75, 84
laparoscopies, 147
Lash, John, 189–90
Laska, Shirley, 122–23
Law, Margaret, 153–54
"Law and Manners" (Moulton), 172
Lederer, Jerome, 171
levees, 123–27
Leveson, Nancy, 167–68, 169, 170
Levine, Peter, 18
Li, Ke Ming, 67
licensing, 102
lighted airways, 58–59, 99
Lilienthal, Michael, 264
limit equilibrium assessment, 126, 127
Lindbergh, Charles "Slim," 46, 51, 65, 93,
    107–8
Lindbergh, Jon, 232
Link, Clayton, 229, 235
Link, Edwin A., Jr.
    overview, 295–97
    about, 17, 60–61
    aerial advertising, 91
    air shows, 90–91
    airplane-repair shop, 90
    aquatic device, *284*
    Clark on, 239
    commemoration, 278, 282, *300–301*
    computers and, 150
    CORDs, 236, 238
    death of, 270
    death of Clayton, 235–36
    decompression chambers, 228–29,
        231–32, *240*
    as diver, 222–27, 229, 234
    as employer, 275

as fisherman, 221–22
Flexman on, 297
flying into Newark, 14–15
flying lessons, 61–62, 64–65
flying school of, 88–89
government regulators and, 149
interships at Harbor Branch, 239
MacInnis on, 239
Marion and, 90
McKee on, 297
ocean engineering at Florida Atlantic
    University, 238–39
on oil industry, 237
organ collecting, 269
philosophy of systems engineering, 190
Pilot Maker, 285–86
simulator companies, 150
Sports Illustrated profile, 221
submersibles, 233–35, 238, *284*
Trainer 50th anniversary, 269
Trainer demo for Army Air Corps, 94
Watson on, 177
work at piano factory, 62
See also Celestial Navigation Trainer;
    flight simulation; Ocean Systems
Link, Edwin A., Sr., 90
Link, George, 90–91
Link, Marilyn, 236
Link, Marion, 90, 91, 94, 222–25, 269
Link Aviation
    Abbey and, 177
    commemoration, 278–79, 282, *300–301*
    formed, 90
    Kelly on, 275
    Lash on, 189
    selling off, 150
    See also Link Trainers; space flight
        simulation
Link Piano and Organ Company, 54–56,
    90, 278–79, 282
Link Trainers
    overview, 4–5, 295–97
    50th anniversary, 269
    Black people as instructors, 103, 104
    Black people training with, 107
    Celestial Navigation Trainer, 107, 108–
        9, *110*, 145
    at circuses, 91
    computers and, 138–39
    Deep Submergence Systems Review
        Group, 231

demo for Army Air Corps, 94
as depth technology, 286
diagrams, *xii*, *xviii*, *20*, *66*, *152*
as entertainment, 88, 90–91, 145
GAT-1, 192–93
government order for, 15
government requirements and, 149
history of, 15–16
instruments, 101
Marilyn and, 236
player pianos and, 183
as psychological training, 138–39
restoring, 188–89, 192, 269–70
selling off, 150
sold to Japan, 94
suspension of disbelief and, 94–95
trusting instruments, 97–98
VAMP, 139
women as instructors, 103
workings of, 88–90
See also Apollo simulator program;
    Mercury simulator
Linkanoe, 221
lions, 51
Locke, John, 207–8
London, UK, 155
Lovell, Jim, 186–87
low-tech similuation, 256–58, 260, 264–65
loyalty, 274, 275, 278
Ludwig, Donald, 253–54
lunar lander, 183

MacInnis, Joe, 231–32, 235–36, 239
Macleod, Joe, 219
maintenance
    overview, 11, 13–14, 192, 200, 216
    costs of, 208
    engineering, 219–20
    hard, 196, 198–99, 200, 202
    in Indian thought, 217–18
    vs. innovation, 199, 217, 219–20
    modernity and, 192
    of pipes, 203–4
    preventative, 204
    soft, 202
    software, 200, 209, 210
    steam power, 196, 198
    themes of, 199
    See also ConOps
maladaption, 119
mañana, 69

Manhattan Project, 163
mappractice, 197–98
maps, 127
Marconi, Guglielmo, 280
marginal utility, 35
The Marriage of Heaven and Hell (Blake), 13
Martin, Roger, 13
mask making, 46
Mazzini, Carl, 189–90
McCarthy, Roger, 131
McKee, Art "Silver Bar," 297
mechanization, 54. *See also* pianos/organs
medicine, 262
Mercury simulator, 177
Merry, Sally Engle, 78
Mesopotamian Empire, 205
Microsoft Flight Simulator, 141
Microsoft Windows, 211
Miliero, 113
military simulation, 265–68
Minard, Charles Joseph, 32–33
Minard Map, 33
Minoan civilization, 205
Minor, C. Sharpe, 55
Mintzberg, Henry, 44
Miss Doran, 49, 50
Mitcham, Carl, 293
Mitchell, Billy, 93
MITRE Corporation, 248
Mock, Ron, 143–44
models, 126–27, 128
modernity, 192
Mohenjo-Daro, 204–5
molasses, 111–16
Monitor, 238
monocausotaxophilia, 12
monotheorism, 12
morality, 208
Moroney, William, 264
Morse code, 60, 232
Morton, James, 133–36
Moselle, 196
Moses, Robert, 117
motion trainers, 63–64, 142–43. *See also* Link Trainers
motivational fidelity, 142
Moulton, John Fletcher, 172
Mount Pinatubo, 82
multilevel thinking, 75–79, 191–92, 252
Murrow, Edward R., 163
music, 54. *See also* pianos/organs

NASA, 175, 178, 183, 185–87. *See also* Apollo simulator program
National Institute of Standards and Technology (NIST), 119
Nautilus, 227
Navier, Claude-Louis, 32, 34
navy (United States), 21–24, 26–28, 163, 230–31, 235–36, 264
Neuberger, Sigmund, 153–54, 157–58
New Orleans, 122–27, 128
A New Theory on Columbus's Voyage... (Link and Link), 224
New York City, 200–204, 207, 212–13
New York City Transit, 209–11, 213–14, 215–16
Newberry, Byron, 45
NIMBY (Not in My Back Yard), 215
No. 1 British Flying Training School, 2–4
Noonan, Fred, 108
normal accidents, 166–69
normal engineering, 182
Notice to Airmen/Missions (NOTAM), 59
nuclear power plants, 158–59, 160, 165, 166, 170–71, 190–91
Nuclear Regulatory Commission, 158–60, 165
nuclear submarines, 164–65, 230, 231, 234
numbers/mathematics, 32, 78, 155, 156

Obedience to the Unenforceable, 172
O'Brien, Karen, 129
Ocean Systems, 230
Ocker, William Charles, 64, 65, 95–96
oil industry, 237
Oklahoma, 48, 49
Olympia, 22
O'Neil, Anne, 209–10, 215–16
Operation CHASE, 237
optimization, 28–29
organs. *See* pianos/organs
Ostendorff, Bill, 158–59, 161, 163, 164–65
Our Future Is in the Air (Picasso), 56
overspecialization, 298

Pabco Pacific Flyer, 48–49, 50
Pan, 245–46, 261
panarchy, 246, 261
paradox of robustness, 122
Parsons, William Barclay, 213–14
Peerless Piano Player Company, 52

penguin system, 64
performance-based design, 174
Perin, Isaac, 196
permanent supportive housing, 251, 252
Perrow, Charles, 166
Perry, John, 233
Peters, David, 146–47
Peterson, Jon, 265
Petrick, Jason, 147–48
petteia, 264
pharmaceuticals, 40–41
physical fidelity, 142
pianos/organs, 52–55, 183, 188, 220, 269. *See also* Link Piano and Organ Company
Picasso, Pablo, 56
Piccard, Auguste Antoine, 227–28
piecemeal approaches, 290–91
Pigott, Peter, 56
Pilot Maker, 285–86
pilots. *See* aviation; flight contests; flight simulation; flight training; Link Trainers; low-tech similuation
Pirtle, Zach, 126, 127
Politics of the Very Worst (Virilio), 165
Popper, Karl, xiii, 290
Port Royal, Jamaica, 224–25
Porter, Theodore, 32
The Positively True Adventures of the Alleged Texas Cheerleader-Murdering Mom (film), 247
Powder Puff Derby, 102
Powell, William Jenifer, 105–6
PowerPoint, 169
prescription drugs, 40–41
pricing, 35
privilege, 36–37
problems
    dissolved, 8
    engineers as solvers of, 6–7
    hard, 7, 8–9, 288
    messy, 8–9, 248, 252, 288
    soft, 7–9, 248, 252, 288
    solving vs. resolved, 7–8
    understanding facets, 12–14
    See also wicked problems; *various ConOps*
Proceedings of the United States Naval Institute (Alger), 22–23
Project Gemini, 177, 180

Project Mercury, 177, 180
projects, 292
public health. *See* wastewater
Public Health Act (Britain), 206
pucker factor, 143
pumping stations, 123
Purity Distilling Company, 111–16

quality-adjusted life years (QALYs), 38–40, 42
Queen Bess, 105

racing boats, 195
radical design, 182–83
radio, 280
radioactive contamination, 160–61
railroads, 41, 211–12, 257
Railtrack, 211–12
Rasmussen, Jens, 76
Reagan, Ronald, 276
Redoubt volcano, 82
rejuvenation, 217–18
reliability, 163, 166–67
rental cars, 167–68
resilience
    overview, 11, 13–14, 262–63
    boats as metaphors, 253–54
    collective, 261–62
    engineering vs. ecological, 244
    hard, 244, 259
    history of word use, 243
    Ludwig on, 253
    messy, 259
    as mindset, 263
    soft, 246, 249, 251–52, 253–54
    as survival, 250–51
    in systems engineering, 262
    Tredgold's timber testing, 242–43, 244
    See also ConOps
responsibility, 164–65
Richards, Jonathan, 148
Rickenbacker, Eddie, 93
Rickover, Hyman, 163–64, 230–31
Ridgway, V.F., 43
risk, 129–31, 164, 165, 166–67, 171
Rittel, Horst, 6, 261
road degradation, 33–34
robustness, 249–50
Roeser, Sabine, 208
Roman Empire, 206
Roosevelt, Eleanor, 106–7

Roosevelt, Franklin Delano, 2, 93–94, 99, 103–4
Roosevelt, Theodore, 27, 264
Rouse, Bill, 74–75, 76–79
Royal Air Force, 1–3
The Royal Air Force in Texas (Killebrew), 3
Royal Navy, 24–26, 27, 28, 43–44, 80, 241
Rubbish! board game, 256–57, 258, 260
Rubick's Cube, 11
Ruggles, William Guy, 64
Russell, Bertrand, 75–76

Sachs, Noah, 258
safety
    overview, 72–73
    complacentcy and, 165
    diversity of opinions, 173–74
    vs. efficacy, 197
    fairness of, 171
    hard, 155, 156–57, 161
    Lederer and, 171–72
    in London, 155–56
    magic numbers, 155, 156
    messy, 168–69, 170
    optimization trap, 172–73
    productivity and, 168
    vs. reliability, 167
    reviews of nuclear power plants, 159
    safe-to-fail systems, 164
    self-regulation, 157
    soft, 159–61, 164
    in steam power, 195–98
    See also airbags; ConOps; maintenance
salamanders, 45
Salas, Eduardo, 150
Salk, Jonas, 263
Sanders, Haydn, 62–63
sanitation. *See* garbage; wastewater
sanitation workers, 258. *See also* garbage
Sawyer, Diane, 163
Say, Jean-Baptiste, 31
Schaff Brothers Piano Com-, 54
Schlesinger, Arthur, Jr., xv
Schluter, Paul, 50, 51
Schneider Trophy, 96
Schultz, Timothy, 95
Scott, Percy, 24–26, 27, 28, 43–44
Scrape project, 183
Scylla, 24–25
Sea Diver, 223, 225, 232
Sea Diver II, 225–26

seafood, 160–61
Seamans, Robert, 175
seat belts, 67, 69, 70–71, 72
segregation, 3, 104
Seiden, Alex, 146
September 11th attacks, 116–20, 122, 244
sewage treatment, 200–204, 205–7
sextants, 107
Shakespeare, William, xiii
Shapin, Steve, 220
shatranj, 264
Shepard, Alan, 177
Sherwood, Susan, 187–89, 192
ships, 21–28, 43–44, 241–42
shipwrecks, 222, 239
The Shock of the Old (Edgerton), 202
Siegel, Greg, 138
sight, 95–96
SIMNET, 266–68
Sims, William Sowden, 23–24, 26–28, 43–44
simulation, 146–47, 150, 177, 190–92, 264, 286. *See also* Apollo simulator program; flight simulation; low-tech similuation
simulator networking, 266–68
Singer Corporation, 150
Sky Chief service, 98–99
sky roadways, 57–58
skyscrapers, 211
slavery, 111–12, 116, 124, 197
Smith, Adam, 30–31
Snook, Mary, 102
social engineering, 290–94
social media, 294–95
societal transitions, 263
software, 200, 209, 210
solutions, 74, 78
Soviet Union, 177
space flight simulation, 177, 180–81. *See also* Apollo simulator program
space programs, 175–77, 180. *See also* NASA
space shuttles, 169–70
space travel ideas, 179–80
Spanish-American War, 22
Spencer, Thomas, 44
Spirit of Peoria, 48
Spirit of St. Louis, 46
Sputnik, 177
Starr, Chauncey, 130–31

steam power, 195–98, 269
Stearman plane, 4
steel, 120
Sténuit, Robert, 229, 232
stimunlators, 264
Stultz, Wilmer, 102
submarines, 227, 230–31. *See also* nuclear
    submarines
Submersible Portable Inflatable Dwellings
    (SPIDs), 229–30, 232–35
subways, 209–10, 213–14. *See also* New
    York City Transit
Sultana, 199
surveillance, 267
"The Survival of the Wisest" (Salk), 263
Swigert, John, 186–87
SWIP project, 183
systems engineering/thinking
    overview, 9–11, 290
    aerospace missions, 184–85
    big pictures, 190
    boats as metaphors, 253–54
    discipline and, 297
    Fukushima, 158–60
    Glazner on, 247, 251
    philosophy, 190
    resilience and, 262
    retiring, 218–19
    technical debt, 209–12
    understanding, 191–92
systems management, 184

technical debt, 209–12
technology (general), 202, 291, 294–95
TechWorks! 188, 192
Telford, Thomas, 270
ten C's, 109
Terrell, Texas, 2–4
Terrible, 26
Thaden, Louise, 102
theatre fires, 154
theory of types, 75–76
THINK, 273–74
Thorpe, Jack, 265–68
Three Mile Island, 165, 190–91
Thresher, 230, 231
timber, 241–43, 244–45
Titanic, 239
toilets, 200–201
tolerance principle, 68
Touch Sanitation Performance (Ukeles), 216

toxic waste, 237, 259
Tozzoli, Guy, 116–18, 122
trade-offs
    economic, 31, 36
    health, 38
    optimization/accuracy, 28
    resilience and, 262
    vulnerabilities and, 121
Transcontinental and Western Air (TWA)
    Flight 6, 98–99
transfer of training, 141
transit, 209–10
Tredgold, Thomas, 242–43, 246–47
tricksters, 261
Trieste, 228, 230
trust, 43, 128, 131, 171, 172
truth, 162
tsunamis, 158–59
Turtle, 227
Tuskegee Airmen, 106–7
Twain, Mark, 198–99
Twin Towers. *See* World Trade Center

Ukeles, Mierle Laderman, 216
underwater capsule, *194*
underwater living, 227–33
United Airlines, 143
United States Industrial Alcohol, 113–14
user testing, 67–68
utility, 32–33, 35, 40

vaccines, 41–42
vagueness
    overview, 11, 13–14
    airbags, 71–72
    collaborative planning and, 86
    defined, 68–69
    hard, 68, 69, 71, 72–73
    of language, 75, 84–86
    messy, 83
    soft, 75
    See also ConOps
values, 8, 36
Variable Anamorphic Motion Picture
    (VAMP), 139
Vaucanson, Jacques de, 53
Verne, Jules, 179–80
vertigo, 142
veterans, 248–53, 254
The Victory at Sea (Sims), 23
Virgin Rail Group, 211–12

Virilio, Paul, 165
virtual environment, 89. *See also* flight
    simulation; low-tech similuation;
    simulation
virtual reality, 89, 286
vision, 95–96
visual effects, 146, 149
visualalization, 76–77
Visualizing Software (Bennett), 191
Vitruvius, Marcus, 244
volcanos, 79–83, 86
Vonnegut, Kurt, 188, 192
vulnerability
    overview, 11, 13–14
    complex adaptive systems,
        118–19
    efficiency and, 46, 160
    hard, 113, 115, 126, 128
    as imprecise, 129
    maladaption, 119
    messy, 116, 122, 123, 128
    vs. risk, 129–31
    as shifting, 128
    soft, 115–16, 119, 121, 122, 128
    trust and, 131
    See also ConOps

Ward, Earlin, 190
Warfield, John, xiii
wargames, 264–68
waste, 254–60
wastewater, 200–204, 205–7
water, 200
Watson, Maurice, 133–36
Watson, Tom, Jr., 177, 275
Watson, Tomas, Sr., 273–75
weather
    airmail and, 92–93
    aviation, 15, 50–51
    Guggenheim fund, 96–98
    Link Trainers, 109
    TWA Flight 6, 98–99
Webb, James, 175–76
Webber, Melvin, 6, 261
Webster, Daniel, 195

Weems, Philip Van Horn, 107–8, 223
welfare programs, 29–30
Wells, H. G., 69
West Coast Main Line, 211–12
Wetmore, Jameson, 71, 72
"What If Hurricane Ivan Had Not Missed
    New Orleans?" (Laska), 123
"When the Chick Hits the Fan" (Downer),
    162–63
Whitney, Joshua, 51
wicked problems
    overview, 288–91, 295
    as collision of problems, 8
    modern, 286–87
    properties of, 6
    as surprising, 261
    See also efficiency; maintenance;
        problems; resilience; vagueness;
        vulnerability
wicked problems defined, xiii, xiv, 6
Wiesner, Jerome, 175
Wilt, Donna, 140–42, 144–45, 146
Wittgenstein, Ludwig, 260
women, 5, 101–4, 105
Women Accepted for Voluntary Emer-
    gency Services (WAVES), 103–4
Women Airforce Service Pilots (WASP),
    104
Woolaroc, 49–50, 108
words as aerosol, xiv
working academy, 270
World Trade Center, 116–20, 122, 130
World War I, 104
World War II, 1–2, 10, 103–4, 106–7
Worster, Donald, 129
Wren, Christopher, 155
Wright, Orville and Wilbur, 56, 62–63, 88,
    295, 296

Young, John, 180
Young, Thomas, 243

Zadeh, Lotfi, 85
Zen, 73, 79, 243
Zhang, Diana, 67